VISCOUS PROFILES AND NUMERICAL METHODS FOR SHOCK WAVES

SIAM PROCEEDINGS SERIES LIST

Neustadt, L.W., Proceedings of the First International Congress on Programming and Control (1966)

Hull, T.E., Studies in Optimization (1970)

Day, R.H. & Robinson, S.M., Mathematical Topics in Economic Theory and Computation (1972)

Proschan, F. & Serfling, R.J., Reliability and Biometry: Statistical Analysis of Lifelength (1974)

Barlow, R.E., Reliability & Fault Tree Analysis: Theoretical & Applied Aspects of System Reliability & Safety Assessment (1975)

Fussell, J.B. & Burdick, G.R., Nuclear Systems Reliability Engineering and Risk Assessment (1977)

Duff, I.S. & Stewart, G.W., Sparse Matrix Proceedings 1978 (1979)

Holmes, P.J., New Approaches to Nonlinear Problems in Dynamics (1980)

Erisman, A.M., Neves, K.W. & Dwarakanath, M.H., Electric Power Problems: The Mathematical Challenge (1981)

Bednar, J.B., Redner, R.,Robinson, E. & Weglein, A., Conference on Inverse Scattering: Theory and Application (1983)

Voigt, R.G., Gottlieb, D. & Hussaini, M. Yousuff, Spectral Methods for Partial Differential Equations (1984)

Chandra, Jagdish, Chaos in Nonlinear Dynamical Systems (1984)

Santosa, Fadil, Symes, William W., Pao, Yih-Hsing & Holland, Charles, Inverse Problems of Acoustic and Elastic Waves (1984)

Gross, Kenneth I., Mathematical Methods in Energy Research (1984)

Babuska, I., Chandra, J. & Flaherty, J., Adaptive Computational Methods for Partial Differential Equations (1984)

Boggs, Paul T., Byrd, Richard H. & Schnabel, Robert B., Numerical Optimization 1984 (1985)

Angrand, F., Dervieux, A., Desideri, J.A. & Glowinski, R., Numerical Methods for Euler Equations of Fluid Dynamics (1985)

Wouk, Arthur, New Computing Environments: Parallel, Vector and Systolic (1986)

Fitzgibbon, William E., Mathematical and Computational Methods in Seismic Exploration and Reservoir Modeling (1986)

Drew, Donald A. & Flaherty, Joseph E., Mathematics Applied to Fluid Mechanics and Stability: Proceedings of a Conference Dedicated to R.C. DiPrima (1986)

Heath, Michael T., Hypercube Multiprocessors 1986 (1986)

Papanicolaou, George, Advances in Multiphase Flow and Related Problems (1987)

Wouk, Arthur, New Computing Environments: Microcomputers in Large-Scale Computing (1987)

Chandra, Jagdish & Srivastav, Ram, Constitutive Models of Deformation (1987)

Heath, Michael T., Hypercube Multiprocessors 1987 (1987)

Glowinski, R., Golub, G.H., Meurant, G.A. & Periaux, J., First International Conference on Domain Decomposition Methods for Partial Differential Equations (1988)

Salam, Fathi M.A. & Levi, Mark L., Dynamical Systems Approaches to Nonlinear Problems in Systems and Circuits (1988)

Datta, B., Johnson, C., Kaashoek, M., Plemmons, R. & Sontag, E., Linear Algebra in Signals, Systems and Control (1988)

Ringeisen, Richard D. & Roberts, Fred S., Applications of Discrete Mathematics (1988)

McKenna, James & Temam, Roger, ICIAM '87–Proceedings of the First International Conference on Industrial and Applied Mathematics (1988)

Rodrigue, Garry, Parallel Processing for Scientific Computing (1989)

Chan, Tony F., Meurant, Gerard, Periaux, Jacques & Widlund, Olof B., Domain Decomposition Methods (1989)

Caflish, Russel E., Mathematical Aspects of Vortex Dynamics (1989)

Wouk, Arthur, Parallel Processing and Medium-Scale Multiprocessors (1989)

Flaherty, Joseph E., Paslow, Pamela J., Shephard, Mark S. & Vasilakis, John D., Adaptive Methods for Partial Differential Equations (1989)

Kohn, Robert V. & Milton, Graeme W., Random Media and Composites (1989)

Mandel, Jan, McCormick, S.F., Dendy, J.E., Jr., Farhat, Charbel, Lonsdale, Guy, Parter, Seymour V., Ruge, John W. & Stüben, Klaus, Proceedings of the Fourth Copper Mountain Conference on Multigrid Methods (1989)

Colton, David, Ewing, Richard & Rundell, William, Inverse Problems in Partial Differential Equations (1990)

Chan, Tony F., Glowinski, Roland, Periaux, Jacques & Widlund, Olof B., Third International Symposium on Domain Decomposition Methods for Partial Differential Equations (1990)

Dongarra, Jack, Messina, Paul, Sorensen, Danny C. & Voigt, Robert G., Proceedings of the Fourth SIAM Conference on Parallel Processing for Scientific Computing (1990)

Glowinski, Roland & Lichnewsky, Alain, Computing Methods in Applied Sciences and Engineering (1990)

Coleman, Thomas F. & Li, Yuying, Large-Scale Numerical Optimization (1990)

Aggarwal, Alok, Borodin, Allan, Gabow, Harold, N., Galil Zvi, Karp, Richard M., Kleitman, Daniel J., Odlyzko, Andrew M., Pulleyblank, William R., Tardos, Éva & Vishkin, Uzi, Proceedings of the Second Annual ACM-SIAM Symposium on Discrete Algorithms (1990)

Cohen, Gary, Halpern, Laurence & Joly, Patrick, Mathematical and Numerical Aspects of Wave Propagation Phenomena (1991)

Gómez, S., Hennart, J. P. , & Tapia, R. A., Advances in Numerical Partial Differential Equations and Optimization: Proceedings of the Fifth Mexico-United States Workshop (1991)

Glowinski, Roland, Kuznetsov, Yuri A., Meurant, Gérard, Périaux, Jacques & Widlund, Olof B., Fourth International Symposium on Domain Decomposition Methods for Partial Differential Equations (1991)

Alavi, Y., Chung, F. R. K., Graham, R. L., Hsu, D. F., Graph Theory, Combinatorics, Algorithms, and Applications (1991)

Wu, Julian J., Ting, T. C. T., Barnett, David M., Modern Theory of Anisotropic Elasticity and Applications (1991)

Shearer, Michael, Viscous Profiles and Numerical Methods for Shock Waves (1991)

VISCOUS PROFILES AND NUMERICAL METHODS FOR SHOCK WAVES

Edited by Michael Shearer
North Carolina State University

siam. Philadelphia

Society for Industrial and Applied Mathematics

VISCOUS PROFILES AND NUMERICAL METHODS FOR SHOCK WAVES

Proceedings of a workshop at North Carolina State University, Raleigh, North Carolina, May 23-25, 1990. This workshop was supported in part by the US Army Research Office Grant No. DAAL03-90-G-0011.

Library of Congress Cataloging-in-Publication Data

Viscous profiles and numerical methods for shock waves/ [edited by]
Michael Shearer.
p. cm.
Proceedings of a workshop at North Carolina State University,
Raleigh, North Carolina, May 23-35, 1990.
Includes bibliographical references.
ISBN 0-89871-283-1
1. Shock waves—Congresses. 2. Viscous flow—Congresses.
3. Differential equations, Hyperbolic—Numerical solutions—
-Congresses. 4. Differential equations, Parabolic—Numerical
solutions—Congresses. I. Shearer, Michael.
QA927.V57 1992
531'.1133—dc20 91-40934

PREFACE

A workshop on shock waves, held at North Carolina State University, May 23-25, 1990, brought together mathematicians interested in the following areas related to shock waves: the theory of hyperbolic conservation laws, numerical methods for hyperbolic and parabolic systems of equations, the theory of traveling waves viscous profiles for parabolic systems, and applications. The workshop was supported in part by a grant from the Army Research Office.

The papers in this volume reflect the considerable breadth of interests in analysis and applications at the workshop. Consequently, it is a rough survey of current mathematical work in the area of shockwaves and provides a rich source of ideas and problems in theoretical and numerical aspects of the field and its many applications. One theme of the meeting strongly represented in the proceedings is the power of ideas from dynamical systems that are being adapted and developed in the context of shockwaves. A related theme,the approximation and stability of shockwaves, is treated in several papers. Applications are another prominent feature of these proceedings. Many of the papers on applications present models that are comparatively new to the mathematics community, with interesting phenomena and ideas for their interpretation.

It is a pleasure to acknowledge the help I received in organizing the workshop. Stephen Schecter deserves special credit for being a constant source of encouragement, good ideas, and hard work. The other members of the organizing committee, Tim Kelley and David Schaeffer, also provided valuable and crucial assistance. Willing and skillful secretarial and administrative services were performed by the NCSU Mathematics Department staff, notably Mary Byrd, Dionne Wilson, and Carolyn Gunton. I am also grateful to Bernadetta DiLisi and Vickie Kearn of SIAM for their professional and helpful involvement at every stage of compiling this volume of papers.

Michael Shearer
North Carolina State University

CONTENTS

Fully-Discrete Methods with Grid Refinement for the Generalized Korteweg-de Vries Equation[1]

Jerry L. Bona†
Vassilios A. Dougalis*
Ohannes A. Karakashian‡
William R. McKinney**

Abstract. A class of fully-discrete schemes for the numerical solution of the periodic initial-value problem for the generalized Korteweg-de Vries equation are presented and tested. These numerical approximations are generated by a Galerkin Finite Element process for the spatial discretization and Implicit Runge-Kutta methods for the temporal discretization. Such schemes possess excellent stability properties, as well as arbitrarily high order rates of convergence in both the spatial and temporal variables. These methods are used with adaptive grid refinement in the spatial and temporal meshes to investigate the stability and instability of a class of travelling-wave solutions known as solitary-waves. The numerical approximations give evidence that solutions of the GKdV equation can develop singularities in finite time.

1. Introduction. In this paper, numerical methods used to generate fully-discrete approximations for the generalized Korteweg-de Vries (GKdV) equation are presented. The GKdV equation is a model used to describe the propagation of nonlinear, dispersive waves and can be written in the form

$$u_t + u^p u_x + \epsilon u_{xxx} = 0, \tag{1.1a}$$

where p is a nonnegative integer and ϵ is a nonzero parameter. Here, the equation is considered with periodic boundary conditions on the spatial interval $x \in [0,1]$ and for

[1] Work supported in part by the Institute of Applied and Computational Mathematics of the Research Center of Crete, Iraklion, Crete, Greece, the National Science Foundation, USA, the Air Force Office for Scientific Research under grant AFOSR-88-0019, and the Science Alliance, University of Tennessee.

†Department of Mathematics and Applied Research Laboratory, The Pennsylvania State University, University Park, PA 16802.

*Department of Mathematics, University of Crete, Iraklion, Crete, Greece.

‡ Department of Mathematics, University of Tennessee, Knoxville, TN 37996.

**Department of Mathematics, North Carolina State University, Raleigh, NC 27695.

finite time intervals $t \in [0, t^*]$, with initial data

$$u(x,0) = u^0(x) \tag{1.1b}$$

where u^0 is a given periodic function of period 1. The value of the solution $u(x,t)$ represents the amplitude of a periodic wave at a point x at time t. The special case $p = 1$ corresponds to the well-known Korteweg-de Vries (KdV) equation.

The global existence of solutions for (1.1), with arbitrary p and large, smooth initial data u^0, is still an open question. Kato **[1]** has shown that if $p < 4$, then the pure initial-value problem for (1.1), wherein u^0 is specified for all x and need not be periodic, does indeed have solutions that exist for all $t > 0$, provided u^0 is sufficiently smooth and decays at infinity. A similar result holds for the periodic case (see for example **[2]** in the case $p = 1$). To investigate numerically the question of global existence, we shall use the family of solitary-wave solutions. A recent theory **[3]** has shown that these special travelling-wave solutions are stable if and only if $p < 4$. However, the theory leaves completely open the manifestation of instability. The numerical simulations given herein indicate that the instability leads to the formation of singularities in finite time.

In Section 2, the numerical schemes used to produce fully-discrete approximations to solutions of the periodic initial-value problem for the GKdV equation are presented. These schemes are a combination of a Galerkin method and Implicit Runge-Kutta methods. In this paper, only a brief description of these methods is given. A more detailed description of these schemes, including proofs of stability and convergence along with an algorithm for their efficient implementation, may be found in **[4]**; for earlier work and results, including a list of numerical techniques for the KdV equation, see **[5]** and the references therein. An important consequence of the methods used here is that they preserve the second invariant of the GKdV equation.

In Section 3, these numerical methods are used with uniform spatial and temporal meshes to simulate the solitary-wave for $p = 5$. With small initial data, the schemes are very effective. However, for larger u^0, it becomes very evident that these methods need modification. This is done in Section 4 where we introduce schemes to refine both the spatial and temporal grids. This adaptive grid algorithm is based on the third invariant of the GKdV equation and a standard inverse property of the finite elements utilized in our scheme.

2. Numerical Approximations.

After introducing some basic notation, the numerical schemes used to generate the fully-discrete approximations are described. These approximations are obtained by a Galerkin Finite Element process for the spatial discretization and conservative Implicit Runge-Kutta (IRK) methods for the time-stepping. These schemes possess excellent stability properties, as well as arbitrarily high order rates of convergence in both the spatial and temporal variable. Thus, they represent an efficient means for highly accurate numerical simulations of solutions of (1.1).

For nonnegative integers m and real p, $1 \le p \le \infty$, let $W^{m,p} = W^{m,p}(0,1)$ denote the usual Sobolev spaces with norm $\|\cdot\|_{m,p}$ consisting of functions whose first m derivatives belong to L^p. For $p = 2$, note that $W^{m,2}$ is the Hilbert space H^m and replace $\|\cdot\|_{m,2}$ with $\|\cdot\|_m$. If in addition $m = 0$, we shall replace $\|\cdot\|_0$ with $\|\cdot\|$, and note that $H^0 = L^2$ with the inner product $(\cdot,\cdot)$.

2.1 Finite Element Spaces. Let r and N be integers with $3 \le r \le N$ and define $h = 1/N$. Let S_h^r denote the space of 1-periodic smooth splines of order r (polynomials of degree less than or equal to $r-1$) defined on a uniform partition of $[0,1]$, $x_i = ih$, $i = 0, \ldots, N$.

It is well-known that if v is a sufficiently smooth periodic function with period 1, then there exists a $\chi \in S_h^r$ such that for $1 \leq m \leq r$,

$$\sum_{j=0}^{m-1} h^j \|v - \chi\|_{j,s} \leq ch^m \|v\|_{m,s}, \quad s = 2 \quad \text{or} \quad s = \infty, \tag{2.1}$$

where c is a constant independent of h, v and χ. In addition to the above *approximation* properties, the finite-dimensional spaces S_h^r also possess the following *inverse* properties. There exists a constant c, independent of h, such that for all $\chi \in S_h^r$,

$$\|\chi\|_\beta \leq ch^{-(\beta-\alpha)}\|\chi\|_\alpha \quad \text{and} \quad \|\chi\|_{\alpha,\infty} \leq ch^{-(\alpha+\frac{1}{2})}\|\chi\|, \tag{2.2}$$

where $0 \leq \alpha \leq \beta \leq r - 1$.

The Galerkin semidiscrete approximation to the solution u(x,t) of (1.1) is defined as a mapping $v_h : [0, t^*] \to S_h^r$ satisfying the relations

$$(v_{ht}, \chi) - (\frac{1}{p+1} v_h^{p+1} + \epsilon v_{hxx}, \chi') = 0, \quad \forall \chi \in S_h^r, \tag{2.3}$$

and

$$v_h(0) = \pi_h u^0, \tag{2.4}$$

where $\pi_h u^0$ denotes any conveniently chosen element of S_h^r (e.g. L^2-projection, interpolant, etc.) such that

$$\|\pi_h u^0 - u^0\| \leq ch^r. \tag{2.5}$$

One may then show that if $u(x,t)$ is sufficiently smooth, then v_h exists uniquely and satisfies the error estimate

$$\max_{0 \leq t \leq t^*} \|v_h(t) - u(\cdot, t)\| \leq ch^r \tag{2.6}$$

where c is a constant independent of h. (See [6] for a proof in the special case $p = 1$ which may easily be extended to larger p). Furthermore, letting $\chi = v_h$ in (2.3), integrating by parts and using periodicity, it is easy to see that

$$\|v_h(t)\| = \|v_h(0)\|, \quad t \geq 0. \tag{2.7}$$

Upon choosing a basis for S_h^r and representing v_h in terms of this basis, (2.3) then becomes a system of N nonlinear ordinary differential equations which is conservative in L^2 by (2.7).

2.2 Implicit Runge-Kutta Methods. To obtain the fully-discrete approximations, we shall discretize the system (2.3) in time by Implicit Runge-Kutta methods. (For details and general remarks concerning Runge-Kutta methods, we refer to [7] and the references therein). For positive integer q, a q-stage IRK method is given as a set of constants arranged in *tableau* form

$$\begin{array}{ccc|c} a_{11} & \cdots & a_{1q} & \tau_1 \\ \vdots & & \vdots & \vdots \\ a_{q1} & \cdots & a_{qq} & \tau_q \\ \hline b_1 & \cdots & b_q & \end{array} .$$

Of particular interest is the family of Gauss-Legendre methods. These methods are appropriate for several reasons. First, they are conservative, a property that is necessary for the existence of a discrete conservation law similar to (2.7). They are algebraically stable, a property that is much stronger than A-stability. Also, this family is optimal in that for a given q, they are the unique Runge-Kutta method of order $2q$ in the context of the approximation of solutions of systems of ordinary differential equations. The tableau for the 2-stage Gauss-Legendre method is given by

$$\begin{array}{cc|c} \frac{1}{4} & \frac{1}{4}-\frac{1}{2\sqrt{3}} & \frac{1}{2}-\frac{1}{2\sqrt{3}} \\ \frac{1}{4}+\frac{1}{2\sqrt{3}} & \frac{1}{4} & \frac{1}{2}+\frac{1}{2\sqrt{3}} \\ \hline \frac{1}{2} & \frac{1}{2} & \end{array},$$

and the tableau for the 3-stage method is

$$\begin{array}{ccc|c} \frac{5}{36} & \frac{80-24\sqrt{15}}{360} & \frac{50-12\sqrt{15}}{360} & \frac{1}{2}-\frac{\sqrt{15}}{10} \\ \frac{50+15\sqrt{15}}{360} & \frac{2}{9} & \frac{50-15\sqrt{15}}{360} & \frac{1}{2} \\ \frac{50+12\sqrt{15}}{360} & \frac{80+24\sqrt{15}}{360} & \frac{5}{36} & \frac{1}{2}+\frac{\sqrt{15}}{10} \\ \hline \frac{5}{18} & \frac{8}{18} & \frac{5}{18} & \end{array}.$$

The fully-discrete approximations are now obtained by applying a q-stage Gauss-Legendre method to the system (2.3). For integer $J > 0$, let $t^n = nk$, $n = 0, 1, \ldots, J$, where the step size is given by $k = t^*/J$. With $u(x,t)$ being the true solution of (1.1), approximations $u_h^n \in S_h^r$ to $u(\cdot, t^n)$, for $n = 0, 1, \ldots, J$ are defined by letting $u_h^0 = \pi_h u^o$, and for $n = 0, 1, \ldots, J-1$,

$$(2.8) \qquad (u_h^{n+1}, \chi) = (u_h^n, \chi) + k \sum_{i=1}^{q} b_i \Big(\frac{1}{p+1}[u_h^{n,i}]^{p+1} + \epsilon u_{hxx}^{n,i}, \chi'\Big), \quad \forall \chi \in S_h^r,$$

where the intermediate values $u_h^{n,i}$, $i = 1, \ldots, q$, are the solutions of the system of q nonlinear equations

$$(2.9) \qquad (u_h^{n,i}, \chi) = (u_h^n, \chi) + k \sum_{j=1}^{q} a_{ij} \Big(\frac{1}{p+1}[u_h^{n,j}]^{p+1} + \epsilon u_{hxx}^{n,j}, \chi'\Big), \quad \forall \chi \in S_h^r.$$

The equations (2.9) are solved by using a Newton iteration as described in [4].

2.3 Stability and Convergence. In this section we summarize results concerning the fully-discrete approximations u_h^n which can be found along with their proofs in [4]. In the following, we shall assume that $u^0(x)$ is sufficiently smooth to guarantee the existence and required smoothness of the solution $u(x,t)$ for $0 \le t \le t^*$ in order that the error estimates hold.

THEOREM 2.1. *For $n = 0, 1, \ldots, J-1$, the equations (2.8) and (2.9) have solutions $u_h^{n+1} \in S_h^r$ and $u_h^{n,i} \in S_h^r$, $i = 1, \ldots, q$. Furthermore,*

$$(2.10) \qquad \|u_h^n\| = \|\pi_h u^0\| \quad \text{for} \quad n = 0, \ldots, J.$$

The above theorem demonstrates that the fully-discrete approximations are conservative in L^2, a discrete analog of the second invariant of the GKdV.

THEOREM 2.2. *Assume that as* $h \to 0$,

(i) *if* $p = 1$, *then* $k = O(h^{\frac{3}{2(q+2)}})$ *for* $q \geq 2$ *or* $k = O(h^{\frac{3}{4}})$ *for* $q = 1$,
(ii) *if* $p = 2$, *then* $kh^{-\frac{1}{2}}$ *is sufficiently small for* $q \geq 2$ *or* $k = O(h^{\frac{3}{4}})$ *for* $q = 1$,
(iii) *if* $p \geq 3$, *then* kh^{-1} *is sufficiently small for all* $q \geq 1$.

If h *is sufficiently small, then for* $n = 0, 1, \dots, J$, *there exists a unique solution* u_h^n *of (2.8) and (2.9) such that*

$$\max_{0 \leq n \leq J} \|u_h^n - u(\cdot, t^n)\| \leq c(k^{2q} + h^r), \quad \text{for} \quad q = 1, 2, \tag{2.11}$$

$$\max_{0 \leq n \leq J} \|u_h^n - u(\cdot, t^n)\| \leq c(k^{q+2} + h^r), \quad \text{for} \quad q \geq 3 \tag{2.12}$$

for some constant c *independent of* h *and* k.

It is worth remark that the above result is sub-optimal if $q \geq 3$. However, in the case of the KdV equation ($p = 1$), it has recently been shown [**8**] that the optimal temporal rate of convergence $2q$ is achieved independent of the number of stages q.

3. Numerical Simulation of Solitary Waves. In this section, the numerical schemes presented in Section 2 are tested. Specifically, we try to simulate the solitary-wave solutions of the GKdV equation. These special, travelling-wave solutions are given by

$$u(x, t) = A\, sech^{2/p}[B(x - x_0 - Ct)], \tag{3.1}$$

where

$$C = \frac{2A^p}{(p+1)(p+2)}, \quad B = \frac{p}{2}\sqrt{\frac{C}{\epsilon}}$$

and A represents the amplitude of the waves. Although (3.1) is an exact solution of the pure initial-value problem for (1.1), if A is large and ϵ is taken sufficiently small then it should also be an approximate solution in the periodic case since the tails of the solitary-wave decay exponentially.

For other numerical results regarding the order of these methods, their efficiency, and the calculation of various errors, see [**4**] and [**5**].

The numerical experiments reported here and in the next section are the results of a FORTRAN program written in double precision and run on both a SUN Sparcstation 1 and an Alliant FX/40 with identical results up to machine roundoff error.

In the following, we have used the 2-stage Gauss-Legendre method ($q = 2$) with cubic splines ($r = 4$). We also have set $p = 5$ and $x_0 = .5$ so that the wave is initially centered at $x = 1/2$. According to the theory in [**3**], $p = 5$ is in the range where solitary-waves are unstable.

In the first test, we took $\epsilon = 1 \cdot 10^{-4}$ and $A = .8$ as the amplitude. For spatial and temporal mesh sizes the values $h = 1/384$ and $k = 1/100$, respectively, were used. The numerical solution was computed up to $t = 18$. In Figure 1 plots of the numerical solution are given at $t = 0, t = 6, t = 12$, and $t = 18$. It is evident from the graphs that the mesh parameters used were sufficient to numerically approximate the travelling-wave solution.

Next, the size of the initial data was increased to $A = 2.0$ for the amplitude. Also, we took $\epsilon = 5 \cdot 10^{-4}$. The step size was reduced to $k = 1/5000$ and the spatial mesh remained at $h = 1/384$. Output of our code in this case are depicted in Figure 2. As can be seen from the graphs, the instability of the solitary-wave required only a small

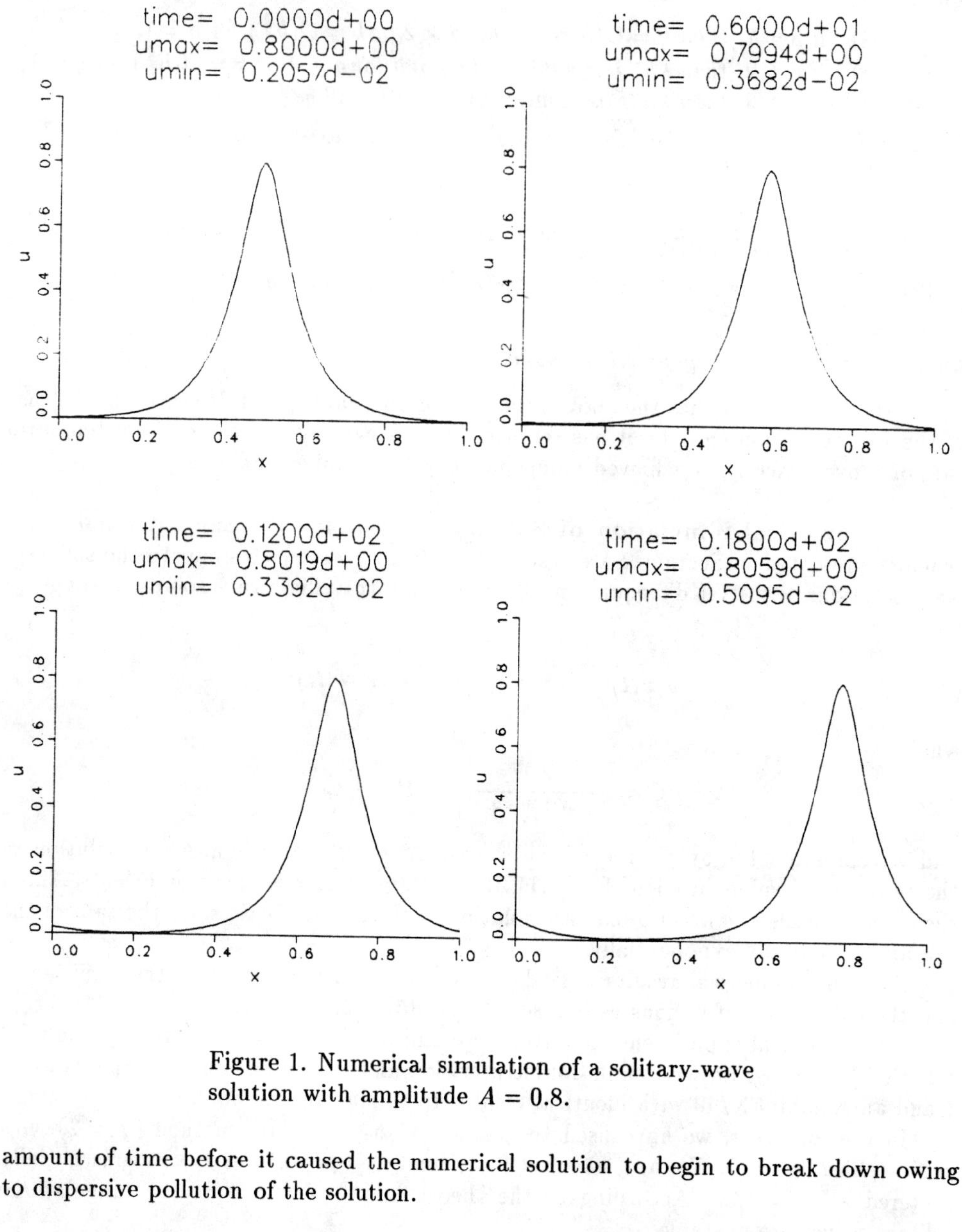

Figure 1. Numerical simulation of a solitary-wave solution with amplitude $A = 0.8$.

amount of time before it caused the numerical solution to begin to break down owing to dispersive pollution of the solution.

4. Grid Refinement. As the results of the previous section indicate, the instability of the solitary-wave solutions leads to substantial numerical errors when using fixed grids. To overcome this difficulty, an algorithm with automatic grid refinement was implemented so that one can approximate a solution $u(x,t)$ of (1.1) that develops a singularity in L^∞ at some point (x^*, t^*). The adaptive mechanism in our code consists of three main parts:

(i) local refinement of the spatial grid,
(ii) selection of a temporal step size k,
(iii) spatial translations of the solution.

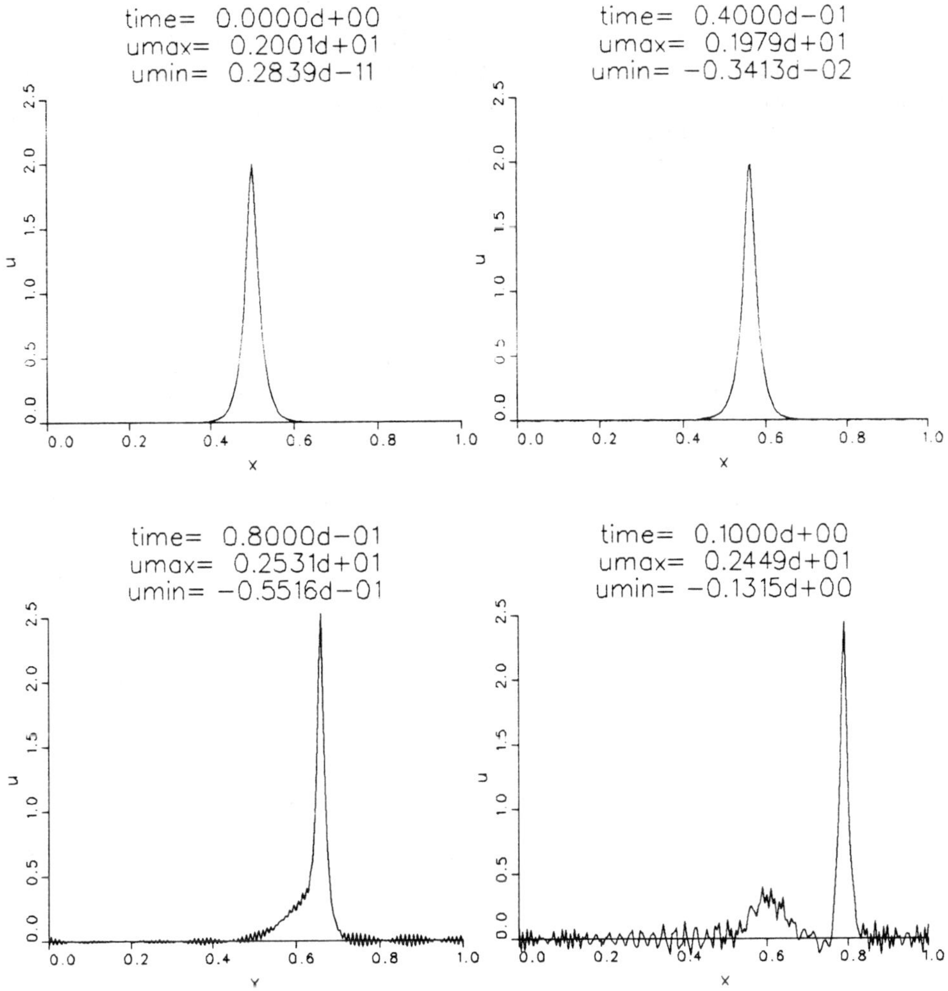

Figure 2. Numerical simulation of a solitary-wave solution with amplitude $A = 2.0$.

Our spatial refinements will consist of adding new nodes, distributed evenly about the midpoint $x = .5$, but in successively smaller neighborhoods of the midpoint. This local refinement is combined with (iii), which takes advantage of the fact that the solution is a travelling-wave, to keep the peak near the midpoint $x = .5$ in the region of highest density of nodes, and away from a region of coarse mesh. This is accomplished by occasionally translating the solution and centering the peak at $x = .5$. (In effect, we translate the solution to conform to fixed grids rather than have a moving grid which conforms to the solution.)

We next describe the automatic implementation of (i).

4.1 Spatial Grid Refinement. Let *NSPLIT* stand for the number of times nodes are to be added and let Ω^* represent the neighborhood of the midpoint $x = .5$ which is the region with the finest grid, with grid size h^*. Each time nodes are added, *NSPLIT* is increased by 1, and both Ω^* and h^* are cut in half. As an example, with *NSPLIT* = 1, we might have $\Omega^* = [0.4, 0.6]$ and $h^* = 1/100$. Then as *NSPLIT* increases to 2, $\Omega^* = [0.45, 0.55]$, $h^* = 1/200$; when *NSPLIT*= 3, $\Omega^* = [0.475, 0.525]$, $h^* = 1/400$, and so on. The effect of the occasional spatial translations is to insure that the peak of the solution remains in the region Ω^*.

From the $L^\infty - L^2$ inverse inequality in (2.2) and the discrete conservation law (2.10), it follows that for a uniform mesh

$$\|u_h^n\|_{0,\infty} \le ch^{-1/2}\|u_h^n\| = ch^{-1/2}\|\pi_h u^0\| \quad \textit{for} \quad n = 0, ..., J. \tag{4.1}$$

Similarly, for a non-uniform grid the minimal grid size h^* must become arbitrarily small if the numerical solution is to develop an arbitrarily large peak. Our particular choice of spatial refinement is based on a local $L^\infty - L^2$ inverse property on Ω^*. Specifically, we use the following test to determine if *NSPLIT* needs to be increased. At each step, compute

$$Z_\infty = \|u_h^n\|_{0,\infty} \quad \text{and} \quad Z_2 = \left[\int_{\Omega^*} (u_h^n)^2\, dx\right]^{1/2}.$$

We then increase *NSPLIT* by 1, add new nodes, and modify Ω^* and h^* if

$$\frac{Z_\infty\sqrt{h^*}}{Z_2} > TOL_1. \tag{4.2}$$

Here TOL_1 must be chosen small enough in accordance with (4.1) to allow for new nodes to be added. In our program, we have used $TOL_1 = 0.2$.

We next describe (ii), the selection of a temporal step size k.

4.2 Temporal Step Size Reduction. The temporal step size is adjusted in an attempt to preserve the third invariant of the GKdV equation. This invariant is defined by

$$I_3(v) = \int_0^1 [v^{p+2} - \frac{(p+1)(p+2)\epsilon}{2}(v_x)^2]\, dx.$$

For exact solutions $u(x,t)$ of (1.1), $I_3(u(\cdot,t)) = I_3(u^0)$, independent of the value of t.

Given u_h^n, a possible u_h^{n+1} is computed using the current step size k. It is accepted if

$$\frac{|I_3(u_h^{n+1}) - I_3(u_h^n)|}{\int_0^1 [(u_h^{n+1})_x]^2\, dx} < TOL_2 \tag{4.3}$$

where TOL_2 is a small parameter. In the results to be presented, the value $TOL_2 = 1\cdot 10^{-5}$ was used. If (4.3) is not satisfied, then k is cut in half and the process is repeated. The denominator in (4.3) is a convenient normalization factor.

4.3 Numerical Solutions with Grid Refinement. The experiment discussed in Section 3 to simulate the solitary-wave solution with large initial data was repeated using (4.2) and (4.3) for spatial and temporal grid refinement, respectively. These experiments showed that the solution was unstable; such instabilities being arguably precipitated by roundoff and truncation errors. To hasten the development of the instability, it was found convenient to use as initial data

$$u^0(x) = 1.01A\, sech^{2/p}[B(x - x_0)]$$

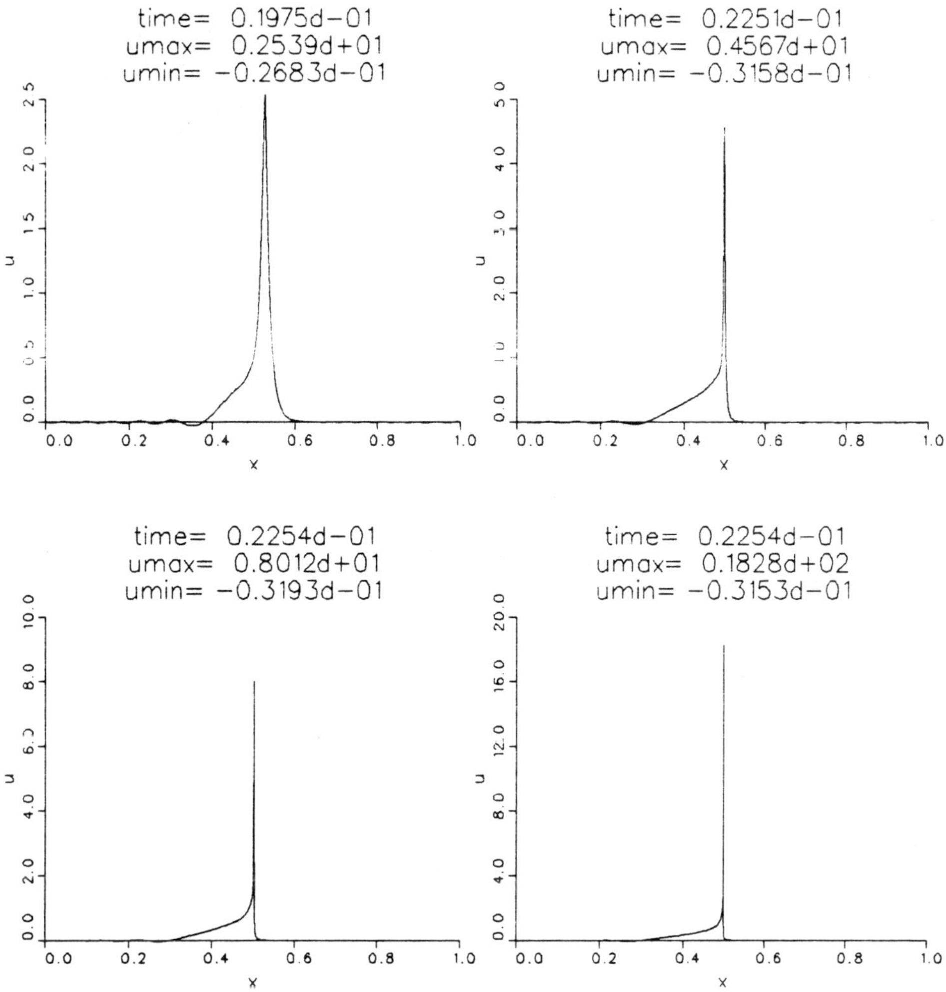

Figure 3. Numerical simulation of a solitary-wave solution with grid refinement, u-axis scaled.

with $A = 2.0, p = 5, \epsilon = 5 \cdot 10^{-4}$, and $x_0 = 1/2$. Again, use was made of the 2-stage Gauss-Legendre method with cubic splines ($q = 2, r = 4$). The initial mesh parameters were $h = 1/192$ and $k = 1/1000$. The graphs in Figure 3 illustrate the growth in the peak of the solution. The plots given are at *NSPLIT*$= 2, 4, 6, 9$. Note that as the peak (*umax* in the plots) significantly increases, we are forced to rescale the vertical axis in order to plot the entire solution. This has the effect of making the numerical solution u_h^n appear to be zero almost everywhere except at the peak, which, due to the translations, is kept near the midpoint $x = 1/2$. Indeed, more plots of this type would convey very little information.

In order to examine the nature of the solution near the peak as the peak continues to grow, one needs to rescale the horizontal axis. Recall that Ω^* is the region of finest grids with center at $x = .5$ and define $\Omega^0 = \Omega^* - 0.5$. In Figure 4 the solution is translated by 0.5 so that the peak is near 0 and then graphed using Ω^0 as the x-axis. The plots given

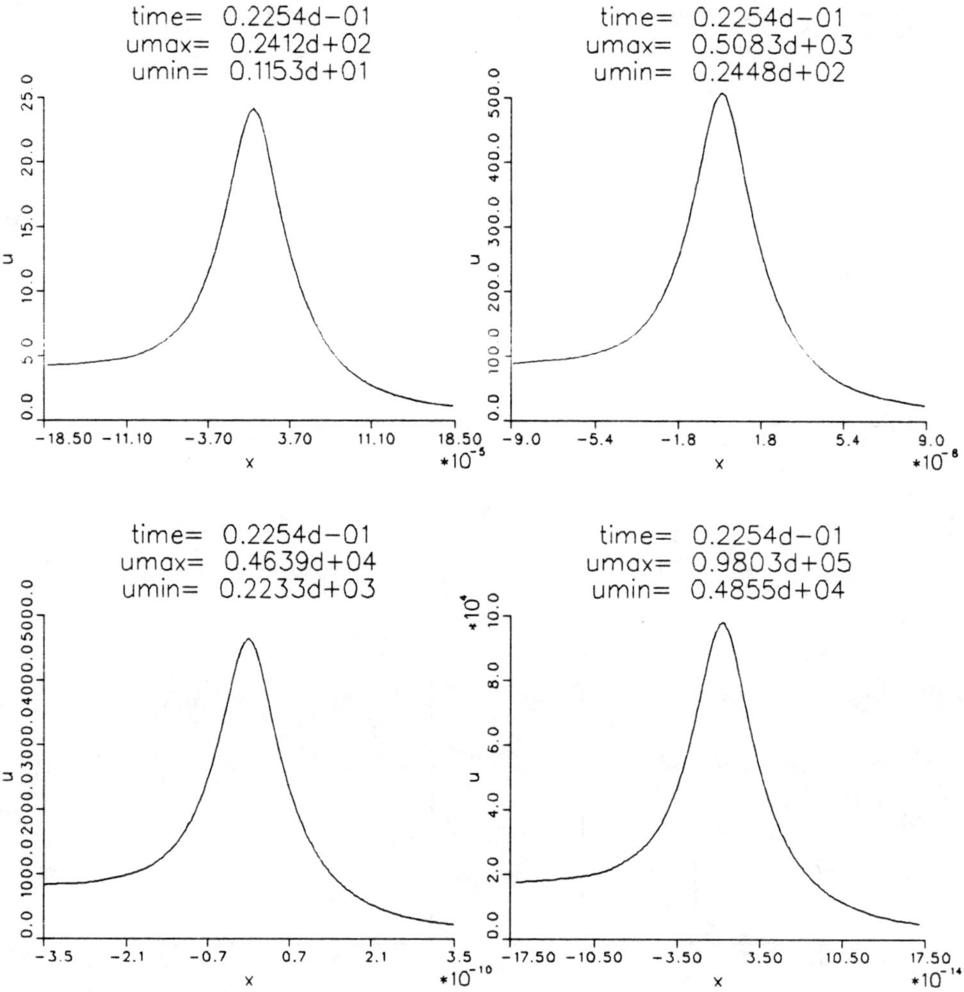

Figure 4. Numerical simulation of a solitary-wave solution with grid refinement, both axes scaled.

are with *NSPLIT*= 10, 21, 29, 40. By the final plot, the spatial and temporal mesh sizes have decreased to approximately $h \approx 10^{-14}$ and $k \approx 10^{-38}$ respectively. Our code is able to follow the peak until it reaches approximately 200,000 while maintaining a smooth profile everywhere including Ω^*. The graphs in Figure 4 suggest that instabilities of solitary-waves can lead to blowup of solutions in L^∞ in finite time.

REFERENCES

1. T. Kato, *On the Cauchy problem for the (generalized) Korteweg-de Vries equation*, Studies in Appl. Math., Advances in Mathematics Supplementary Studies, Academic Press, NY **8** (1983), 93-130.

2. J.L. Bona and R. Smith, *The initial-value problem for the Korteweg-de Vries equation*, Philos. Trans. Roy. Soc. London Ser. A **278** (1975), 555-604.

3. J.L. Bona, P.E. Souganidis and W.A. Strauss, *Stability and instability of solitary waves of KdV type*, Proc. Royal Soc. London **A 411** (1987), 395-412.

4. J.L.Bona, V.A. Dougalis, O.A. Karakashian and W.R. McKinney, *Conservative high order schemes for the Generalized Korteweg-de Vries equation*, submitted.

5. J.L.Bona, V.A. Dougalis and O.A. Karakashian, *Fully-discrete Galerkin methods for the Korteweg-de Vries equation*, Comp. & Maths. with Appls. **12A No. 7** (1986), 859-884.

6. G.A. Baker, V.A. Dougalis and O.A. Karakashian, *Convergence of Galerkin approximations for the Korteweg-de Vries equation*, Math. Comp. **40** (1983), 419-433.

7. K. Dekker and J.G. Verwer, *Stability of Runge-Kutta methods for stiff nonlinear differential equations, CWI monographs*, North Holland, Amsterdam (1984).

8. O.A. Karakashian and W.R. McKinney, *On optimal high order in time approximations for the Korteweg-de Vries equation*, Math. Comp., to appear.

CHAPTER 2

Regularizations of the Inviscid Burgers Equation*

M. Brio**
M. Temple-Raston†

Abstract
The inviscid Burgers equation with $\sin x$ initial-values has infinite derivatives at time equal to 1. We show that the Kepler problem and the Burgers equation have a surprising amount in common. Their similarities are exploited to give two regularizations of the inviscid Burgers equation, one 'elastic' and the other 'inelastic'.

Introduction

In continuum mechanics the inviscid Burgers equation is usually derived by a multi-scaling argument, which to first order of approximation describes the motion on the advection scale. By the method of characteristics, the initial-value problem, $u(0,x) = f(x)$, for the inviscid Burgers equation

$$u_t - uu_x = 0, \tag{1}$$

has an analytic solution of the form:

$$u(x,t) = f(x+ut) \equiv f(\phi). \tag{2}$$

ϕ is the characteristic (Lagrangian) variable. One can verify that the derivatives of $u(x,t)$ appearing in the Burgers equation (1), however, are infinite when $t = t_b \equiv 1/\max(f'(\phi))$. The break-time, t_b, identifies a weak discontinuity in the inviscid Burgers equation, that is, a discontinuity in the derivatives. At the break-time the inviscid Burgers equation (1)

*M. B. acknowledges partially supported from an NSF grant DMS-8810782.
**Department of Mathematics, University of Arizona, Tucson, Arizona, 85705 USA
†Department of Mathematics and Statistics, Concordia University, 7141 Sherbrooke Street West, Montréal, Québec, Canada H4B 1R6

is clearly meaningless. To extend solutions to the inviscid Burgers equation beyond t_b regularizations are introduced. Regularizations are designed either to prevent the infinite derivatives from occurring, or to prescribe the continuation of the solution through the discontinuities. We review briefly the known methods of regularization.

One approach used to prevent the steepening of the waveform is to argue physically that dispersion, diffusion or dissipation play a greater rôle at steep regions. At some point in the evolution, one adds to the inviscid Burgers equation some of these terms, or all of them, to give an equation which has the property that the waveform never reaches a discontinuity. Alternatively, one may permit the discontinuity. In this case, the regularization involves rewriting the inviscid Burgers equation as an integral equation. The inviscid Burgers equation in integral form is better suited to infinite derivatives, because it is not sensitive to questions of continuity or differentiability. Solutions to the integral equation are called 'weak solutions'.

Our interest in this paper is in attainable weak discontinuities, typified by sound waves, stress waves, electromagnetic waves, and water waves. Strong discontinuities, discontinuities in the solution as well as in the derivative, typical in gas dynamics shall not be considered here. Therefore we restrict our study to those situations for which dispersive, diffusive or dissipative effects are found not to contribute significantly to the prevention of infinite derivatives. This is a preliminary report.

In the next section we consider the Bessel-Fubini solution to the inviscid Burgers equation with sine initial-values, as a generalized solution for all time, t. We examine the evolution numerically. We also extend the form of solution to a larger class of initial-values. Experimental comparison will appear elsewhere. This is the first of two regularizations we propose in this paper. Our second regularization will not assume that the waveform is expressible in a Fourier series. We are interested here in non-linear waveforms. Our approach in this section is to make use of a close relationship between the Burgers equation and the Kepler problem (the classical problem of two massive body dynamics for a $1/r^2$ force field). This relationship and the known regularization of the Kepler problem suggest that there are regularizations which are complex analytic continuations of time.

1. The Bessel-Fubini solution

The inviscid Burgers equation (1) with initial-values $u(0,x) = \sin x$ has an explicit Fourier series solution, the Bessel-Fubini solution (e.g., [1]),

$$u(x,t) = 2\sum_{n=1}^{\infty} \frac{J_n(nt)}{nt} \sin nx. \tag{3}$$

The $J_n(x)$ are the Bessel functions. It is interesting that Bessel invented the $J_n(x)$ and found the series (3) as a solution to a another problem: the Kepler problem.

A number of things are easily deduced about the series (3). Most importantly, this series is convergent not only for $t < 1 = t_b$, but also beyond the break-time. For $t \leq 1$, the following inequalities hold

$$J_n(nt) < \begin{cases} n^{-1/2}\,(1-t^2)^{-1/4}, & t<1, \\ n^{-1/3}, & t=1. \end{cases} \tag{4}$$

For $t > 1$,

$$J_n(nt) \sim n^{-1/2}(t^2-1)^{-1/4}(\pi/2)^{1/2}\cos\left[n(\sqrt{t^2-1} - \cos^{-1}(1/t)) - \pi/4\right], \tag{5}$$

for $n \to \infty$ [2]. From (4) and (5), the series solution (3) can be seen to be bounded from above by a convergent series which is independent of x, so that (3) is uniformly convergent in x for arbitrary t. In addition, therefore, $u(x,t)$ is continuous in x.

We propose that (3) is a generalized solution to the inviscid Burgers equation with sine initial data. This is analogous to the definition of a generalized solution to the wave equation by a convergent Fourier series. In the case of the wave equation, the generalized solution is the same as the unique weak solution in the distributional sense [3]. It is easily seen that our generalized solution is not a weak solution.

The Bessel-Fubini series in (3) can be truncated to give a good approximation of the evolution of $\sin x$. In Figure 1 we have summed the first 400 terms for representative values of time. Our results are not sensitive to the number of terms taken in the sum. Note the attenuation.

A similar approach for a generalized solution can be used for an arbitrary periodic or non-periodic initial data. For simplicity, we assume an odd initial function $f(x)$. The formulæ are as follows (see appendix),

$$u(x,t) = \sum_{n=1}^{\infty} \frac{F_n(nt)}{nt} \sin nx \tag{6}$$

where $F_n(z)$ is defined as

$$F_n(z) = \frac{1}{\pi}\int_0^{\pi} \cos(nx - zf(x))dx. \tag{7}$$

In order to conclude the convergence of (6), we rewrite $F_n(nt)$ as

$$F_n(nt) = \frac{1}{\pi} Re \int_0^\pi e^{in(x-tf(x))} dx = \frac{1}{\pi} Re \int_0^\pi e^{in\tau(x)}\, dx. \tag{7'}$$

If the interval $(0, \pi)$ can be split into a finite number of intervals where $\tau(x)$ is smooth and monotonic, then [4]

$$F_n(nt) \sim \frac{1}{n^{1/p}}, \qquad p > 0, \tag{10}$$

where p is related to the order of the first nonvanishing derivative at the points of non-differentiability, and is one at the discontinuities of the initial function $f(x)$. In any case, if $f(x)$ is such that (7′) exists, for example $f(x)$ is L_1, then the series (6) is convergent to some L_2 function since the squares of the coefficients of (6) form a convergent sequence. Convergence is in a mean-square sense and is also almost everywhere pointwise [5].

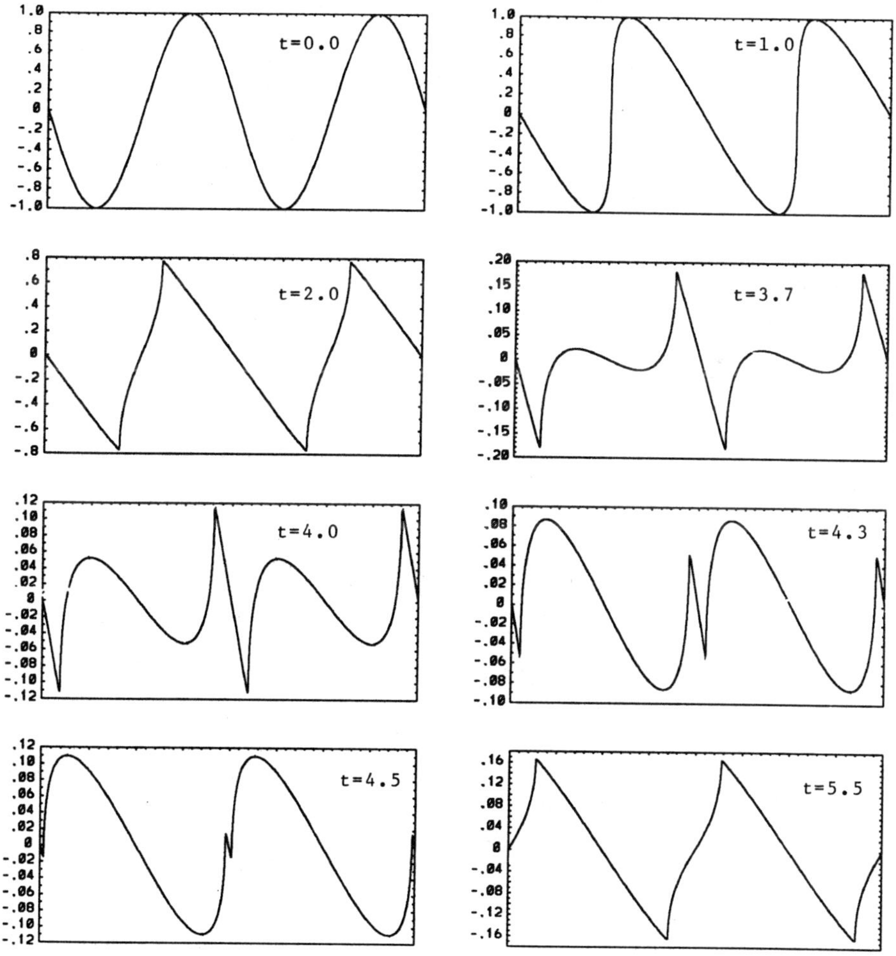

Fig. 1 The evolution of $\sin x$ by the inviscid Burgers equation at times t =0.0, 1.0, 2.0, 3.7, 4.0, 4.3, 4.5, 5.5. The break-time is $t_b = 1.0$.

2. The Kepler problem and the Burgers equation

In the previous section we remarked that the Bessel-Fubini series is a solution to both the Burgers and Kepler problems. The coincidence in solution is due to an algebraic equation, the Kepler equation, which is found to be fundamental to both. We shall show that the Kepler equation is the characteristic equation for the inviscid Burgers equation with sine initial-value. This explains why the solutions to both problems are the same. It is clear that algebraic equations do not require the differentiability that differential equations do, so we shall assume that the algebraic equation contains the information for the continuation of the solution through the infinite derivatives. Indeed, the Kepler equation has been used to continue Kepler motions through particle collisions [7,8]. In a similar manner, we propose to use the Kepler equation to regularize the Burgers equation. In other words, the geometry of the Kepler problem prompts the regularization presented below.

The Kepler problem, which we review here, describes the motion of two massive bodies governed by an attractive force, $F = -k/r^2$, where r is the spatial distance between the masses. We derive the Kepler equation from the differential equations of motion. Integrating the equations of motion for time, t, (e.g., see [6]) we find that

$$t = \sqrt{\frac{m}{2}} \int \frac{dr}{\sqrt{E + k/r - l^2/2mr^2}}. \tag{11}$$

The angular momentum, l, is related to an expression involving the energy, E, the eccentricity, e, and the mass, m: $l^2/2m = k^2(e^2-1)/4E$. The Kepler equation follows from the change of variable,

$$r = -\frac{k}{2E}(1 - e\sqrt{1-u^2}). \tag{12}$$

Equation (12) is physically sensible when $E < 0$ and $e \leq 1$. Equation (11) is thus integrated to give

$$t = \frac{k\sqrt{m}}{(-2E)^{\frac{3}{2}}}(\arcsin u - ue),$$

or,

$$\kappa = \arcsin u - ue, \tag{13}$$

where we have defined $\kappa \equiv t(-2E)^{3/2}/k\sqrt{m}$. Equation (13) is the Kepler equation. Once k and E are specified, solving the Kepler equation for the intermediate variable u is equivalent to knowing r. If there is conservation of orbital angular momentum, then the dynamics can be shown to be elliptical motion, possibly degenerate.

Rewriting equation (13) to read as

$$u = \sin(ue + \kappa) \equiv \sin\phi, \tag{14}$$

we observe that the Kepler equation is also the general form of solution to the inviscid Burgers equation, $u_e - uu_\kappa = 0$, with initial-values $\sin\kappa$ (see (2)). For Kepler motion, the eccentricity e is fixed and κ, which is proportional to time, monotonically increases. But when $u(e,\kappa)$ is interpreted as a velocity distribution, as it is for the Burgers equation, the characteristic equations

$$\frac{du}{dt} = 0, \quad \frac{de}{dt} = 1, \quad \frac{d\kappa}{dt} = -u,$$

tell us that the eccentricity is the time variable ($e = t$), and that κ is the spatial variable ($\kappa = x$).

In the Kepler problem particle collisions can occur when $e = 1$. The regularization through collisions is called the 'reflection principle' and can be understood from the geometrical picture of Kepler motion as first formulated by Hamilton [7], and recently updated by Milnor [8]. Having fixed the eccentricity, the Kepler equation correctly reproduces the 'reflection principle'. The inviscid Burgers equation with initial-values $\sin\kappa$, likewise, defines a dynamical flow in the space of Kepler motions where the flow is parametrized by the eccentricity, e. The infinite derivatives in the inviscid Burgers equation develop when $e = t_b = 1$ and $\kappa = 0$, but the algebraic equation is well-defined. When $e > 1$, the algebraic equation has triple-valued solutions for u. While triple-valuedness is not forbidden, the relationship to the Kepler problem is lost. However, it is well-known that when $e > 1$ the Keplerian orbits are no longer ellipses, but hyperbolæ. Thus we examine the hyperbolic Kepler equation.

The Kepler equation for hyperbolic orbits ($e > 1$) is somewhat different, but is derived in a manner similar to the elliptic Kepler equation (13). In this case, the change of variables is $r = k(1 + e\sqrt{1+v^2})/2E$, and the integral in (11) is evaluated to give

$$\alpha = ve - \operatorname{arcsinh} v, \tag{15}$$

where $\alpha = t(2E)^{3/2}/k\sqrt{m}$. This is the hyperbolic Kepler equation.

The hyperbolic Kepler equation (15) gives the characteristic equation, $v = \sinh(ve - \alpha)$, for the inviscid Burgers equation

$$v_e + vv_\alpha = 0, \tag{16}$$

with $\sinh\alpha$ initial-values. The initial-values are not important here, but the differential equation (16) which defines the evolution for $e > 1$ is useful. Comparing (1) and (17) we conclude that the continuation which maintains the relationship to the Kepler problem is equivalent to time reversal of the development prior to $e = 1$. Thus the regularization prevents further distortion of the sine initial-value problem and returns the system back to $\sin x$. The motion can be extended by making e cyclic with period 2.

The elliptic and hyperbolic Kepler equations are the same equation defined over $\mathbf{C}^2 \times \mathbf{R}$. That is, the hyperbolic and elliptic variables are related through the equations $v = iu$, $\alpha = -i\kappa$, and $e = e$. Therefore the regularization suggested in this section is the proposal that one use the characteristic equations to solve for time as a function of u and x, complex analytically continue time, and restrict to evolutions which keep time real and monotonically increasing.

3. Conclusion

A more complete account of this work shall follow, but we should like to say a few words about experiment. The similarities between the Burgers equation and the general theory of surface waves can form the basis of a comparison to both numerical and laboratory experiment. Waveform evolution for nonlinear elastic surface acoustic waves (SAWs) have been studied in [9], most of these articles use the method of multiple scales. Numerical methods applied to the truncated ODEs resulting from the integral representation of the surface acoustical equations, steepen in the same way as for the Burgers equation [10]. Moreover, Parker in [10] has argued that the equations will indeed have a finite break-time. Therefore the regularizations in this paper may have application, for instance, to the theory of surface acoustic waves.

Acknowledgements

We would like to thank John Hunter for bringing to our attention the question of the meaning of the Bessel-Fubini series beyond the break-time. M.B. acknowledges partially supported from an NSF grant DMS-8810782.

Appendix

We repeat the derivation of the Bessel-Fubini solution by replacing the sine initial-values with a C^1, odd, 2π-periodic function $f(x)$, with $f(x)$ vanishing on the boundary. The same steps can be applied to a full Fourier series or transform. Recall the characteristic equations (5)

$$x(\xi,t) = \xi - ut,$$
$$u(x,t) = f(\xi),$$

and assume that

$$u(x,t) = \sum_{n=1}^{\infty} b_n(t) \sin nx,$$

where x is real. Then,

$$b_n(t) = \frac{2}{\pi} \int_0^\pi u(x,t) \sin nx \, dx = \frac{2}{\pi} \int_0^\pi f(\xi) \sin[n\, x(\xi,t)] \frac{\partial x}{\partial \xi} \, d\xi.$$

Integrating by parts gives

$$b_n(t) = -\frac{2}{\pi n t} \int_0^\pi (1 - tf'(\xi)) \cos[n\, x(\xi,t)] \, d\xi + \frac{2}{\pi n t} \int_0^\pi \cos[n\, x(\xi,t)] \, d\xi.$$

The first of these integrals vanishes. By analogy with the definition of the Bessel function, let

$$F_n(y) = \frac{1}{\pi} \int_0^\pi \cos[n\xi - yf(\xi)] \, d\xi,$$

so that

$$u(x,t) = 2 \sum_{n=1}^{\infty} \frac{F_n(nt)}{nt} \sin nx.$$

References

[1] D.T. Blackstock, *Propagation of Plane Sound Waves of Finite Amplitude in Non-dissipative Fluids*, J. Acoust. Soc. Am., 34 (1962), pp.9-30.

[2]G.N. Watson, *A Treatise on the Theory of Bessel Functions*, 2nd ed., (Cambridge, 1944).

[3]S.G. Mikhlin, *Mathematical Physics; an Advanced Course*, North Holland, (Amsterdam, 1971).

[4] L. Sirovich, *Techniques of Asymptotic Analysis*, Springer-Verlag, (New York, 1971).

[5]R.E. Edwards, *Fourier Series*, 2nd ed., Springer-Verlag, (New York, 1982).

[6] H. Goldstein, Classical Mechanics, 2nd ed., Addison-Wesley, (Reading, 1980).

[7] W.R. Hamilton, *Proc. Roy. Irish Acad.*, **3** (1846) pp.344-353 (Math. Papers v.2, 287-294, Cambridge University Press, 1940).

[8] J. Milnor, *On the geometry of the Kepler problem*, Amer. Math. Monthly, **90** (1983), pp.353-365.

[9] R.W. Lardner, *Nonlinear surface waves on an elastic solid*, Int. J. Eng. Sci., 21 (1983), pp.1331-1342.

[10] D.F. Parker, *Waveform evolution for nonlinear surface acoustic waves*, Int. J. Eng. Sci., **26**, No. 1 (1988)pp.59-75.

CHAPTER 3

Upwind Simulation of a Steady-State Electron Shock Wave in a Submicron Semiconductor Device

Carl L. Gardner*

Abstract

The hydrodynamic model treats electron flow in a semiconductor device through the Euler equations of gas dynamics, with the addition of a heat conduction term. Thus the hydrodynamic model PDEs have hyperbolic, parabolic, and elliptic modes.

The nonlinear hyperbolic modes support shock waves. Numerical simulations of a steady-state electron shock wave in a submicron semiconductor device are presented, using a steady-state second upwind upwind method. For the $n^+ - n - n^+$ diode (which models the channel of a MOSFET), the shock wave is fully developed in Si (with a 1 volt bias) at 300 K for a 0.1 micron channel and at 77 K for a 1.0 micron channel.

1 Introduction

The hydrodynamic model equations consist of a set of nonlinear conservation laws for particle number, momentum, and energy, coupled to Poisson's equation for the electric potential. The nonlinear conservation laws are just the Euler equations of gas dynamics for a gas of charged particles in an electric field, with the addition of a heat conduction term [1]. Thus the hydrodynamic model PDEs have hyperbolic, parabolic, and elliptic modes.

The nonlinear hyperbolic modes support shock waves. Electron shock waves should occur in silicon devices at short length scales or at low temperatures. Steady-state second upwind simulations of a steady-state electron shock wave in a submicron semiconductor device will be presented, using the hydrodynamic model. The simulations were performed for the $n^+ - n - n^+$ diode (which models the channel of a MOSFET). With a 1 volt bias across the diode, the shock wave is fully developed in Si at 300 K for a 0.1 micron channel and at 77 K for a 1.0 micron channel [2].

I give a brief introduction to the hydrodynamic model in section 2, and describe the nonlinear waves that the model supports. In section 3, I describe the one-dimensional

*Department of Computer Science, Duke University, Durham, NC 27706. Research supported in part by NSF grant DMS-8905872.

$n^+ - n - n^+$ diode problem, and discuss the second upwind discretization of the 1D steady-state hydrodynamic equations. In section 4, I compare central difference and second upwind computations for steady-state subsonic flow of electrons in the $n^+ - n - n^+$ diode. In section 5, I discuss transonic electron flow in the $n^+ - n - n^+$ diode using the second upwind method. Section 5.1 presents a steady-state Mach 1.6 electron shock wave in Si at 300 K, and section 5.2 presents a steady-state Mach 4.1 electron shock wave in Si at 77 K. These electron shock waves are analogous to the shock wave that forms in a converging-diverging Laval nozzle for transonic gas flow.

2 The hydrodynamic model

A fairly general set of transport equations for device simulation has been derived by Bløtekjær [3] from the Boltzmann equation:

$$\frac{\partial n}{\partial t} + \nabla\cdot(n\mathbf{v}) = 0 \tag{1}$$

$$\frac{\partial \mathbf{p}}{\partial t} + \mathbf{v}\nabla\cdot\mathbf{p} + \mathbf{p}\cdot\nabla\mathbf{v} = -en\mathbf{E} - \nabla(nT) + \left(\frac{\partial \mathbf{p}}{\partial t}\right)_c \tag{2}$$

$$\frac{\partial W}{\partial t} + \nabla\cdot(\mathbf{v}W) = -en\mathbf{v}\cdot\mathbf{E} - \nabla\cdot(\mathbf{v}nT) + \nabla\cdot(\kappa\nabla T) + \left(\frac{\partial W}{\partial t}\right)_c \tag{3}$$

where n is the electron density, $\mathbf{v}$ is the velocity, $\mathbf{p}$ is the momentum density, e (> 0) is the electronic charge, $\mathbf{E}$ is the electric field, T is the temperature in energy units, W is the energy density, κ is the heat conduction coefficient, and the subscript c indicates collision terms. I will assume that the energy bands are parabolic:

$$\mathbf{p} = mn\mathbf{v}, \quad W = \frac{3}{2}nT + \frac{1}{2}mnv^2 \tag{4}$$

where m is the effective electron mass.

In addition to the transport equations (1)–(3), we have Poisson's equation for the electric field

$$\nabla\cdot(\epsilon\nabla\phi) = -e(N_D - N_A - n), \quad \mathbf{E} = -\nabla\phi \tag{5}$$

where N_D is the density of donors and N_A is the density of acceptors. Eqs. (1)–(5) determine the variables n, $\mathbf{p}$, W, and ϕ.

Eqs. (1)–(3) are in conservation form, and may be written in terms of the variables n, $\mathbf{v}$, T, and ϕ. These variables represent the simplest choice for upwind methods and for the $n^+ - n - n^+$ diode boundary conditions.

If κ is set to zero in Eq. (3), then Eqs. (1)–(3) are the Euler equations with source terms due to the collision terms and the electric field. (The polytropic gas constant $\gamma = \frac{5}{3}$, as is appropriate for a monatomic gas.) These equations then are hyperbolic, with five nonlinear waves [4, 5]:

nonlinear wave	*characteristic speed*	*discontinuous variables*
2 shocks	$\lambda_\pm = v \pm c$	v, n, T
3 contacts	$\lambda_0 = v$	$\mathbf{v}_T$, n, T

v is the particle velocity normal to the wave, and $\mathbf{v}_T$ is the particle velocity tangential to the wave. The soundspeed $c = \sqrt{\gamma T/m}$.

If now we let κ be nonzero, Eqs. (1)–(3) become hyperbolic (4 modes) plus parabolic (1 mode) [6]. There are now only four nonlinear hyperbolic waves:

nonlinear wave	*characteristic speed*	*discontinuous variables*
2 shocks	$\lambda_\pm = v \pm c$	v, n
2 contacts	$\lambda_0 = v$	$\mathbf{v}_T$

The contact wave corresponding to a jump in T has disappeared due to the $\nabla\cdot(\kappa\nabla T)$ term in Eq. (3), and the soundspeed $c = \sqrt{T/m}$. For $T = \alpha T_0$ and $T_0 = 300$ K, $c = \sqrt{\alpha}\, 1.3 \times 10^7$ cm/s, while for $T_0 = 77$ K, $c = \sqrt{\alpha}\, 6.6 \times 10^6$ cm/s.

3 The $n^+ - n - n^+$ diode problem

As a model problem, I simulate the flow of electrons in a submicron $n^+ - n - n^+$ silicon diode. This device models the channel in a MOSFET, and clearly exhibits hot carrier effects at submicron scales. The diode begins with an n^+ "source" region, is followed by an n "channel" region, and ends with an n^+ "drain" region.

Following Baccarani and Wordeman [7], I take

$$\left(\frac{\partial \mathbf{p}}{\partial t}\right)_c = -\frac{\mathbf{p}}{\tau_p}, \quad \tau_p = m\frac{\mu_{n0}}{e}\frac{T_0}{T} \tag{6}$$

$$\left(\frac{\partial W}{\partial t}\right)_c = \frac{-\left(W - \frac{3}{2}nT_0\right)}{\tau_w}, \quad \tau_w = \frac{m}{2}\frac{\mu_{n0}}{e}\frac{T_0}{T} + \frac{3}{2}\frac{\mu_{n0}}{ev_s^2}\frac{TT_0}{T + T_0} \tag{7}$$

$$\kappa = \kappa_0\frac{\mu_{n0}}{e}nT_0 \tag{8}$$

where T_0 is the ambient temperature, $\mu_{n0} = \mu_{n0}(T_0, N_D + N_A)$ is the low field electron mobility, and $v_s = v_s(T_0)$ is the saturation velocity.

The simplest boundary conditions for the simulations are $n = N$ and $T = T_0$ at x_{min} and x_{max}, and $e\phi(x_{min}) = T\ln(n/n_i)$ and $e\phi(x_{max}) = T\ln(n/n_i) + eV$, where V is the bias across the diode and n_i is the intrinsic electron concentration.

The steady-state computations use the second upwind method [2], a conservative discretization which captures the essential physical transport property of supersonic flow, that advected quantities directly influence the solution only downstream.

The 1D steady-state hydrodynamic equations have the form

$$\begin{bmatrix} f_n \\ f_v \\ f_T \\ f_\phi \end{bmatrix} = \frac{d}{dx}\begin{bmatrix} vg_n \\ vg_v \\ vg_T \\ 0 \end{bmatrix} + \begin{bmatrix} 0 \\ h_v \\ h_T \\ h_\phi \end{bmatrix} + \begin{bmatrix} 0 \\ s_v \\ s_T \\ s_\phi \end{bmatrix} = 0 \tag{9}$$

where

$$g_n = n, \quad g_v = mnv, \quad g_T = \frac{5}{2}nT + \frac{1}{2}mnv^2 - en\phi \tag{10}$$

$$h_v = -en\frac{d\phi}{dx} + \frac{d}{dx}(nT), \quad h_T = -\frac{d}{dx}\left(\kappa\frac{dT}{dx}\right), \quad h_\phi = \epsilon\frac{d^2\phi}{dx^2} \tag{11}$$

$$s_v = \frac{mnv}{\tau_p}, \quad s_T = \frac{\frac{1}{2}mnv^2 + \frac{3}{2}n(T - T_0)}{\tau_w}, \quad s_\phi = e(N - n). \tag{12}$$

In the second upwind method, the advection terms $d(vg)/dx$ in Eq. (9) are discretized using second upwind differences

$$\frac{d}{dx}(vg)_i \approx (v_{i+\frac{1}{2}}g_R - v_{i-\frac{1}{2}}g_L)/\Delta x \tag{13}$$

where

$$g_R = \begin{cases} g_i & (v_{i+\frac{1}{2}} > 0) \\ g_{i+1} & (v_{i+\frac{1}{2}} < 0) \end{cases}, \quad g_L = \begin{cases} g_{i-1} & (v_{i-\frac{1}{2}} > 0) \\ g_i & (v_{i-\frac{1}{2}} < 0) \end{cases} \tag{14}$$

and central differences are used for h_v, h_T, and h_ϕ.

I use Newton's method to linearize the discretized version of Eq. (9):

$$J \begin{bmatrix} \delta n \\ \delta v \\ \delta T \\ \delta \phi \end{bmatrix} = - \begin{bmatrix} f_n \\ f_v \\ f_T \\ f_\phi \end{bmatrix} = -f, \tag{15}$$

where J is the Jacobian and f is the 4x1 residual. The new solution is obtained by setting

$$\begin{bmatrix} n \\ v \\ T \\ \phi \end{bmatrix} \leftarrow \begin{bmatrix} n \\ v \\ T \\ \phi \end{bmatrix} + t \begin{bmatrix} \delta n \\ \delta v \\ \delta T \\ \delta \phi \end{bmatrix} \tag{16}$$

where t is a damping factor [8] between 0 and 1, chosen to insure that the norm of the residual f decreases monotonically. Newton's equation may be solved directly by sparse matrix techniques, or by block iterative methods [9].

The steady-state shock simulations presented here are reproduced in Ref. [10] using a time-dependent "essentially non-oscillatory" (ENO) upwind scheme [11]. The ENO method uses an approximate Riemann solver to evolve the solution preceded by an adaptive non-oscillatory reconstruction of the solution, and is capable of resolving shocks over 2-3 grid points. The upwind stencil is chosen to minimize the likelihood of crossing a discontinuity. The scheme used in Ref. [10] is sixth-order accurate in space in regions of smooth flow. Steady-state equilibrium is achieved in a few thousand timesteps.

4 Computations for subsonic flow

For the subsonic computations, I take a diode consisting of a 0.1 micron source, a 0.4 micron channel, and a 0.1 micron drain. In the n^+ region, the doping density $N = 5 \times 10^{17}$ cm^{-3}, while in the n region $N = 2 \times 10^{15}$ cm^{-3}. The ambient device temperature $T_0 = 300$ K $= 0.0259$ eV. The saturation velocity $v_s = 1.03 \times 10^7$ cm/s. Note that $v_s < c(T_0) = 1.3 \times 10^7$ cm/s.

To validate the second upwind method for subsonic problems, I performed simulations (with 120 grid intervals) of the $n^+ - n - n^+$ diode at $V = 1$ volt using the hydrodynamic model. The simulations are in good agreement with other hydrodynamic and Monte Carlo simulations of the same device [12, 13, 14, 1], although there are differences due to the different numerical methods and the different models for τ_p, τ_w, and κ.

Fig. 1 compares the temperature of the electron gas for the central difference and second upwind methods. Note that the temperature increases as the electrons flow from left to right across the channel. The maximum temperature is five times higher than ambient. Recall that the drift-diffusion model assumes the electron temperature $T = T_0$ throughout the device. Also note the slight cooling of the electron gas as the electrons enter the channel. This cooling results from the electrons having to overcome a small potential barrier at the junction.

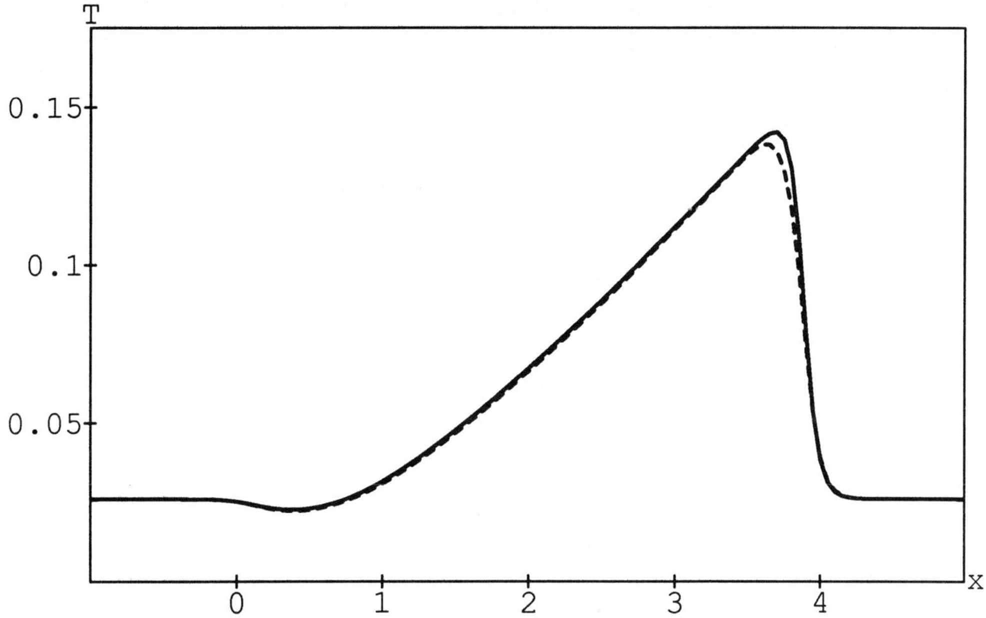

Figure 1: Electron temperature in eV for V = 1 volt, 0.4 micron channel, 300 K: central difference (solid) vs. second upwind (dotted). x is in 0.1 microns.

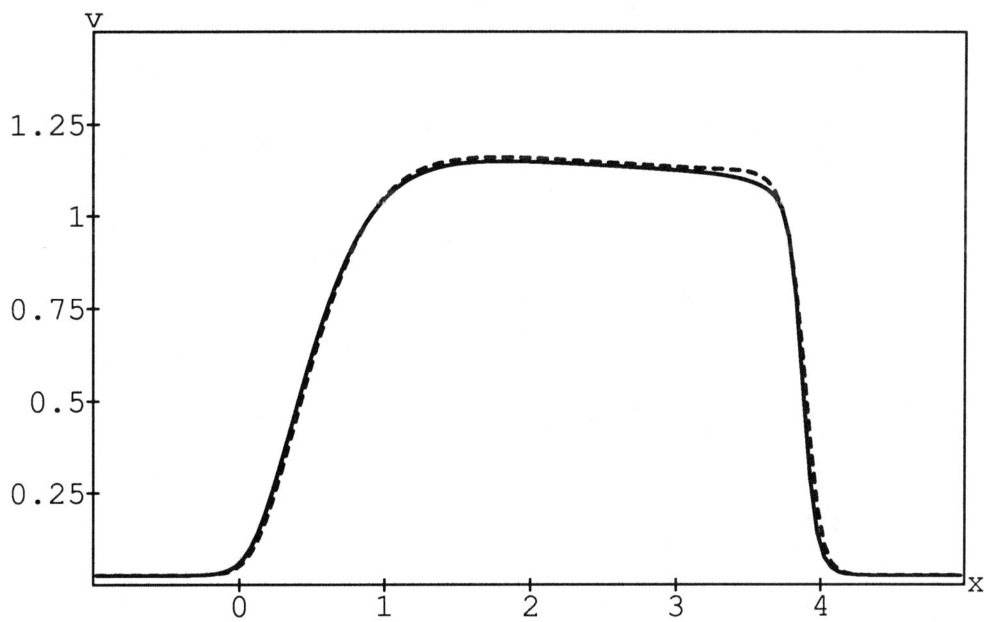

Figure 2: Electron velocity in 10^7 cm/s for V = 1 volt, 0.4 micron channel, 300 K: central difference (solid) vs. second upwind (dotted). x is in 0.1 microns.

In Fig. 2 I compare the electron velocity for the central difference and second upwind methods. Note the velocity overshoot phenomenon $v > v_s$ throughout most of the channel. Typical Mach numbers for these flows are $\lesssim 0.7$.

5 Computations for transonic flow

Higher electron velocities can be obtained by increasing the bias V across a device or by making the active device length shorter. Since the electron soundspeed $c = \sqrt{T/m}$, higher electron Mach numbers may also be obtained by lowering the ambient temperature T_0.

In this investigation, I present two parameter regimes for the $n^+ - n - n^+$ diode in which there is a transition from subsonic to supersonic electron flow just to the right of the $n^+ - n$ junction and from supersonic to subsonic flow to the left of the $n - n^+$ junction. In analogy with gas dynamical flow in a Laval nozzle[1], a shock wave develops at the transition from supersonic to subsonic flow. The $n^+ - n - n^+$ doping of the diode corresponds to the converging-diverging geometry of the Laval nozzle. For further analysis of the shock waves, see Ref. [2].

Since the $n^+ - n - n^+$ diode models the channel of a MOSFET, the development of this shock wave should have implications for MOSFET performance, perhaps as a threshold effect in the IV curves or switching time.

The hydrodynamic shocks simulated below have a more complicated structure than shocks supported by the Euler equations of gas dynamics, due to the heat conduction term $\nabla\cdot(\kappa\nabla T)$ in Eq. (3), the relaxation time source terms in Eqs. (2) and (3), and the coupling of the electron gas to the electric field. For the (inviscid) Euler equations, a shock wave is an exact discontinuity with a step function jump in n, v, and T, where v is the normal velocity of the gas [4]. In gas dynamics with heat conduction and viscosity, a shock wave varies rapidly but smoothly over a short viscous length scale [4, 16, 17]. (The hydrodynamic model neglects the effects of viscosity in the electron gas.) If there is heat conduction, but no viscosity, the situation is more complex [16, 17]. In the frame of the shock, the flow ahead of the shock is supersonic and behind the shock is subsonic. If the Mach number M of the flow ahead of the shock is greater than one but less than a critical Mach number M_c, the shock is a rapidly varying smooth profile. If $M > M_c$, then there is an outer profile to the shock wave with an inner exact discontinuity. ($M_c = v_c/c \approx 1.74$ for a polytropic gas with $\gamma = 5/3$, where $c = \sqrt{T/m}$.) In simulations with numerical viscosity, the inner discontinuity will be spread out over several grid intervals.

The transonic computations use the second upwind method. (The central difference method becomes unstable just as the maximum Mach number of the flow exceeds 1.)

5.1 Electron shock wave at 300 K

For the transonic computations at 300 K, I take a diode consisting of a 0.1 micron source, a 0.1 micron channel, and a 0.1 micron drain. In the n^+ region, the doping density $N = 5 \times 10^{17}$ cm^{-3}, while in the n region $N = 2 \times 10^{15}$ cm^{-3}.

Figs. 3–5 present a simulation of a Mach 1.6 electron shock wave at $V = 1$ volt (with 120 grid intervals). Fig. 3 shows the doping profile and the electron density. The electron density in the channel is almost two orders of magnitude greater than the background doping density for this very short device. The shock profile develops at $x_s \approx 0.035$ microns (see

[1] See Ref. [15] for a theoretical and numerical investigation of a steady-state shock wave in a Laval nozzle.

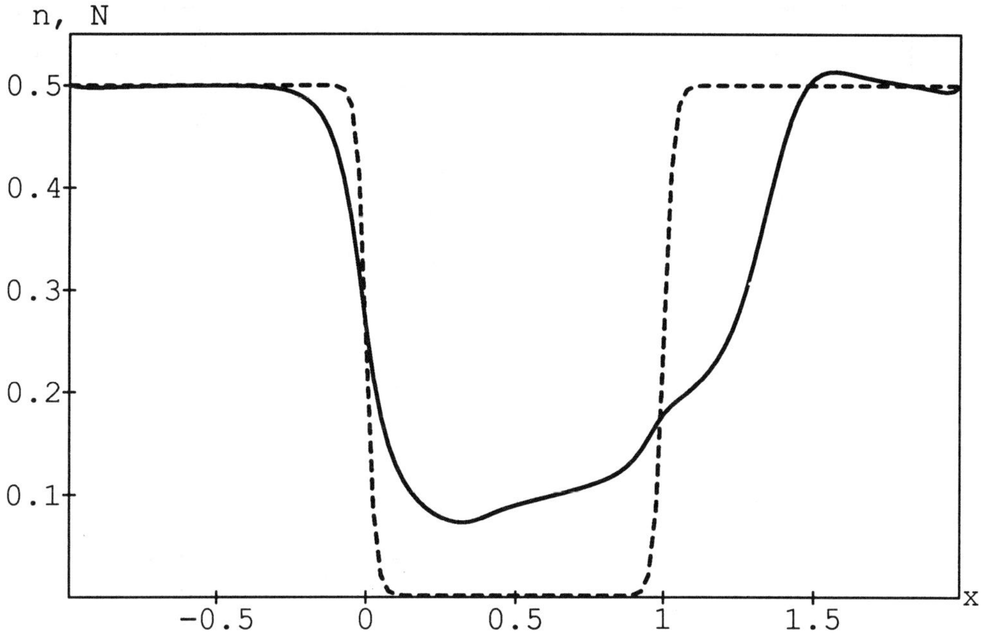

Figure 3: Doping profile (dotted) and electron density (solid) in 10^{18} cm^{-3} at V = 1 volt for 0.1 micron channel device at 300 K. x is in 0.1 microns.

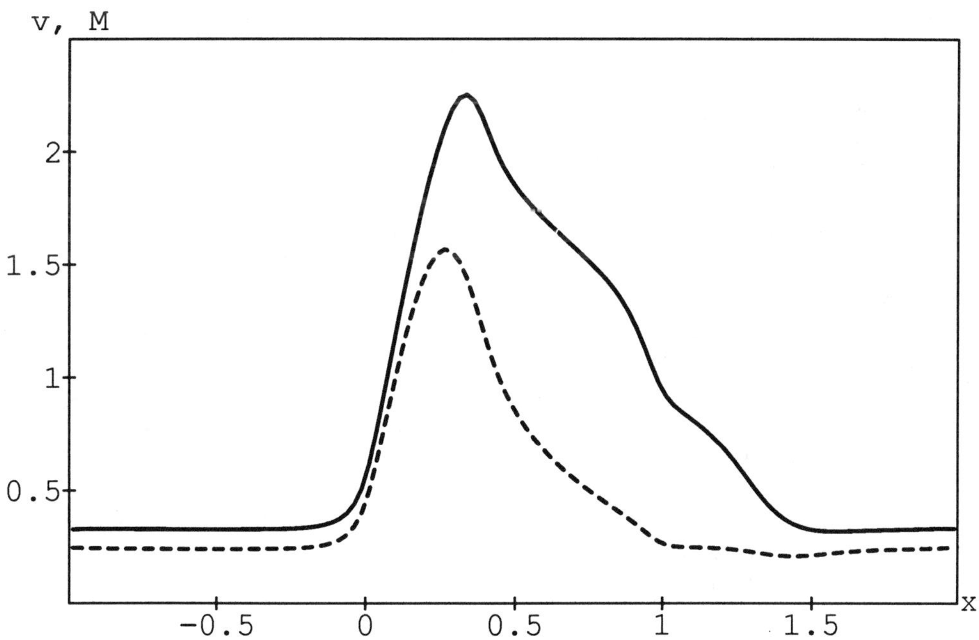

Figure 4: Electron velocity in 10^7 cm/s and electron Mach number (dotted) for V = 1 volt, 0.1 micron channel, 300 K. x is in 0.1 microns.

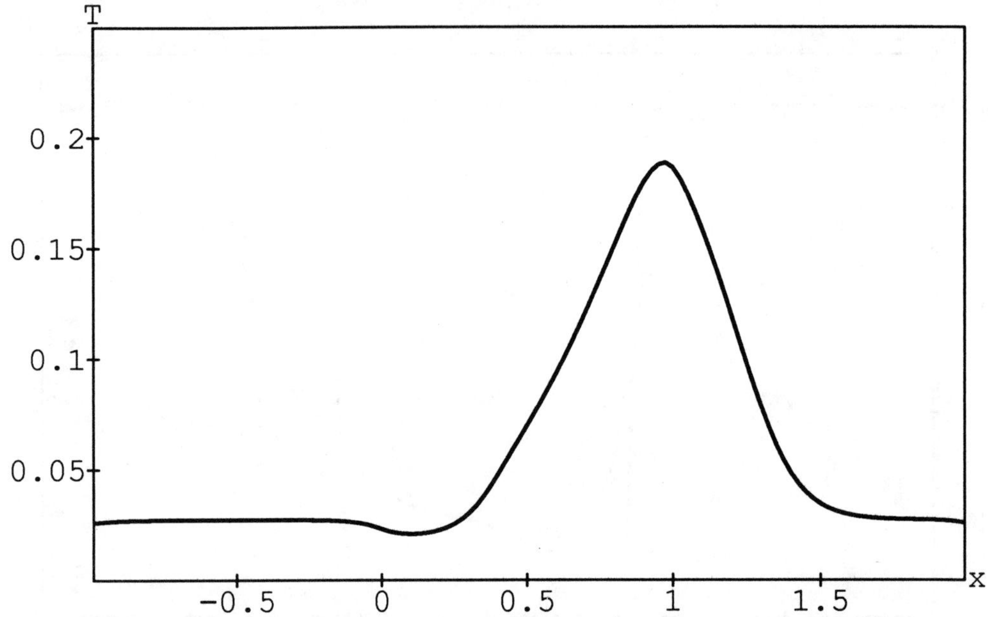

Figure 5: Electron temperature in eV for $V = 1$ volt, 0.1 micron channel, 300 K. x is in 0.1 microns.

Fig. 4) and is spread over about $6\Delta x$. The electron flow is subsonic behind the shock profile and supersonic ahead of the shock profile. Note that the shock wave is a sharp profile in v and M. The electron temperature (Fig. 5) is elevated well into the drain.

5.2 Electron shock wave at 77 K

For the transonic computations at 77 K, I take a diode consisting of a 0.1 micron source, a 1.0 micron channel, and a 0.1 micron drain. In the n^+ region, the doping density $N = 10^{18}$ cm^{-3} (at 77 K), while in the n region $N = 10^{15}$ cm^{-3} (at 77 K). The ambient device temperature $T_0 = 77$ K $= 0.00665$ eV. The saturation velocity $v_s = 1.26 \times 10^7$ cm/s. Note that $v_s > c(T_0) = 0.66 \times 10^7$ cm/s.

Figs. 6–8 present a simulation of a Mach 4.1 electron shock wave at $V = 1$ volt (with 240 grid intervals). Fig. 6 shows the doping profile and the electron density. The shock profile develops at $x_s \approx 0.3$ microns (see Fig. 7) and is spread over about $10\Delta x$. Note the dramatic cooling (Fig. 8) of the electron gas as the electrons enter the channel, due to the potential barrier at the junction.

The Mach 4.1 shock wave is a much sharper profile in v and M than the Mach 1.6 shock wave. The high-order non-oscillatory scheme in Ref. [10] does resolve the shock more sharply than the second upwind method, and will be important in resolving strong shocks and two-dimensional shocks.

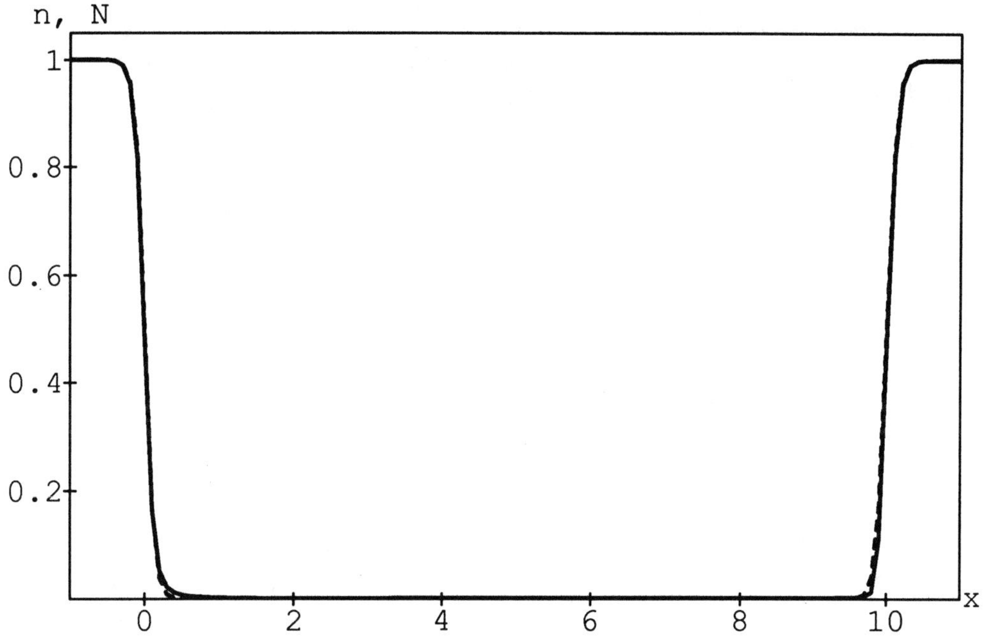

Figure 6: Doping profile (dotted) and electron density (solid) in 10^{18} cm^{-3} at $V = 1$ volt for 1 micron channel device at 77 K. x is in 0.1 microns.

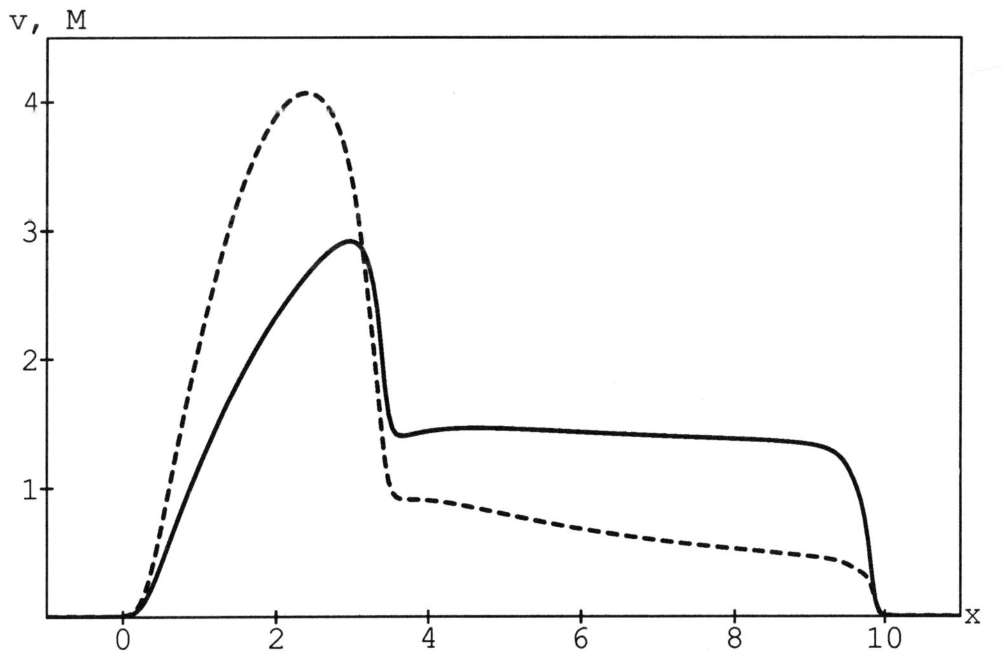

Figure 7: Electron velocity in 10^7 cm/s and electron Mach number (dotted) for $V = 1$ volt, 1 micron channel, 77 K. x is in 0.1 microns.

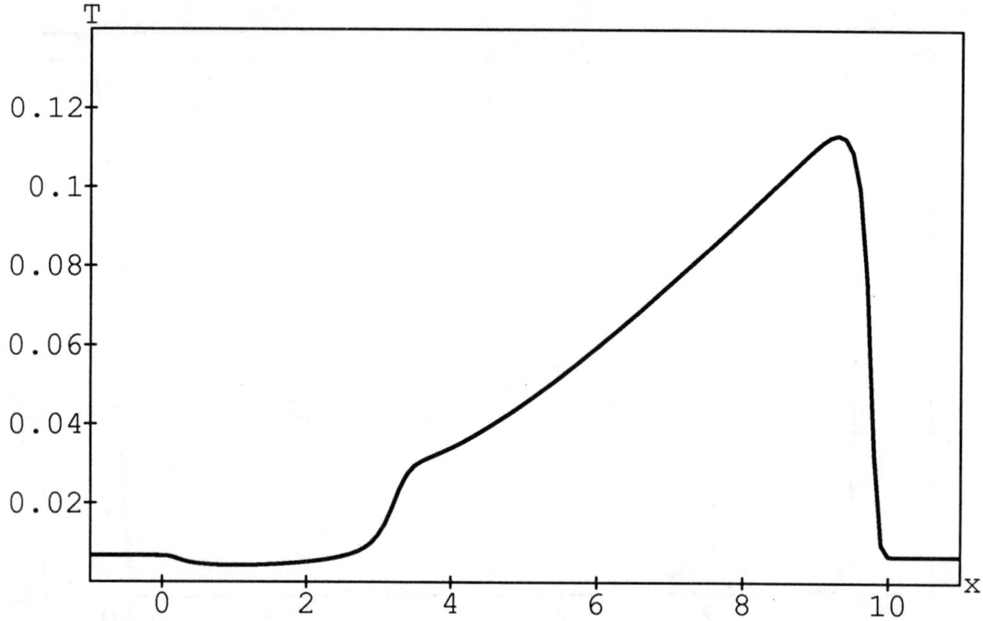

Figure 8: Electron temperature in eV for $V = 1$ volt, 1 micron channel, 77 K. x is in 0.1 microns.

References

[1] C. L. Gardner, J. W. Jerome, and D. J. Rose, "Numerical methods for the hydrodynamic device model: Subsonic flow," *IEEE Transactions on Computer-Aided Design of Integrated Circuits and Systems*, vol. 8, pp. 501–507, 1989.

[2] C. L. Gardner, "Numerical simulation of a steady-state electron shock wave in a submicrometer semiconductor device," *IEEE Transactions on Electron Devices*, vol. 38, pp. 392–398, 1991.

[3] K. Bløtekjær, "Transport equations for electrons in two-valley semiconductors," *IEEE Transactions on Electron Devices*, vol. ED-17, pp. 38–47, 1970.

[4] R. Courant and K. O. Friedrichs, *Supersonic Flow and Shock Waves.* New York: Springer-Verlag, 1948.

[5] P. D. Lax, *Hyperbolic Systems of Conservation Laws and the Mathematical Theory of Shock Waves.* Philadelphia: SIAM, 1972.

[6] C. L. Gardner, "Nonlinear hyperbolic waves in the hydrodynamic model for semiconductor devices," to appear.

[7] G. Baccarani and M. R. Wordeman, "An investigation of steady-state velocity overshoot effects in Si and GaAs devices," *Solid State Electronics*, vol. 28, pp. 407–416, 1985.

[8] R. E. Bank and D. J. Rose, "Global approximate Newton methods," *Numerische Mathematik*, vol. 37, pp. 279–295, 1981.

[9] C. L. Gardner, P. J. Lanzkron, and D. J. Rose, "A parallel block iterative method for the hydrodynamic device model," *IEEE Transactions on Computer-Aided Design of Integrated Circuits and Systems*, to appear in 1991.

[10] E. Fatemi, C. L. Gardner, J. W. Jerome, S. Osher, and D. J. Rose, "Simulation of a steady-state electron shock wave in a submicron semiconductor device using high-order upwind methods," in *Computational Electronics: Semiconductor Transport and Device Simulation*, Boston: Kluwer Academic Publishers, 1991.

[11] E. Fatemi, J. W. Jerome, and S. Osher, "Solution of the hydrodynamic device model using high-order non-oscillatory shock capturing algorithms," *IEEE Transactions on Computer-Aided Design of Integrated Circuits and Systems*, vol. 10, pp. 232–244, 1991.

[12] A. Gnudi, F. Odeh, and M. Rudan, "Investigation of non-local transport phenomena in small semiconductor devices," to appear.

[13] F. Odeh, M. Rudan, and J. White, "Numerical solution of the hydrodynamic model for a one-dimensional semiconductor device," *COMPEL*, vol. 6, pp. 151–170, 1987.

[14] M. V. Fischetti and M. Rudan, "Hydrodynamic and Monte Carlo simulation of an $n^+ - n - n^+$ submicron device," to appear.

[15] J. Glimm, G. Marshall, and B. Plohr, "A generalized Riemann problem for quasi-one-dimensional gas flows," *Advances in Applied Mathematics*, vol. 5, pp. 1–30, 1984.

[16] Ya. B. Zel'dovich and Yu. P. Raizer, *Physics of Shock Waves and High-Temperature Hydrodynamic Phenomena*. New York: Academic Press, 1966.

[17] R. Menikoff and B. J. Plohr, "The Riemann problem for fluid flow of real materials," *Reviews of Modern Physics*, vol. 61, pp. 75–130, 1989.

CHAPTER 4

Stability of One-Dimensional Waves in Weak and Singular Limits

R. A. Gardner*
C. K. R. T. Jones**

Abstract. Methods for determining the stability of travelling waves are discussed. An index which counts the eigenvalues in the right half-plane is described and its particularly effective use in singular limits is outlined in an example of a diffusive phase-field model. The problem of the stability of weak shocks in viscous conservation laws is considered and the absence of unstable eigenvalues is shown for scalar and 2×2 systems. Weak "shock-like" solutions also appear in Ginsburg-Landau equations and recent results, due to Kapitula, which give algebraic decay in algebraically weighted norms for perturbations of the wave, are explained in the context of this theory.

1 INTRODUCTION.

Nonlinear parabolic systems of the form

$$u_t = Du_{xx} + f(u, u_x) \qquad (u \in \mathbb{R}^n, \ x \in \mathbb{R}^1) \tag{1}$$

frequently admit a physically important class of self-similar wave solutions, $u(x,t) = V(\xi)$ where $\xi = x - \theta t$, which satisfy associated systems of o.d.e.'s

$$-\theta V' = DV'' + f(V, V'). \tag{2}$$

A typical problem is to locate solutions of (2) which tend to limits $V_\pm$ as $\xi \to \pm\infty$. A central issue is to determine when such solutions are asymptotically stable solutions. We shall consider them as steady states of

$$u_t = Du_{\xi\xi} + \theta u_\xi + f(u, u_\xi), \tag{1$'$}$$

the equation obtained from (1) under the change of variables $(x,t) \mapsto (\xi, t)$. In addition to the obvious physical significance of stability, this question also arises in other connections, such as the numerical approximation of solutions of conservation laws, and the construction and validity of matched asymptotic expansions for curved, multidimensional fronts in

*Dept. of Mathematics, University of Massachusetts, Amherst, MA 01003. Supported by NSF Grant # DMS-8922384

** Div. of Applied Mathematics, Brown University, Providence, R. I. 02912. Supported by NSF Grant # DMS-8801627

solutions of singularly perturbed reaction-diffusion systems. The purpose of this paper is to describe some recently developed geometric tools for analyzing the stability of travelling wave solutions, and also, to remark upon their application to the stability of viscous shock profiles.

To begin with, we say that a travelling wave $V(\xi)$ is a *stable* solution of $(1)'$ in a prescribed norm $\|\cdot\|$, if there exists $\delta > 0$ so that, whenever $\|u(\xi,0) - V(\xi)\| < \delta$, there is a $k \in \mathbb{R}$ for which $\|u(\xi,t) - V(\xi+k)\| \to 0$ as $t \to +\infty$. The choice of norm can be a crucial issue. For bistable reaction-diffusion systems, the sup norm usually suffices, while for viscous shock profiles a weighted norm is more appropriate. The choice of norm turns out to be closely associated with the properties of the spectrum $\sigma(L)$ of the linear operator L, obtained by linearizing $(1)'$ about $V(\xi)$,

$$Lp = Dp'' + (\theta + a(\xi))p' + b(\xi)p, \tag{3}$$

where $a(\xi) = d_2 f(V(\xi), V'(\xi))$ and $b(\xi) = d_1 f(V(\xi), V'(\xi))$, and its associated eigenvalue problem,

$$Lp = \lambda p. \tag{4}$$

Since the domain is unbounded, L can have both point spectrum, $\sigma_p(L)$ and essential spectrum $\sigma_e(L)$, either of which can cause an instability in the underlying wave. In particular, we distinguish between the two distinct situations depicted in Figure 1.

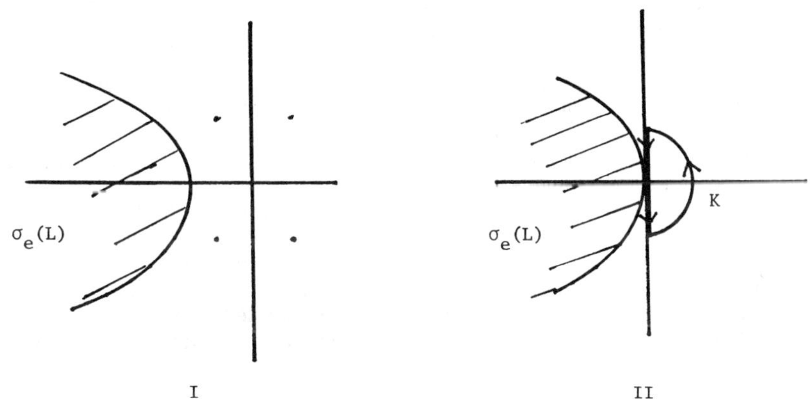

Figure 1. The configuration of essential spectrum

In the first case, which is typical of waves for bistable reaction-diffusion systems, the essential spectrum of L satisfies $Re\ \lambda < -\beta$ for some $\beta > 0$ and all $\lambda \in \sigma_e(L)$. In the second case, $\sigma_e(L)$ is tangent to the imaginary axis at $\lambda = 0$; this is the typical situation for viscous shock profiles for conservation laws and for the Ginzburg -Landau system. Each case has its own subtleties. In order to better appreciate this, we mention two well-known general theorems.

Theorem 1 (Weyl's lemma). Let $L_\pm$ be the constant coefficient operators obtained from L by setting $a = a_\pm, b = b_\pm$ where $a_\pm, b_\pm$ are the limiting values of $a(\xi), b(\xi)$ at $\xi = \pm\infty$; then $\sigma(L_-) \bigcup \sigma(L_+)$ divides $\mathbb{C}$ into components, and the component containing $Re\lambda > 0$ is free from essential spectrum.

Since the operators $L_\pm$ have constant coefficients, their spectra, and hence $\sigma_e(L)$, are easily computed (see § 2. A).

Theorem 2 (Linearized Stability). Suppose that L is a sectorial operator and that $\sigma_e(L)$ is as in case I. If in addition $\sigma_p(L) \setminus \{0\} \subset \{Re\ \lambda < 0\}$ and $\lambda = 0$ is a simple eigenvalue of L, then $V(\xi)$ is (nonlinearly) stable; more precisely, for some $c, M > 0$ and $k \in \mathbb{R}$, we have that

$$\|V(\xi + k) - u(\xi,t)\|_\infty \le M\|V(\xi) - u(\xi,0)\|_\infty e^{-ct}$$

(see e.g. Henry [12], Ball [2], Bates and Jones [3]). The latter result draws upon the theory of analytic semigroups and is therefore available in great generality. Regarding its application, the difficult aspect is to verify the conditions concerning the point spectrum of L; this usually depends sensitively upon the internal structure the underlying wave. In the next section, we shall describe some geometric machinery for finding the location and multiplicities of eigenvalues λ of L. There are two principal constructions in the theory:

* An analytic function $D(\lambda)$ with domain $\mathbb{C}\backslash\sigma_e(L)$ whose roots (counting order) coincide with the eigenvalues of L (counting multiplicity).

* A complex vector bundle $\mathcal{E}(K)$, where K is a simple closed curve in $\mathbb{C}$ with $K \cap \sigma(L) = \emptyset$.

Together, they provide a vehicle for counting eigenvalues of L interior to K. In regard to checking the hypothesis of Theorem 2, it is frequently possible to bound the modulus $|\lambda|$ of any unstable eigenvalue λ a priori; hence K is usually taken to be a curve enclosing $\lambda = 0$, which has a sufficiently large portion of $Re\ \lambda \geq 0$ in its interior. A key ingredient in the construction of and computation with $D(\lambda)$ and $\mathcal{E}(K)$ is the flow induced by the linearized equations on the space of complex lines in $\mathbb{C}^N, \mathbb{C}P^{N-1}$, or more generally, the Grassmannian $G_k(\mathbb{C}^N)$ of k-planes in $\mathbb{C}^N$. We call this associated flow the *projectivized flow*; it will be described in some detail in the next section.

There are several difficulties with extending this procedure to problems which are of type II. The most formidable problem is that $\lambda = 0$ is not a spectral set since it is embedded in $\sigma_e(L)$; thus Theorem 2 is not available in this case. One does not expect exponential decay here.

Another difficulty is that our procedure for counting unstable eigenvalues requires $D(\lambda)$ to be defined inside K and hence, in a full neighborhood of $\lambda = 0$. However, the general construction of $D(\lambda)$ and $\mathcal{E}(K)$ requires that $K \subset \mathbb{C} \setminus \sigma_e(L)$. It turns out that in many cases of interest, $D(\lambda)$ can be analytically continued into the essential spectrum. This enters crucially into making an appropriate choice of weighted norm.

The stability index for travelling waves was first developed by Alexander, Gardner and Jones in [1]; in addition to the application to the FitzHugh-Nagumo systems in [1], applications to some other diffusive waves were given by Gardner and Jones [8,10]. A parallel construction for solutions of boundary value problems was developed by Gardner and Jones in [9], and subsequently applied to some predator-prey systems by Gardner [7]. It is also used by Kapitula in the context of Ginsburg-Landau equations, see [14] or section 3C below.

2 THE SEARCH FOR $\sigma_p(L)$.

A. Preliminaries. Express the linearized eigenvalue equations (4) as an equivalent first order system:

$$p' = q$$

$$q' = D^{-1}[-(\theta + a(\xi))q + (\lambda I - b(\xi))p],$$

or in abbreviated form, as

$$Y' = A(\xi, \lambda)Y \tag{5}$$

where $Y = (p, q) \in \mathbb{C}^N, N = 2n$, and $A(\xi, \lambda)$ is the appropriate $N \times N$ coefficient matrix. Since the underlying wave $U(\xi)$ tends to limits $V_\pm$ as $\xi \to \pm\infty$, the matrices $A(\xi, \lambda)$ have limits $A_\pm(\lambda)$ as $\xi \to \pm\infty$.

Definition. $\Omega \subset \mathbb{C}$ has *consistent splitting* if there exists k such that $A_\pm(\lambda)$ both have k eigenvalues with positive real part and $N - k$ eigenvalues with negative real part for all $\lambda \in \Omega$.

Remarks:

(1) In the cases we consider, there exists Ω with consistent splitting containing $\{Re\lambda > 0\}$; in these cases we have $k = n = N/2$.

(2) The curves in $\mathbb{C}$ where $A_{\pm}(\lambda)$ have pure imaginary eigenvalues define the boundary of $\sigma_e(L)$; in fact $\sigma_e(L)$ is contained in the region to the left of the union of these curves.

B. The Evans Function. If Ω has consistent splitting, then it is not difficult to see that, for each $\lambda \in \Omega$, there exist k independent solutions of (5), $Y_i(\xi,\lambda)$ $1 \le i \le k$, which decay to zero as $\xi \to -\infty$, and $N-k$ independent solutions $Y_i(\xi,\lambda), k+1 \le i \le N$, which decay to zero as $\xi \to +\infty$. (A rigorous construction is given in the next section). In view of the hyperbolicity of $A_{\pm}(\lambda)$, λ is an eigenvalue of L if and only if (5) possesses a nontrivial solution which decays to zero at *both* $-\infty$ and $+\infty$. Hence the span of either set satisfies half of the conditions at $\pm\infty$ required of eigenfunctions. Clearly, both conditions are satisfied if and only if these two subspaces intersect nontrivially. This suggests introducing the N-form

$$\eta(\xi,\lambda) = Y_1(\xi,\lambda) \wedge Y_2(\xi,\lambda) \wedge \cdots \wedge Y_N(\xi,\lambda).$$

Clearly, η vanishes for all ξ if and only if λ is an eigenvalue; furthermore, we have by Abel's formula that

$$\eta' = \text{trace } A(\xi,\lambda)\eta.$$

Definition. The Evans function $D(\lambda)$ is

$$D(\lambda) = e^{-\int_\xi \text{trace } A(s,\lambda)ds}\eta(\xi,\lambda).$$

The following properties of $D(\lambda)$ are proved in [1].
1. $D(\lambda)$ is analytic in λ for $\lambda \in \Omega$, and independent of ξ.
2. $D(\lambda_0) = 0$ if and only if $\lambda_0 \in \sigma_p(L)$.
3. The order of the root of D at $\lambda = \lambda_0$ equals the algebraic multiplicity of λ_0 as an eigenvalue of L.
The last property was proved in [1] by a rather complicated perturbation procedure. A simpler, more direct argument was recently given in [10].

C. The Augmented Unstable Bundle. The previous construction has a geometric character in that it specifies how the eigenvalues of L are determined by the intersection of certain distinguished subspaces of solutions of (5), namely

$$Y_-(\xi,\lambda) = \text{span }\{Y_1(\xi,\lambda),\cdots,Y_k(\xi,\lambda)\}$$
$$Y_+(\xi,\lambda) = \text{span }\{Y_{k+1}(\xi,\lambda),\cdots,Y_N(\xi,\lambda)\}.$$

The connection with geometry can be carried substantially further; this is the goal of the next construction.

The set $Y_-(\xi,\lambda)$ is a family of k-dimensional subspaces of $\mathbb{C}^N$ which depends continuously on parameters (ξ,λ); in other words, it has the structure of a complex k-plane bundle over the base space $\mathbb{R}\times\Omega$. It will be convenient to compactify the base space by reparametrizing ξ with a parameter $\tau \in [-1,1]$, namely

$$\xi = \frac{1}{2\kappa} ln\left(\frac{1+\tau}{1-\tau}\right).$$

We can then express (5) as an equivalent, augmented, autonomous system of the form

$$Y' = A(\tau,\lambda)Y$$
$$\tau' = \kappa(1-\tau^2) \tag{6}$$

where $A(\tau,\lambda)$ is obtained from $A(\xi,\lambda)$ by reparametrization by τ. If $\kappa > 0$ is not too large and if the wave decays exponentially to $V_\pm$ at $\xi = \pm\infty$, then $A(\tau,\lambda)$ is C^1 in its arguments.

The compactified system (6) is very useful for constructing subspaces of solutions of (5) with specified behavior at $\xi = \pm\infty$. The idea is to apply the stable/unstable manifold theorem to the rest points of (6), namely $(\vec{0},-1)$ and $(\vec{0},+1)$. More precisely, since $A(\pm 1,\lambda) = A_\pm(\lambda)$ have consistent splitting for $\lambda \in \Omega$, there exist k dimensional (resp. $N-k$ dimensional) subspaces $U_\pm(\lambda)$ (resp. $S_\pm(\lambda)$) associated with the portion of the spectra of $A_\pm(\lambda)$ with positive (resp. negative) real part. The rest point $(\vec{0},-1)$ of (6) is therefore hyperbolic with a k (complex) $+1$ (real) dimensional unstable subspace; (the unstable real direction is the τ-direction). The unstable manifold theorem provides a $2k+1$ (real) dimensional manifold $\mathcal{M}^u$ of solutions of (6) which tend to $(\vec{0},-1)$ in backward time. By the linear structure of the Y-components, $\mathcal{M}^u \cap \mathbb{C}^N \times \{\tau\}$ is a k-dimensional vector space. Put

$$Y_-(\tau,\lambda) = \mathcal{M}^u \cap \mathbb{C}^N \times \{\tau\};$$

then $Y_-(\tau,\lambda)$ "tends" to the unstable subspace $V_-(\lambda)$ of $A_-(\lambda)$ as $\tau \to -1$ (see section C, below). We therefore have constructed a k-plane bundle over the base space $[-1,1) \times \{\lambda\}$ for each $\lambda \in \Omega$.

The key idea is to compactify the base space by somehow tracking the behavior of $Y_-(\tau,\lambda)$ as $\tau \to +1$. In general it is not always possible to characterize this limit; in fact, this occurs precisely when λ is an eigenvalue. However, we have the following result, whose proof can be found in [1].

Lemma: If $\lambda \in \Omega\backslash\sigma_p(L)$ then $Y_-(\tau,\lambda)$ tends to $V_+(\lambda)$ (in the Grassmanian) as $\tau \to +1$. The situation is depicted in Figure 2. Evidently, for $\lambda \notin \sigma_p(L)$, $Y_-(\tau,\lambda)$ forms a k-plane bundle over the base space $[-1,1] \times \{\lambda\}$.

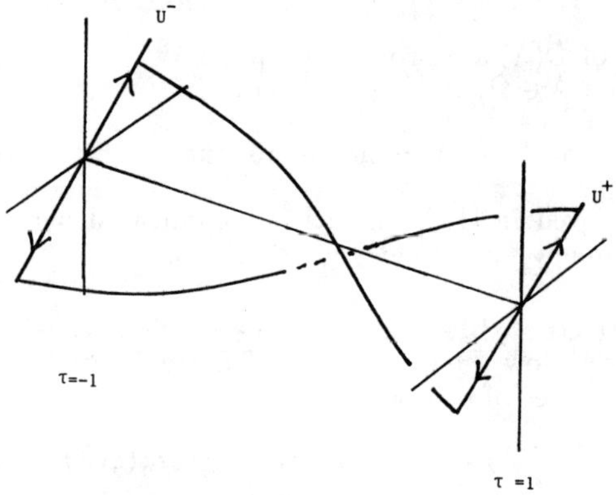

Figure 2. λ is not an Eigenvalue

Let K be a simple closed curve in Ω with $K \cap \sigma_p(L) = \emptyset$. We assert that the twisting of the fibers $Y_-(\tau,\lambda)$ for $(\tau,\lambda) \in [-1,1] \times K$ determines the number of eigenvalues of L interior to K. More precisely, let $\mathcal{E}(K)$ be the bundle $\pi : \mathcal{E}(K) \to B$ over base space

$$B = \{-1\} \times K^0 \cup [-1,1] \times K \cup \{+1\} \times K^0$$

when K^0 is the interior of K, with total space $\mathcal{E}(K)$ defined by

$$\pi^{-1}(\lambda,\tau) = \begin{cases} V_-(\lambda) & \text{for } \tau = -1,\ \lambda \in K^0 \\ Y_-(\tau,\lambda) & \text{for } \tau \in [-1,1],\ \lambda \in K \\ V_+(\lambda) & \text{for } \tau = +1,\ \lambda \in K^0. \end{cases}$$

By the defining property of Y_- at $\tau = -1$, together with the above lemma concerning the behavior of Y_- at $\{1\} \times K$, it follows that $\mathcal{E}(K)$ is a bundle. In particular, the fibers match up continuously where the caps $\{\pm 1\} \times K^0$ are glued onto the sides $[-1,1] \times K$ of the base space. The resulting bundle is called the *augmented unstable bundle.*

The above construction provides us with a bundle over a 2 -sphere B. If K^0 does not contain eigenvalues of L this bundle is easily seen to be trivial; in that case $Y_-(1,\lambda)$ is defined for *all* $\lambda \in K^0$ and equals $V_+(\lambda)$, so that $\mathcal{E}(K)$ is a retract of a bundle over a solid sphere, which is contractible. Thus the eigenvalue count of L inside K is related to the extent to which $\mathcal{E}(K)$ differs from a trivial bundle. This is measured by a topological (integer) invariant $c_1(\mathcal{E}(K))$ associated with the bundle called the *first Chern number* of $\mathcal{E}(K)$. The following theorem, whose proof can be found in [1], makes this statement precise.

Theorem 3. Suppose that $K \subset \Omega \backslash \sigma_p(L)$; then the following three quantities are equal:

(i) $c_1(\mathcal{E}(K))$;

(ii) The winding number of the curve $D(K)$ with respect to the origin;

(iii) The number of eigenvalues of L interior to K counting algebraic multiplicities.

We shall not elaborate here upon the definition of c_1; indeed, much of the topology entering into the proof of Theorem 3 does not figure prominently in the applications. However, c_1 does provide some powerful computational machinery for counting eigenvalues; we illustrate this later in the paper.

We finally remark that if K contains $\lambda = 0$ in its interior and if K contains a sufficiently large portion of $Re\ \lambda \geq 0$, then $c_1(\mathcal{E}(K)) = 1$ implies that the hypotheses of Theorem 2 are satisfied, so that the underlying wave is stable. It therefore seems appropriate to call $c_1(\mathcal{E}(K))$ the *stability index*, keeping in mind however, that it provides only a crude measure of stability.

D. Projectivized Flows and Grassmannians. In order to define $\mathcal{E}(K)$ and to compute $c_1(\mathcal{E}(K))$ we require some apparatus for studying the global behavior of the vector space $Y_-(\xi,\lambda)$. A useful conceptual tool in this regard is the Grassmannian $G_k(\mathbb{C}^N)$ of k-planes in $\mathbb{C}^N$, topologized so that two elements in G_k are close whenever they have bases which are close in the Euclidean topology. In particular, given a linear system $Y' = AY$ with $Y \in \mathbb{C}^N$, there is an associated flow on $G_k(\mathbb{C}^N)$ which we denote by $\hat{Y}' = A_k(\hat{Y})$, where $\hat{Y}$ is the generic point in $G_k(\mathbb{C}^N)$ and A_k is the vector field on $G_k(\mathbb{C}^N)$ induced by A. The latter system is called the *projectivized flow*; its solutions are obtained from the linear flow by forming the span of k independent solutions. The special case when $k = 1$ is of particular importance; here $G_1(\mathbb{C}^N) = \mathbb{C}P^{N-1}$ is complex projective space, the space of complex lines in $\mathbb{C}^N$.

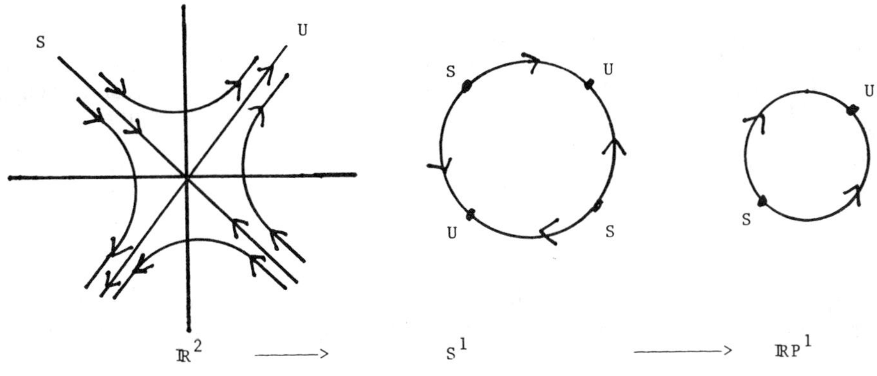

Figure 3. A saddle in the Plane

An important observation, of which we shall make repeated use, is that a k-dimensional eigenspace E of A is A-invariant, and hence, E is a critical point of A_k. Furthermore, if A is hyperbolic with k (resp. $N-k$) eigenvalues of positive (resp. negative) real part, and if E is the eigenspace associated with the unstable eigenvalues, then E is an *attracting* rest point for the projectivized flow on $G_k(\mathbb{C}^N)$. This is most easily visualized in the case of a saddle in $\mathbb{R}^2$, as depicted in Figure 3.

Rays through the origin behave under the linear flow as indicated in the middle figure after being identified to points on S^1; identification of antipodal points in S^1 yields the projectivized flow on $\mathbb{R}P^1$. In regard to the asymptotic matrices $A_\pm(\lambda)$ of $A(\xi,\lambda)$, the unstable subspaces $U^\pm(\lambda)$ are attractors for the flow $A_\pm(\lambda)$ induces on $G_k(\mathbb{C}^N)$ and the stable subspaces $S^\pm(\lambda)$ are repellers for the induced flow on $G_{N-k}(\mathbb{C}^N)$. If we now projectivize the augmented system (6), the flow on $G_k(\mathbb{C}^N)\times[-1,1]$,

$$\hat{Y}' = A_k(\tau,\lambda;\hat{Y}) \qquad (7)$$

$$\tau' = \kappa(1-\tau^2)$$

is qualitatively as depicted in Figure 4; i.e., $\hat{Y}_-(\tau,\lambda)$ is an orbit of (7) which connects the rest point $(U^-(\lambda),-1)$ to $(U^+(\lambda),+1)$, provided of course that $\lambda \notin \sigma_p(L)$.

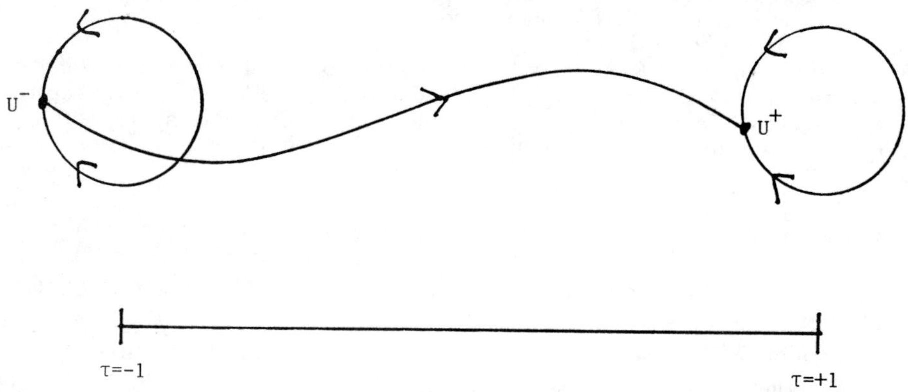

Figure 4. The Flow on Projective space

In this manner one obtains a continuous map $\Phi : B \to G_k(\mathbb{C}^N)$ called the *classifying map* of $\mathcal{E}(K)$. The specification of the map Φ is equivalent to the specification of the bundle, and in particular, $c_1(\mathcal{E}(K))$ is determined by the homotopy type of Φ.

E. General Remarks about Applications to Fast-Slow Systems.

The stability index is well suited to singular perturbation problems, wherein the underlying equations have a small parameter ε such as the shock strength, or a diffusion coefficient which is tending to zero. In such situations, the phase space for the underlying wave can be decomposed into manifolds of slowly varying solutions away from which solutions behave on a rapid time scale. This structure is inherited by the linearized equations. In terms of the augmented unstable bundle, it is manifested as a Whitney sum decomposition,

$$\mathcal{E}(K,\varepsilon) = \oplus_{i=1}^{\ell}\mathcal{E}_i(K,\varepsilon)$$

of lower dimensional subbundles whose fibers are determined by the spans of certain distinguished solutions, which exhibit behavior on one of the characteristic time scales. A crucial property of the first Chern number is that it is additive on the summands in a Whitney sum,

$$c_1(\mathcal{E}(K,\varepsilon)) = \sum_{i=1}^{\ell} c_1(\mathcal{E}_i(K,\varepsilon)). \tag{8}$$

A second crucial property of c_1 is that it is topological, so that the stability index is invariant under homotopies of the equations which preserve the bundle. This suggests that, in order to compute c_1 of the summands, we allow the small parameter ε to tend to zero and compute the limits $\mathcal{E}_{iR}(K)$ of each of the perturbed summands. In practice, the reduced summands occur as the augmented unstable bundles for certain (formal) reduced eigenvalue problems obtained directly from the equations in which ε has been set equal to zero. (In a singular problem, there can be several distinct reduced systems, depending on the scaling of the equations that is being used). The reduced eigenvalue problems are typically much simpler than the perturbed problem; in particular, it is frequently possible to perform an eigenvalue count directly via Sturm-Liouville theory. By Theorem 3, this can be used to compute the Chern number of the reduced, and also by homotopy invariance, the perturbed summand. Finally, (8) enables this information to be amassed into a statement about $c_1(\mathcal{E}(K,\varepsilon))$.

F. Geometric Singular Perturbation Theory. Problems with fast-slow dynamics have a convenient expression in terms of invariant manifolds. In an appropriate scaling, involving the small parameter ε, the (nonlinear) travelling wave equations can be expressed in the form

$$x' = G(x,\varepsilon) \tag{9}$$

wherein $G(x,0)$ admits a manifold $\mathcal{M}_0$ of rest points. Since (9) is the first order reduction of a second order system (2), it is usually the case that the dimension of $\mathcal{M}_0$ is even, say $\dim \mathcal{M}_0 = 2r$, with $1 \leq r < n$. Under certain conditions, Fenichel [5] proved the existence of an invariant manifold $\mathcal{M}_\varepsilon$ of (9) for small $\varepsilon > 0$ near $\mathcal{M}_0$. The main condition for determining when $\mathcal{M}_0$ perturbs smoothly is *normal hyperbolicity*, which requires that the eigenvalues μ associated with eigenvectors of $dG(x,0)$ for $x \in \mathcal{M}_0$, which are transverse to $T_x\mathcal{M}_0$, satisfy $|Re\,\mu| > \beta$, for some $\beta > 0$, and all $x \in \mathcal{M}_0$. Typically, $\mathcal{M}_0$ is a graph over a slow subspace; in other words, the state vector $x = (x_S, x_F)$ splits into a $2r$-dimensional slow subspace x_S, and the remaining fast components x_F are given as a function $x_F = \gamma(x_S)$ of the slow components. Fenichel's theorem also provides equations for the perturbed flow on $\mathcal{M}_\varepsilon$

$$\dot{x}_S = P_S G(x_S, \gamma(x_S), \varepsilon) + O(\varepsilon), \tag{10}$$

when P_S is projection on the slow subspace and the independent variable has been rescaled to $s = \varepsilon^{-1}\xi$. It is frequently the case that the reduced problem (10) at $\varepsilon = 0$ can be explicitly analyzed, and that travelling waves occur as a transverse intersection of invariant manifolds of rest points inside $\mathcal{M}_0$. The persistence of such connections for $\varepsilon > 0$ follows immediately from Fenichel's theorem and transversality. This provides an efficient machine for producing travelling wave solutions of the perturbed problem.

This viewpoint is also useful in studying the linearized problem,

$$\begin{aligned} Y' &= A(\tau,\lambda,\varepsilon)Y, \\ \tau' &= \varepsilon\kappa(1-\tau^2). \end{aligned} \tag{11}$$

In most situations, the domain Ω in which $A_\pm(\lambda,\varepsilon)$ have consistent splitting contains $Re\lambda \geq 0$, in which case $k = n = N/2$. Note that $A(\tau,0,\varepsilon) = dG(x(\tau),\varepsilon)$, so that by normal hyperbolicity, $A(\tau,0,\varepsilon)$ has $2r$ slow eigenvalues and $N-2r$ fast eigenvalues. It follows that the n eigenvalues of $A_\pm(\tau,0,\varepsilon)$ with positive real part split into r slow eigenvalues and $n-r$ fast eigenvalues. It turns out that this splitting frequently persists for λ in compact subsets of Ω. This suggests that Whitney sum decompositions of $\mathcal{E}(K,\varepsilon)$ can be found of the form

$$\mathcal{E}(K,\varepsilon) = \mathcal{E}_S(K,\varepsilon) \oplus \mathcal{E}_F(K,\varepsilon)$$

where $\mathcal{E}_S$ is r-dimensional and $\mathcal{E}_F$ is $n-r$ dimensional. The fast summand is determined by considering the projectivized flow on $G_{n-r}(\mathbb{C}^N)$,

$$\hat{Y}' = A_{n-r}(\hat{Y}, \tau, \lambda, \varepsilon) \tag{12}$$
$$\tau' = \varepsilon\kappa(1-\tau^2).$$

Note that at $\varepsilon = 0$, τ appears as a parameter in the equations. Since $A(\tau,\lambda,0)$ has $n-r$ fast eigenvalues of positive real part, it follows from our remarks in the previous section that this matrix admits an $n-r$ dimensional subspace $U_F(\tau,\lambda,0)$ which is an attracting rest point of (12) when $\varepsilon = 0$ for each fixed τ, or equivalently, $(U_F(\tau,\lambda,0),\tau), -1 \le \tau \le 1$ is an attractor for the parametrized flow. Since attractors perturb to nearby flows, it follows that (12) admits an attractor $(\hat{Y}_F(\tau,\lambda,\varepsilon),\tau), -1 \le \tau \le 1$, for small $\varepsilon > 0$. By construction, $\hat{Y}_F(\tau,\lambda,\varepsilon)$ remains near the fast unstable subspace of $A(\tau,\lambda,\varepsilon)$; in particular, this is true at $\tau = -1, +1$. Thus no solution in $Y_F(\tau,\lambda,\varepsilon)$ forms an eigenfunction; in other words, $\hat{Y}_F$ provides an orbit connecting $U_F(-1,\lambda,\varepsilon)$ to $U_F(+1,\lambda,\varepsilon)$, and it can therefore be used to construct a fast subbundle $\mathcal{E}_F$ of $\mathcal{E}$ by the usual capping procedure:

$$\pi^{-1}(\tau,\lambda) = \begin{cases} U_F^-(-1,\lambda,\varepsilon) & \text{for } \tau = -1, \ \lambda \in K^0 \\ \hat{Y}_F(\tau,\lambda,\varepsilon) & \text{for } -1 \le \tau \le 1, \ \lambda \in K \\ U_F^+(+1,\lambda,\varepsilon) & \text{for } \tau = +1, \ \lambda \in K^0. \end{cases}$$

Since the fast-slow splitting is valid for all λ in compact subsets of Ω, it follows that the connection $\hat{Y}_F$ from U_F^- to U_F^+ exists for all $\lambda \in K \cup K^0$; it therefore follows that $\mathcal{E}_F(K,\varepsilon)$ is trivial, so that

$$c_1(\mathcal{E}_F(K,\varepsilon)) = 0.$$

Evidently, all the action in this setting is in the slow summand $\mathcal{E}_S(K,\varepsilon)$. In order to define $\mathcal{E}_S$ we decompose $\mathbb{C}^N$ into fast and slow components, $Y = (Y_S, Y_F)$, where Y_S lies in the $2r$-dimensional slow subspace and $Y_F = \Gamma(\tau,\lambda,\varepsilon)Y_S$ for an appropriate matrix Γ. When $\varepsilon = 0$, the set of vectors $(Y_S, \Gamma Y_S)$ forms a $2r$-dimensional subspace $S_0(\tau,\lambda)$ of rest points of (11) for each $\tau \in [-1,1]$. If we consider the projectivized flow induced by (11) on $\mathbb{C}P^{N-1} \times [-1,1]$,

$$\hat{Y}' = A_1(\hat{Y}, \tau, \lambda, \varepsilon) \tag{13}$$
$$\tau' = \varepsilon\kappa(1-\tau^2)$$

it follows that $(\hat{S}_0(\tau,\lambda),\tau), \tau \in [-1,1]$ forms a normally hyperbolic manifold of rest points of (13) at $\varepsilon = 0$. Fenichel's theorem [5] then implies that (13) admits a smooth invariant manifold $(\hat{S}_\varepsilon(\tau,\lambda),\tau)$, for $\tau \in [-1,1]$ and small $\varepsilon > 0$. The flow on this manifold is given (with a rescaled independent variable) by the projectivization on $\mathbb{C}P^{r-1}$ of

$$Y_S' = A^S(\tau,\lambda,\varepsilon)Y_S, \tag{14}$$
$$\tau' = \varepsilon\kappa(1-\tau^2)$$
$$A^S(\tau,\lambda,\varepsilon)Y_S = P_S A(\tau,\lambda,\varepsilon)(Y_S, \Gamma(\tau,\lambda,\varepsilon)Y_S).$$

In many applications $r = 2$, so that (14) can be analyzed via Sturm-Liouville theory. Assuming that the eigenvalues Σ_S of (14) in Ω can be found at $\varepsilon = 0$, the curve K is chosen in the usual manner so that $K \subset \Omega \backslash \Sigma_S$. It then follows that for $\lambda \in K$, an r-dimensional subbundle $\mathcal{E}_S(K,\varepsilon)$ of $\mathcal{E}(K,\varepsilon)$ can be defined by the usual capping procedure, so that

$$\mathcal{E}(K,\varepsilon) = \mathcal{E}_F(K,\varepsilon) \oplus \mathcal{E}_S(K,\varepsilon)),$$

and by previous remarks,

$$\begin{aligned} c_1(\mathcal{E}(K,\varepsilon)) &= c_1(\mathcal{E}_S(K,\varepsilon)) \\ &= \text{number of eigenvalues of } \textstyle\sum_s \text{ in} \\ &\quad K^0 \text{ (counting multiplicities).} \end{aligned}$$

3 APPLICATIONS.

A. Waves in Phase-Field Models. We briefly sketch the implementation of the methods of the previous section to travelling wave solutions of the equation

$$u_t = \varepsilon^4 \partial^6 u/\partial x^6 + A\varepsilon^2 \partial^4 u/\partial x^4 + B\partial^2 u/\partial x^2 + f(u)$$

where $f(u)$ is a qualitative bistable cubic with stable roots at $u = 0$ and $u = 1$; (see [4] for a discussion of the physical origins of the model). The travelling wave equations can be written as

$$u_1' = \varepsilon u_2 \quad u_3' = u_4$$

$$u_2' = \varepsilon u_3 \quad u_4' = u_5$$

$$u_5' = u_6$$

$$u_6' = -Au_5 - bu_3 - \theta u_2 - f(u_1)$$

where $\xi = (x - \theta t)/\varepsilon$ and "prime"$= d/d\xi$, or $x' = G(x,\varepsilon)$ where $x = (u_1, \cdots, u_6)$. Clearly x splits into fast components $x_F = (u_3, u_4, u_5, u_6)$ and slow components $x_S = (u_1, u_2)$. The slow manifold $\mathcal{M}_0$ at $\varepsilon = 0$ is defined by

$$x_F = \gamma(x_s) = (-\theta B^{-1} u_2 - B^{-1} f(u_1), 0, 0, 0),$$

which is easily seen to be normally hyperbolic. The slow flow (10) (with $s = \varepsilon^{-1}\xi$) is

$$\begin{aligned} \dot{u}_1 &= u_2 \\ B\dot{u}_2 &= -\theta u_2 - f(u_1) + O(\varepsilon), \end{aligned} \tag{15}$$

which is just the travelling wave system for the bistable reaction-diffusion equation. Existence of the wave connecting $u = 0$ to $u = 1$ follows by appending $\theta' = 0$ in the usual manner to obtain transversality.

In order to implement the linearized stability analysis we need to check (in the notation of the previous section) that $A_\pm(\lambda,\varepsilon)$ have consistent splitting in $\Omega = \{Re\,\lambda > \beta\}$ for some $\beta < 0$. It is easily seen that there is a $3-3$ splitting for $Re\,\lambda > \max\{f'(0), f'(1)\} = \beta$ for all sufficiently small ε, so that $\mathcal{E}(K,\varepsilon)$ is 3-dimensional, for all appropriate $K \subset \Omega$. Thus if $\mu_i^\pm(\lambda,\varepsilon)$ are the eigenvalues of $A_\pm(\lambda,\varepsilon)$, then

$$Re\,\mu_i^\pm > 0 \quad (i = 1,2,3) \qquad\qquad Re\,\mu_i^\pm < 0 \quad (i = 4,5,6).$$

Furthermore, for λ in compact subsets of Ω the fast-slow splitting of $Y = (Y_S, Y_F)$ is $Y_S = (Y_1, Y_2), Y_F = (Y_3, \cdots, Y_6)$, so that $r = 2$. Thus the eigenvalues of $A_\pm$ satisfy

$$Re\,\mu_i^\pm = O(1) \text{ for } i = 1,2,5,6$$

$$Re\,\mu_i^\pm = O(\varepsilon) \text{ for } i = 3,4.$$

We can therefore expect a fast-slow decomposition of $\mathcal{E} = \mathcal{E}_S \oplus \mathcal{E}_F$ with $\dim \mathcal{E}_S = 1$ and $\dim \mathcal{E}_F = 2$. The general analysis of the fast summand via the projectivized flow (12) on $G_2(\mathbb{C}^6)$ is applicable here, showing that $\mathcal{E}_F(K,\varepsilon)$ exists and that $c_1(\mathcal{E}_F(K,\varepsilon)) = 0$ for all sufficiently small ε.

We next need to define the slow summand $\mathcal{E}_S(K,\varepsilon)$. The general prescription in section 2F for passing from the full system (13) to the reduced slow linearized system (14) is applicable here. In fact, a simple computation shows that (14), at $\varepsilon = 0$, is just

$$\begin{aligned} \dot{Y}_1 &= Y_2 \\ B\dot{Y}_2 &= -\theta Y_2 - f'(u_R(\xi))Y_1, \end{aligned} \tag{16}$$

where $u_R(\xi)$ is the connecting solution of (15); this problem is just the linearization of (15). The stability of $u_R(\xi)$ was proved by Fife and McLeod [6]; it follows from this that for $Re\,\lambda \geq \beta$, the only eigenvalue λ of (16) is at $\lambda = 0$ and that $\lambda = 0$ is simple. Hence if K is any simple closed curve in Ω containing $\lambda = 0$ in its interior, $\mathcal{E}_S(K,\varepsilon)$, and hence $\mathcal{E}(K,\varepsilon)$, can be defined, and

$$c_1(\mathcal{E}(K,\varepsilon)) = c_1(\mathcal{E}_S(K,\varepsilon)) = 1$$

for all sufficiently small ε. This proves stability of the wave. The full proof can be found in [10].

B. Conservation Laws. The stability of viscous profiles of shock wave solutions of conservation laws has been significantly advanced in recent years by several investigators (see Liu [16], Goodman [11], Kawashima [15]). This work has primarily been based upon p.d.e. methods. Analyses based upon spectral analysis of the linearized equations have only been partially successful. In particular, Sattinger's approach [19] of using exponentially weighted spaces works for the scalar equation, but places restrictions on the initial perturbation which are stronger than necessary. Furthermore, as the work of Pego [18] shows, this approach does not generalize to give rigorous results for systems. One of the main difficulties with this approach is that the essential spectrum is as in case II in Figure 1, so that Theorem 2 is unavailable. The iteration scheme described below for the Ginsburg-Landau problem should be applicable here too in order to obtain nonlinear stability. This is a very difficult issue that we shall not explore further here. Rather, we shall discuss the eigenvalue problem in $Re\,\lambda \geq 0$ in the case of systems of $n = 1$ and $n = 2$ conservation laws.

In view of the stability theorems already in the literature (see [16]) these results are not new. Our goal here is to illustrate how the methods of section II can be applied to conservation laws, with the hope that this will be of use in the stability analysis of waves arising in nonclassical settings where the stability question is still open.

Consider then travelling wave $u(\xi)$ solutions of

$$u_t + f(u)_x = u_{xx}$$

connecting states V_L to V_R. The linearized problem

$$Lp = p'' + \theta p' + [df(u(\xi))p]' = \lambda p$$

always has the translation eigenvalue at $\lambda = 0$, and $\lambda = 0$ is embedded in $\sigma_e(L)$. Thus, it is not possible to form $\mathcal{E}(K)$ if $\lambda = 0$ is interior to K, since in that case, $A^{\pm}(\lambda)$ do not have consistent splitting for all $\lambda \in K$.

This problem is bypassed by working with the integrated variable r where $r' = p$, which satisfies

$$L_I r = r'' + [\theta - df(u(\xi))]r' = \lambda r.$$

Although $\lambda = 0$ is an eigenvalue of L_I with eigenfunction $u(\xi) - V_L$, it can be shown that $D(0) \neq 0$; this will enable us to define the bundle $\mathcal{E}(K)$ where K is the semicircle depicted in case II of Figure 1.

In order to see how this is accomplished, consider first the case of the scalar conservation law. The eigenvalue problem $L_I r = \lambda r$ can be expressed as a two-dimensional system

$$Y' = A(\xi,\lambda)Y \tag{17}$$

with $Y = (r, r')$. The eigenvalues of $A_{\pm}(\lambda)$ are

$$\mu_{\pm}(\lambda) = \frac{1}{2}(-\theta + \sigma) \pm \frac{1}{2}[1 - \theta + \sigma)^2 + 4\lambda]^{\frac{1}{2}},$$

where $\sigma = f'(V_L)$ or $f'(V_R)$. In order to define $\mathcal{E}(K)$, we need to track the solution $\hat{Y}_-(\xi, \lambda)$ of the projectivization of (17) on $\mathbb{C}P^1$,

$$\hat{Y}' = A_1(\hat{Y}, \xi, \lambda). \tag{18}$$

From the above expression for $\mu_-(\lambda)$ and $\mu_+(\lambda)$ the real parts of $\mu_{\pm}(\lambda)$ at $\xi = \pm\infty$ are as depicted in Figure 5.

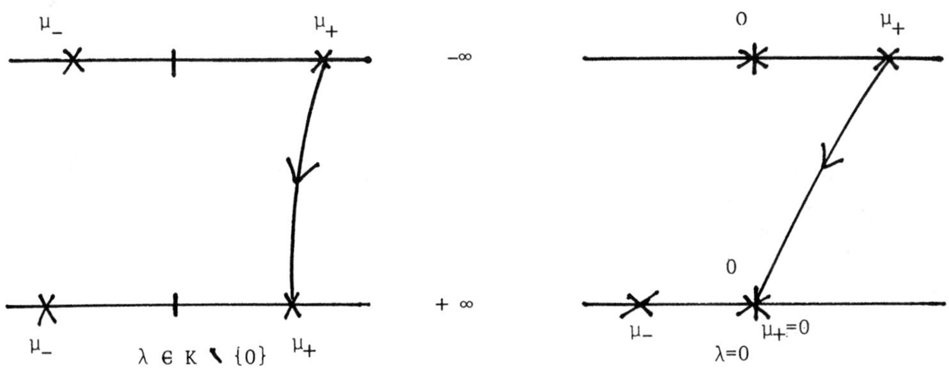

Figure 5. Real Parts of Eigenvalues–Scalar Case

Note that consistent splitting of $A_{\pm}(\lambda)$ breaks down at $\lambda = 0 \in K$. For $\lambda \in K \backslash \{0\}$, the (unique) solution of (18) which tends to the eigendirection associated with $\mu_+(\lambda)$ at $-\infty$ necessarily tends to $\mu_+(\lambda)$ at $+\infty$; this is because $\lambda \in K$ is not an eigenvalue, as was proved in [19]; this can also be seen directly from Sturm-Liouville theory. When $\lambda = 0$, the connection persists. Indeed the eigenfunction is known explicitly to be $u(\xi) - V_L$. This means that even though $\lambda = 0 \in K$ is an eigenvalue, the bundle $\mathcal{E}(K)$ still can be formed, and, by filling in the interior of K, we see in the usual way that $c_1(\mathcal{E}(K)) = 0$. This is hardly surprising because we knew that there were no unstable eigenvalues to begin with. The point here is that $\mathcal{E}(K)$ still makes sense even in this marginal situation where the general constructions of section 2 are not valid.

We next consider the case of weak 1-shocks for strictly hyperbolic, genuinely nonlinear 2×2 systems. If ε is the shock strength, the viscous profile can be constructed by geometric singular perturbation theory with ε as small parameter along the lines presented in §II.F (see Majda and Pego [17]). In particular, the wave is contained in a normally hyperbolic 2-dimensional center manifold. The (integrated) linearized system,

$$Y' = A(\xi, \lambda, \varepsilon) Y \tag{19}$$

with $Y \in \mathbb{C}^4$, has a similar fast-slow structure (see e.g. Pego [18]). The eigenvalues of $A_{\pm}(\lambda, \varepsilon)$ are

$$\mu_2^{\pm}(\lambda, \varepsilon) = \frac{1}{2}(-\theta + \sigma_i) \pm \frac{1}{2}[(-\theta + \sigma_i)^2 + 4\lambda]^{\frac{1}{2}}$$

$$\sigma_i = \text{ eigenvalue of } df(u) \text{ at } u = u_L \text{ or } u = u_R,$$

for $i = 1, 2$. By genuine nonlinearity, the real parts of the $\mu_i^{\pm}$'s are as depicted in Figure 6.

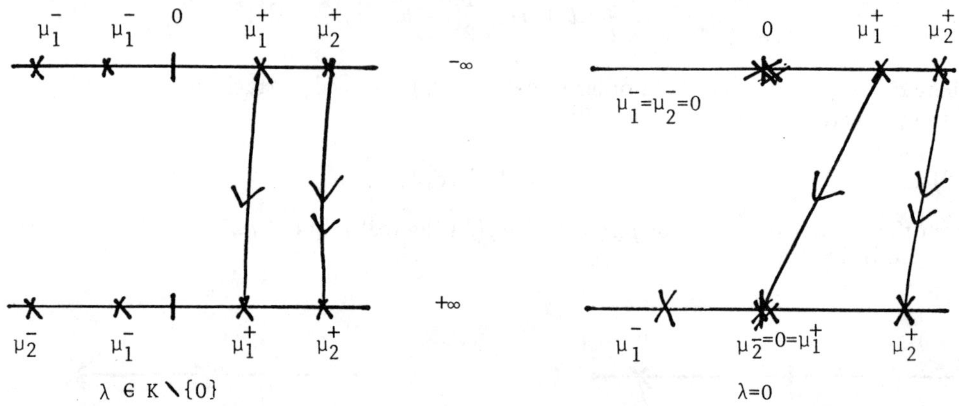

Figure 6. Real Parts of Eigenvalues–2 x 2 System

Since we are considering a weak 1-shock, $\mu_1^{\pm}(\lambda,\varepsilon)$ are of order ε and $\mu_2^{\pm}(\lambda,\varepsilon)$ are of order (1) at both $\xi = \pm\infty$, *provided that $\lambda \in K$ is bounded away from zero.* However, this splitting breaks down at $\lambda = 0$ since $\mu_2^-(0) = 0$. In regard to the projectivized flow induced by (19) on $\mathbb{C}P^3 \times [-1,1]$

$$\hat{Y}' = A_1(\hat{Y}, \tau, \lambda, \varepsilon), \quad (20)$$

$$\tau' = \varepsilon\kappa(1-\tau^2)$$

we need connecting orbits $\hat{Y}_1$ and $\hat{Y}_2$ between the eigendirections associated with μ_1^+ and μ_2^+ at $\pm\infty$. By the usual capping procedure, these solutions will provide the slow and fast summand $\mathcal{E}_S$ and $\mathcal{E}_F$ of $\mathcal{E}$. The fast connection $\hat{Y}_2$ can be handled by the techniques described in section 2F for all $\lambda \in K$, since μ_2^+ is a uniformly fast eigenvalue for all such λ. For $\lambda \in K$ and bounded away from zero, the fast-slow splitting persists and we have that (20) is uniformly approximated by the reduced system (14). A computation shows that (14) in this example is the projectivization of the linearized scalar equation (18), which has been analyzed above. For λ bounded away from zero we therefore obtain a connecting orbit $\hat{Y}_1$ of (20) between the slow unstable eigendirections associated with $\mu_1^+(\lambda)$.

For λ near zero, the fast-slow splitting of the projectivized system breaks down, and we require a different argument. The saving grace is that at $\lambda = 0$, we know a priori that there is a "μ_1^+ to μ_1^+" connection, namely $u(\xi) - V_L$ where $u(\xi)$ is the wave itself. This is because the wave decays at the slow rates μ_1^+ (at $\pm\infty$) to its limits V_L, V_R, which means that $u(\xi) - V_L$ necessarily connects the desired rest points of the projectivized flow. An argument is then required to show that this connection persists for small λ with $Re\,\lambda \geq 0$. We shall not pursue this further here; however it is quite plausible that this should be the case because $Re\,\mu_2^-(\lambda)$ becomes negative for $Re\,\lambda \geq 0$ and $\lambda \neq 0$. Assuming this point, the slow connection $\hat{Y}_1(\xi,\lambda)$ then exists for all $\lambda \in K$, enabling us to construct the slow summand $\mathcal{E}_S(K,\varepsilon)$, and hence the entire bundle,

$$\mathcal{E}(K,\varepsilon) = \mathcal{E}_S(K,\varepsilon) \oplus \mathcal{E}_F(K,\varepsilon).$$

Furthermore, as already noted c_1 of the fast summand is zero. Also, c_1 of $\mathcal{E}_S(K,\varepsilon)$, by homotopy invariance, equals c_1 of the bundle $\mathcal{E}(K)$ constructed for the scalar problem (17), and hence is zero too. Thus $c_1(\mathcal{E}(K,\varepsilon)) = 0$, so that (19) has no unstable eigenvalues.

C. Ginsburg-Landau Equations. Ginsburg-Landau equations arise in fluid dynamics and pattern formation. They are a caricature of the behavior of the full partial differential equation near a bifurcation. The equations govern a complex variable B;

$$B_t = \mu B_{xx} + f(|B|)B, \tag{21}$$

where μ is a complex variable and f takes complex values. The cubic Ginsburg-Landau equation (often called the Ginsburg-Landau equation) is derived from (21) by setting $\mu = 1 + i\sigma$ and $f(|B|) = (1 + i\nu)|B|^2$. The so-called "$\lambda - \omega$" system is a special case of (21) for which $\mu = 1$ and f satisfies the conditions

$$\frac{\partial}{\partial r} Ref(r) < 0, \qquad \frac{\partial}{\partial r} Imf(r) > 0.$$

In the usual notation: $\lambda(r) = Ref(r)$ and $\omega(r) = Imf(r)$.

It should be noted that equation (21) is invariant under both translation and phase change. Travelling waves are solutions of (21) that are functions of the single variable $\xi = x - \theta t$ and hence satisfy

$$A_t = A_{\xi\xi} + \theta A_\xi + f(|A|)A. \tag{22}$$

Plane waves are solutions of (22) of the form $r_0 e^{i\alpha_0 \xi}$. It is easily checked that a condition for plane waves is:

$$Ref(r_0) = \alpha_0^2 \text{ and } Imf(r_0) = -\theta\alpha_0.$$

We shall assume that such a plane wave exists and further, that the spectrum of the linearization of (22), about the plane wave, lies in the closed left half plane, see Howard and Kopell [13]. A travelling wave can be written as $A(\xi) = R(\xi)\exp\{i\int_0^\xi \alpha(s)ds\}$ and it is natural to seek waves that have plane waves as asymptotic states, i.e. $R(\xi) \to r_0$ and $\alpha(\xi) \to \alpha_0$. In [13], Howard and Kopell find travelling waves that connect plane waves which are close to each other. In fact, they prove the following result.

Theorem (Howard and Kopell). For each fixed plane wave $r_- \exp\{i\alpha_- \xi\}$, satisfying the above conditions, there is a family of plane waves $r_+(\varepsilon)\exp\{i\alpha_+(\varepsilon)\xi\}$, for $0 < \varepsilon \ll 1$, and a family of travelling waves $R(\xi)\exp\{i\int_0^\xi \alpha(s)ds\}$, so that
1. $R(\xi) \to r_\pm, \alpha(\xi) \to \alpha_\pm$ as $\xi \to \pm\infty$,
2. $r_+(\varepsilon) \to r_-, \alpha(\varepsilon) \to \alpha_-$ as $\varepsilon \to 0$.

These waves are called "weak" shocks as they connect asymptotic states that lie close to each other, and they enjoy a kind of entropy condition, see Howard and Kopell [13]. Of interest, then, is the stability of these "weak" travelling waves. Before linearizing, we perform a transformation that depends on the wave. Let $A(\xi, t) = V(\xi, t)\exp\{i\int_0^\xi \alpha(s)ds\}$, then V satisfies

$$V_t = V_{\xi\xi} + (2i\alpha + \theta)V_\xi + (i\alpha_\xi + \theta i\alpha - \alpha^2)V + f(|V|)V. \tag{23}$$

We now linearize the right hand side at the wave $V = R(\xi)$ and denote the resulting operator L. We rewrite the eigenvalue problem $LP = \lambda P$, in the usual way, as

$$Y' = A(\lambda, \xi)Y. \tag{24}$$

The asymptotic matrices, $A_\pm(\lambda)$, correspond to the eigenvalue problem for the linearization about the plane wave states in their own rotating frames. A computation shows that the set $\{\lambda : Re\lambda \geq 0, \lambda \neq 0\}$ gives consistent splitting, with $A_\pm(\lambda)$ having 2 eigenvalues of positive real part $\rho_i^\pm(\lambda)(i = 1,2)$ and 2 of negative real part, $\eta_i^\pm(\lambda)(i = 1,2)$. The essential spectrum passing through the origin prevents this splitting from being extended into the left half plane. However, in a similar fashion to the last section, one sees that for $A_-(\lambda)$ it is one of the η_i^- that changes the sign of its real part as λ goes through 0. In forming the bundle $\mathcal{E}(K)$ we use the unstable directions at $-\infty$, and this is still possible if K passes to the left of 0. It should be noted that it is exactly because of the entropy condition that the eigenvalues move in this way. Going to the weak limit, i.e. $\varepsilon \to 0$, we see that $c_1(\mathcal{E}(K)) = 0$ since in the limiting problem (the plane wave state) there is no spectrum in the right half plane. It therefore also holds for $0 < \varepsilon \ll 1$, and there are no unstable eigenvalues.

The above argument shows that these weak waves have no unstable spectrum but the usual invariant manifold theorems do not apply as the spectrum in the left half plane cannot be separated from that on the imaginary axis, see Figure 1, case II. The work of Kapitula [14], shows how this situation can be analyzed using algebraically weighted norms. We need some definitions. Let $w(\xi)$ be a function whose reciprocal is L^1, for instance $w(\xi) = 1+|\xi|^\alpha$, with $\alpha > 1$. We set

$$\|f\|_k = \|w^k(\xi)f(\xi)\|_\infty$$

and

$$\|f\|_{k,2} = \|f\|_k + \|\partial f\|_k + \|\partial^2 f\|.$$

In the following theorem we write a general solution of (23) as $V(\xi,t) = R(\xi) + u(\xi,t)$.

Theorem (Kapitula) If the above wave is sufficiently weak and $\|u_0\|_{4,2}$ is sufficiently small then

$$\|u(\cdot,t)\|_{2,2} \leq \frac{C}{1+t}\|u_0(\cdot)\|_{4,2}$$

where $u(\xi,0) = u_0(\xi)$.

Kapitula's theorem then gives an algebraic rate of decay, but only in an algebraically weighted space. Note also that the weight on the initial data is two powers stronger than on the solution. The proof is quite difficult and will not be presented here but we shall make some remarks about the structure of the proof.

The semigroup generated by L, the linearization of (23) about the wave, can be expressed as a contour integral involving the resolvent

$$S(t) = \int_\Gamma e^{\lambda t} R(\lambda, L) d\lambda$$

where Γ encloses all the spectrum, see Figure 7.

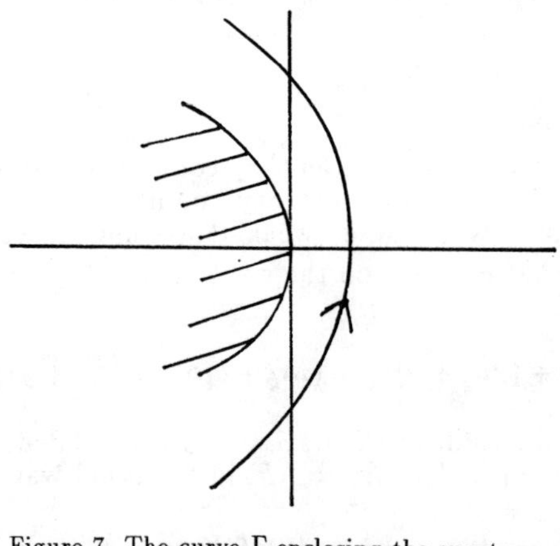

Figure 7. The curve Γ enclosing the spectrum

If it could be shifted into the left half plane, an exponential decay rate on $S(t)$ would be possible. The marginal case of shifting it to the origin, giving a parabolic touching with the imaginary axis can result in some subtle algebraic decay rates.

The curve cannot be shifted in general but it turns out that an estimate on $R(0,L)u$ can be obtained if $\|u\|_3 < \infty$, namely

$$\|R(0,L)u\|_2 \leq c\|u\|_3. \tag{25}$$

The derivation of (25) involves delicate analysis of pieces of the Green's function. The key ingredient is to notice that the singularity of the resolvent at 0 is determined by $D(\lambda)$. In the same way that $\mathcal{E}(K)$ can be defined with K in the left half plane, it is also true that $D(\lambda)$ can be analytically continued into the left half plane, including 0. Moreover, the surprising fact is that $D(0) \neq 0$!

An estimate on the semigroup can be derived from (25). This is then put together in an iteration scheme so as to control the higher order terms. The whole procedure is quite involved but appears to give the correct interpretation of the effect of essential spectrum passing through the origin.

References

[1] Alexander, J., Gardner, R. and Jones, C., A topological invariant arising in the stability analysis of travelling waves, Journal für die Reine und Angewandte Mathematik, 410 (1990), pp167-212.

[2] Ball, J., Saddle point analysis for an ordinary differential equation in a Banach space, and an application to dynamic buckling of a beam, in Nonlinear Elasticity, R.W. Dickey, ed., Academic Press, NY (1973) pp.93-160.

[3] Bates, P. and Jones, C., Invariant manifolds for semilinear partial differential equations, Dynamics Reported, 2 (1988), pp. 1-38.

[4] Caginalp, C. and Fife, P., Higher order phase field models and detailed anisotropy, Phys. Rev. B. 34 (1986), pp. 4940-4943.

[5] Fenichel, N., Geometric singular perturbation theory for ordinary differential equations, JDE, 31 (1979), pp. 193-226.

[6] Fife, P. and McLeod, J.B., The approach of solutions of nonlinear diffusion equations to travelling front solutions, Arch. Rat. Mech. Anal. 65 (1977), pp. 335-361.

[7] Gardner, R.A., Stability and Hopf bifurcation of steady state solution of a singularly perturbed reaction diffusion system, submitted to SIAM Journal of Mathematical Analysis.

[8] Gardner, R. and Jones, C., Stability of travelling wave solutions of diffusive predator-prey systems, to appear in Trans. AMS.

[9] Gardner, R. and Jones, C., A stability index for steady state solutions of boundary value problems for parabolic systems, to appear in JDE.

[10] Gardner, R. and Jones, C., Travelling waves of a perturbed diffusion equation arising in a phase field model, to appear in Indiana Journal of Mathematics.

[11] Goodman, J., Nonlinear asymptotic stability of viscous shock profiles for conservation laws, Arch. Rat. Mech. Anal. 95 (1986).

[12] Henry, D., The Geometric Theory of Semilinear Parabolic Equations. Lecture Notes in Math. #840, Springer-Verlag, 1981.

[13] Howard, L. and Kopell, N., Slowly varying waves and shock structures in reaction-diffusion equations, Studies in Appl. Math., 56 (1977).

[14] Kapitula, T., Stability of travelling waves with applications to Ginsburg-Landau equations, Ph.D thesis, U. of Maryland, 1991.

[15] Kawashima, S., Large-time behavior to hyperbolic - parabolic systems of conservation laws and applications, Proc. Roy. Soc. Ed. 106A (1987).

[16] Liu, T. -P., Nonlinear stability of shock waves for viscous conservation laws, Memoirs AMS, Vol. 56 #328 (1985).

[17] Majda, A. and Pego, R., Stable viscosity matrices for systems of conservation laws, JDE 56 (1985) pp. 229-262.

[18] Pego, R., Linearized stability of extreme shock profiles for systems of conservation laws with viscosity, MRC Technical Summary Report #2457, Madison, WI, 1982.

[19] Sattinger, D., Weighted norms for the stability of travelling waves, JDE 25 (1977).

CHAPTER 5

A Quantitative Theory of Fluid Chaos

James Glimm*, †, **
Qiang Zhang*, **

Abstract. A quantitatively correct theory has been developed for the chaotic mixing of acceleration driven fluid mixing processes. The theory is in agreement with both laboratory experiment and direct simulation of fluid mixing by the front tracking method.

The theory starts from the microphysics of the full two-fluid compressible Euler equations, and proceeds to the experimentally (and computationally) correct value for the growth rate of the mixing layer for a Rayleigh-Taylor unstable fluid interface. The theory is based on the renormalization group.

The key parameters in the theory are set consistently from (a) experimental data and (b) computational data based on the front tracking method. These two sources of data were shown to agree, in a new analysis of the experimental data.

Direct simulation of chaotic mixing rates also produces results which agree with experiment, within the limits of experimental error, for times up to $1\frac{1}{2}$ generations of bubble merger. At later times, three-dimensional effects become important, and cause the exactly two dimensional computations to diverge from the experiments. These computational results are a unique capability of the front tracking method, in terms of their resolution and their agreement with laboratory experiments over an extended time domain.

The results were obtained in collaboration with a number of coworkers, including X. L. Li, R. Menikoff and D. H. Sharp.

* Department of Applied Mathematics and Statistics, State University of New York at Stony Brook, Stony Brook, NY 11794-3600
†Supported in part by the National Science Foundation, grant DMS-8901884 and by the Army Research Office, grant DAAL03-K-0017
** Supported in part by the Applied Mathematical Sciences subprogram of the Office of Energy Research, U. S. Department of Energy, under contract DE-FG02-90ER25084

I. Introduction. The purpose of our study is to obtain a quantitative theory which predicts the statistical behavior of the chaotic mixing of acceleration driven fluid mixing processes. An interface between distinctive fluids can be stable or unstable. In the stable case, a small scale perturbation at the interface will not grow in magnitude, or will diminish if dissipation is present in the system. However, in the unstable case, a small perturbation at the interface will grow dynamically. A classical example is the interface between two fluids of different density under the influence of gravity or external forces. For compressible, inviscid fluids, the motion of the fluids is governed by the Euler equations,

$$\rho_t + \nabla\cdot(\rho\vec{v}) = 0,$$

$$(\rho\vec{v})_t + \nabla\cdot(\rho\vec{v}\times\vec{v}) + \nabla P = \rho\vec{g},$$

$$[\rho(\frac{1}{2}q^2 + e)]_t + \nabla\cdot[\rho\vec{v}(\frac{1}{2}q^2 + i)] + = \rho\vec{v}\cdot\vec{g}, \tag{1}$$

where ρ is the density, P is the pressure, $\vec{v}$ is the velocity, q is the magnitude of $\vec{v}$, e is the specific internal energy and

$$i = e + \frac{P}{\rho}$$

is the specific enthalpy. The thermodynamic quantities pressure and density are related by the equation of state,

$$e = e(P, \rho).$$

When gravity points from the light fluid to the heavy fluid the system is stable. If the direction of gravity is reversed, the system becomes unstable. The light (heavy) fluid starts to penetrate into the heavy (light) fluid and the Rayleigh-Taylor instability develops at the interface. The portions of the light fluid penetrating into the heavy fluid are called bubbles and the portions of the heavy fluid penetrating into the light fluid are called spikes. If the initial interface contains only one Fourier mode, the interface at late times will contain one bubble and one spike in each periodic interval. We denote such systems single-bubble (or single-mode) systems. When the initial interface contains multiple modes, the interface at late times will contain several bubbles and spikes. We denote such systems multi-bubble (or multi-mode) systems. In multi-bubble systems, different bubbles interact with each other. During the process of the interactions, small bubbles are swept downstream and large bubbles expand, leading to a process at the bubble interface called merger. At the end of each generation of bubble merger, the number of bubbles on the interface is reduced by a factor of two. For an ensemble of bubbles, a chaotic mixing layer is formed. The collective behavior of the ensemble leads to the phenomenon that the width of the chaotic mixing layer expands with a constant acceleration.

The Rayleigh-Taylor mixing layer can be analyzed in terms of various regimes. We will be interested in the bubble regime, located at the edge of the mixing layer closest to the undisturbed heavy fluid. The chaotic mixing layer involves multiple length scales. The dominant behavior of the chaotic mixing layer in the bubble regime is characterized by the large scale structures, i.e. the envelope formed by the tips of the bubbles. Since the length scale changes dynamically in the chaotic mixing layer, a renormalization group method is applied to the statistical behavior of the chaotic mixing layer. The prediction of the growth rate of the chaotic mixing layer is the goal of our study. The studies of the motion of single-bubble systems and the interaction between bubbles provide the basis for the prediction. The dynamics of the single-bubble Rayleigh-Taylor instability is modeled by an ODE [1] and the interaction between the bubbles has been approximated by a superposition theory [2]. Based on a modification of the Sharp-Wheeler model [3] and the superposition theory, the renormalization group model expresses the time evolution of average physical quantities in terms of the contributions due to the motion of the individual bubbles, the interactions between bubbles and the conservation of probability [4]. There are three characteristic parameters associated with the chaotic mixing of a

Rayleigh-Taylor unstable interface: the asymptotic velocity of a single bubble, the correlation in the vertical distance between different bubbles and a dimensionless acceleration constant of an ensemble of interacting bubbles. These three parameters are the key parameters of the single-bubble instability, the pair-interaction between bubbles and the statistical behavior of an ensemble of bubbles respectively. The renormalization group fixed point model has provided a new way of analyzing these key parameters and has established a relationship among them. Our quantitative study of the renormalization group fixed point model shows that the agreement between the experimental data and the theoretical results is within the accuracy of the experiments and the theory [5].

The problem of the mixing of two fluids under the influence of external forces was first investigated by Rayleigh [6] and later by Taylor [7]. Since then, various methods have been used to study this classical problem, including nonlinear integral equations [7,8], boundary integral techniques [10,11], conformal mapping [12], modeling the statistical behavior of the Rayleigh-Taylor instability by a set of equations which contain a few characteristic parameters [2-4,13], vortex-in-cell methods [14,15], high order Godunov methods [16], and front tracking [1,2,17,18]. Much of this work has been carried out in the limit of incompressible fluids or in the limit of one-fluid systems. The terminal bubble velocity in an incompressible system is well established from these studies. We study systems containing two compressible fluids. Experimental data on studies of the Rayleigh-Taylor instability are also available [19,20,21]. For a review of the Rayleigh-Taylor instability and its applications to science and engineering, see reference [22].

The front tracking method is used for numerical simulations of the motion of the single- and multi-bubble unstable interfaces. The front tracking method provides a unique capability of elimination of numerical diffusion and high resolution near discontinuous data.

In Section II, we describe the numerical method used in our computations of Rayleigh-Taylor unstable interfaces. In Section III, we introduce a model for the single-bubble Rayleigh-Taylor instability. In Section IV, we discuss the superposition hypothesis for the interaction between bubbles. In Section V, we present the renormalization group fixed point theory which predicts the growth rate of the Rayleigh-Taylor chaotic mixing layer. The conclusions from our study of the chaotic mixing layer are given in Section VI.

II. The Numerical Method. An important issue in the numerical study of a system with discontinuous solutions is the elimination of numerical diffusion. In the numerical study of the Rayleigh-Taylor instability the discontinuity appears in the fluid density at the interface between two fluids. The front tracking method is a suitable method for studying systems containing discontinuities. It provides high resolution near and on the interface between the mixing fluids. In the front tracking method, a lower dimensional moving grid is embedded in the rectangular computational grid to fit the boundaries of the discontinuity interface dynamically. The solution varies smoothly in each connected region bounded by the interfaces, and varies abruptly across interfaces. For two-dimensional computations, an interface contains curves and a curve is formed by a set of doubly linked linear bonds. Curves intersect only at their end points which are called nodes. For three-dimensional computations, an interface contains surfaces and a surface is formed by a set of connected triangles. A surface is bounded by the curves which define the boundaries of the surface. Different surfaces are only connected along curves. There are two states associated with each point on the interface, one for each side of the interface. Therefore the discontinuity can be preserved in the data structure for the interfaces. See reference [23] for a detailed description of the data structures used in the front tracking method. The propagation of the front consists of two steps. First all points on the interface are propagated along the direction normal to the interface by solving a non-local Riemann problem using the data on the interface and the data in the nearby interior region on each side of the interface. Then corrections to the states are made to include the contributions of the waves tangential to the interface. The regular interior points, i.e. the points at the center of each mesh block, are propagated by applying a high order finite difference scheme. Each distinctive flow region is labeled by a different component. The interior computation always uses the data of the same component. A front-adapted grid is generated each timestep for the purpose of interpolation of numerical data. For two-dimensional computations, the interpolation grid contains a set of triangles and rectangles. The edges of the triangles fit the boundaries of the discontinuity interface. Therefore the data used for interpolation never cross the discontinuity interface. Linear interpolation is performed for triangles. In the domain where the solution varies smoothly, a rectangle is used. Bilinear interpolation is used for rectangles. Curves may intersect during the process of propagation. When curves intersect, one first identifies which sector is unphysical. Then depending upon the physical problem, an appropriate bifurcation algorithm is applied in untangling the interface. Occasionally, the untangle procedure may not be successful, e.g. if there is no physical sector identified by the algorithm. In this case the front is repropagated with a smaller timestep. A detailed description of the front tracking method for gas dynamics is given in references [24,25]. The front is regridded occasionally to achieve a uniform distribution of points on each curve or each surface (for three dimensional computations). The regridding also serves as a numerical surface tension to regularize the ill-posedness of the Euler equations. Extensive validation tests applying the front tracking method to the study of the single-bubble Rayleigh-Taylor instability have been performed [17]. The effects of grid size, the remeshing frequency, the amplitude of the initial perturbation and the boundaries at the top and bottom of the computational domain have been tested and studied [17]. The studies show that numerical solutions converge under mesh finement. Numerical surface tension has negligible effects on the solutions unless the interface has been regridded very frequently, e.g. at every timestep [17]. See reference [17] for the details of these studies. A recent study of multi-bubble interactions in the Rayleigh-Taylor chaotic mixing layer, using the front tracking method, also shows that the large structures are convergent under mesh refinement [2].

Our numerical simulations are performed in two space dimensions. The computational domain is initialized with a small amplitude, single- or multi-mode sinusoidal interface. The computational domain has periodic boundaries on the left and right and reflective boundaries on the top and bottom. The states in the system, i.e. the states on each side of the interface and the states on mesh points, are initialized by the analytic solution of the linearized Euler equations for isothermal systems [17]. The analytic solution is an exponentially stratified distribution of density and pressure along the direction of gravity. Gravity is pointing upwards and the light fluid lies on top of the heavy fluid. This configuration is consistent with the experimental setup [19]. The full Euler equations are used with the front tracking method to update the evolution of the system. The polytropic equation of state $e = \dfrac{(\gamma - 1)P}{\rho}$, with the ratio of specific heats $\gamma = 1.4$, is used in our numerical simulations.

III. The Single-Bubble Theory. The study of the single-bubble Rayleigh-Taylor instability [1] provides a basis for the study of the interactions between bubbles [2] and the study of the chaotic mixing layer [4,5]. The motion of the tips of bubble and spike represents the dominant behavior of the development of the single-bubble Rayleigh-Taylor instability. Both experiments and numerical simulations show that the motion of the tips of a bubble and spike follows successively the stages of linear growth, constant acceleration (free fall) and asymptotic terminal velocity. In the linear regime, the magnitude of the perturbation is small but it grows exponentially with time. The growth rate is determined by the linearized Euler equations [17,26]. In the free fall regime, the velocity of the bubble or spike grows linearly with time and the magnitude of the acceleration reaches its maximum value. Eventually, the velocity of the spike or bubble approaches a limiting value v_∞ in the terminal velocity regime, for systems with finite density ratio. For one-fluid systems, the asymptotic behavior of the spike is free fall, while the asymptotic behavior of the bubble is still motion with a terminal velocity. The two-fluid system can be specified by three dimensionless parameters: the density ratio of the heavy fluid to the light fluid

$D = \frac{\rho_h}{\rho_l}$, or equivalently the Atwood number $A = \frac{\rho_h - \rho_l}{\rho_h + \rho_l}$, the dimensionless compressibility $M^2 = \frac{\lambda g}{c_h^2}$ and the ratio of the specific heats γ. Here λ is a wave length and c_h is the speed of sound in the heavy fluid near the interface. Four parameters will be sufficient to describe the entire development process of the single-bubble Rayleigh-Taylor instability: the linear growth rate σ for the linear regime, renormalized gravity g_R for the constant acceleration regime, the terminal velocity v_∞ for the asymptotic regime and the decay constant b, i.e. the rate at which the system approaches the asymptotic regime. An example of the interface for the single-bubble Rayleigh-Taylor instability for systems with $A = 1/5$, $M^2 = 0.5$ is shown in Figure 1. The results were obtained using the front tracking method.

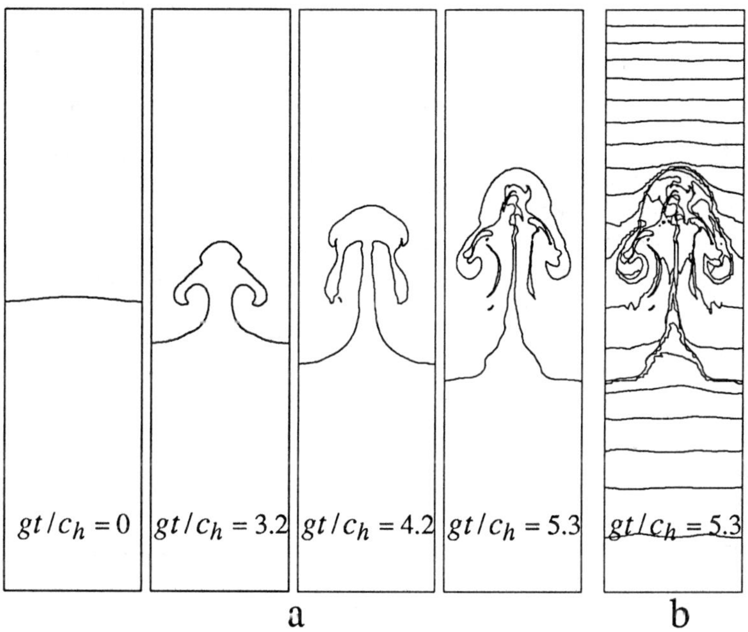

Figure 1. Plots of the interface position and density contours for $A = 1/5$, $M^2 = 0.5$, $\gamma = 1.4$ in a computational domain 1×6 with a 40 by 240 grid. Only the upper two thirds of the computational region is shown in the plot because nothing of interest occurs in the remainder of the region. (*a*) The interface position for successive timesteps. (*b*) Density contours. In this figure and in Figure 4, the low density fluid is above the high density fluid and gravity is directed upward. A bubble of light fluid moves downward and a spike of heavy fluid moves upward.

The dynamics of the single-bubble Rayleigh-Taylor instability has been modeled by an ODE which is expressed in terms of the four parameters introduced above [1],

$$\frac{dv}{dt} = \frac{\sigma v(1-\frac{v}{v_\infty})}{\frac{\sigma}{b}\frac{v}{v_\infty} + (1-\frac{v}{v_\infty}) + [\frac{\sigma v_\infty}{g_R} - (1+\sqrt{\frac{\sigma}{b}})^2]\frac{v}{v_\infty}(1-\frac{v}{v_\infty})}. \tag{2}$$

The solution of this dynamical model is

$$t - t_0 = \frac{1}{\sigma}\ln(\frac{v_t}{v_0}) + [\frac{1}{g_R} - (\frac{1}{\sqrt{\sigma}} + \frac{1}{\sqrt{b}})^2\frac{1}{v_\infty}](v_t - v_0) - \frac{1}{b}\ln(\frac{v_\infty - v_t}{v_\infty - v_0}), \tag{3}$$

where v_0 is the velocity of a spike or a bubble at t_0. Each term on the right hand side of the above expression has a clear physical meaning. The first term is the contribution from the linear regime, the second term is from the free fall regime and the third term is from the asymptotic regime.

Eq. (2) can also be solved for the positions of spikes or bubbles as a function of the velocity,

$$h - h_0 = (\frac{1}{\sigma} - \frac{1}{b})(v_t - v_0) + [\frac{1}{g_R} - (\frac{1}{\sqrt{\sigma}} + \frac{1}{\sqrt{b}})^2\frac{1}{v_\infty}]\frac{v_t^2 - v_0^2}{2} - \frac{v_\infty}{b}\ln(\frac{v_\infty - v_t}{v_\infty - v_0}), \tag{4}$$

where h_0 is the initial position of the tip of the spike or bubble. Extensive numerical simulations show a good agreement between the analytic results of the dynamical model and the numerical solutions of the full Euler equations. The physical parameters of the model which characterize the behavior of bubbles and spikes in each distinctive regime have been set by: the properties of the single bubble or single spike in various limits, well established results for incompressible fluids, and numerical solutions of the full Euler equations obtained with the front tracking method. The growth rate of the linear stage can be determined by linearizing the Euler equations. Phenomenological formulas for the dependence of these physical parameters on the Atwood number and the dimensionless compressibility have been established for the parameter range being studied [1]. For example, the asymptotic velocity of a bubble is given approximately by

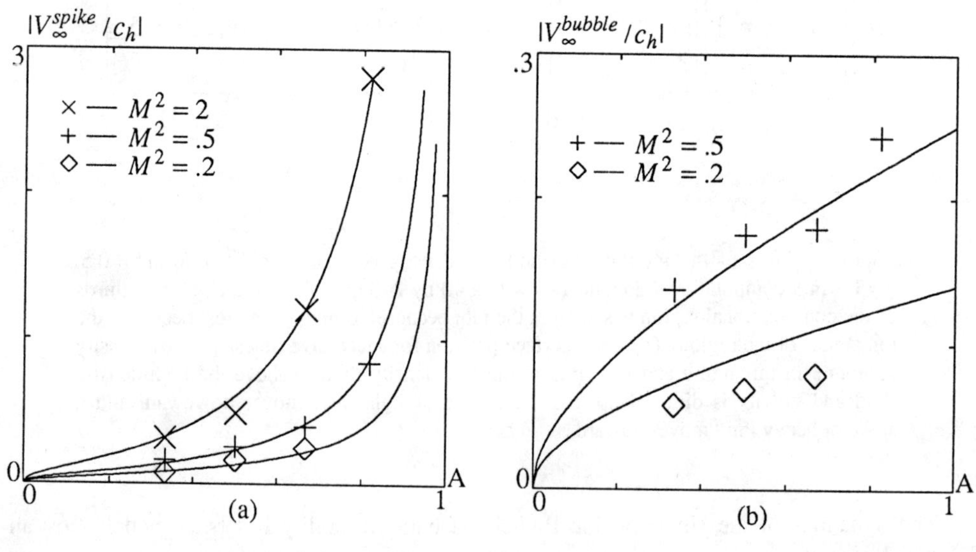

Figure 2. Plots of the ratio of the terminal velocity to the sound speed of the heavy gas of (*a*) spike and (*b*) bubble versus the Atwood ratio. The solid curves in (a) and (b) are the fitting results of phenomenological formulas.

$$\frac{v_\infty}{c_h} = 0.15M^2A + 0.23MA^{\frac{1}{2}}. \tag{5}$$

The comparison of the results of Eq. (5) and the results of numerical simulation is shown in Figure 2.

By using Eq. (3) and the phenomenological formulas [1], we can evaluate the approximate velocity of the spikes and bubbles. The error between the numerical results and the values of these formulas lies in the range of 2-15% for the spikes and in the range of 5-30% for the bubbles. An example is given in Figure 3. In Figure 3 the less smooth solid curves are the numerical data obtained by applying the front tracking method to the full two dimensional Euler equations. The smooth solid curves are the results of fitting numerical data to the solution of the dynamic model, i.e. Eq. (3). The dashed curves are the results of Eq. (3) using phenomenological formulas [1].

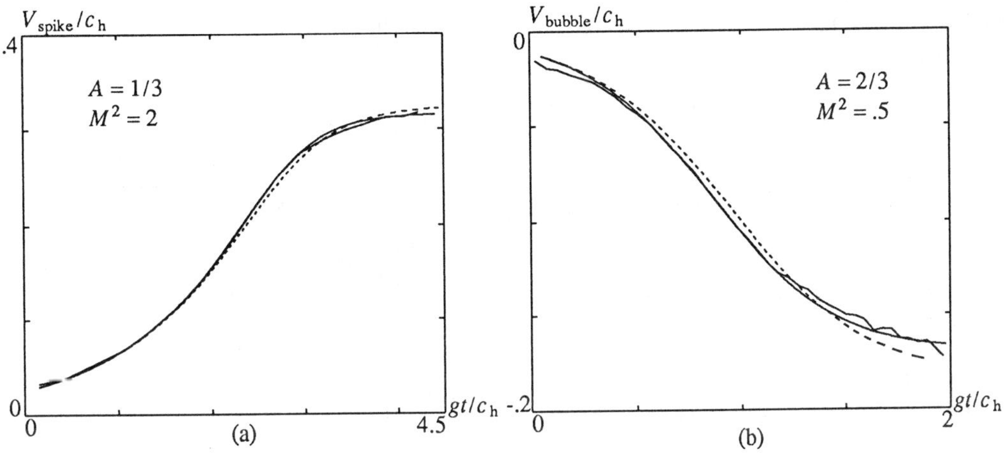

Figure 3. A comparison of numerical results obtained from the front tracking method with the results of single-bubble theory for (a) spike, and (b) bubble. The less smooth solid curves are the numerical data for the full two dimensional Euler equations obtained with the front tracking method. The smooth solid curves are the results of fitting numerical data to Eq. (3). The dashed curves are the results of Eq. (3) using phenomenological formulas. The parameters used in the simulations are $A = 1/3$, $M^2 = 2$ in (a), and $A = 2/3$, $M^2 = 0.5$ in (b).

On dimensional grounds, the terminal velocity of the bubble should be proportional to $\sqrt{\lambda g}$, i.e. $v_\infty = c_1\sqrt{\lambda g}$. Here c_1 is the constant of proportionality and it is a function of the dimensionless parameters A, M and γ only. From Eq. (5), we have,

$$c_1 = 0.15MA + 0.23A^{\frac{1}{2}}. \tag{6}$$

Since a constant value for γ was used in the study of the single bubble theory, the dependence of c_1 on γ was not explored.

IV. The Bubble Interactions. When more than one bubble is present in the system, the bubbles interact with each other. The dominant mechanism of bubble interaction is merger, whereby bubbles of different radii propagate with different velocities, and the leading bubbles grow in size at the expense of their neighbors. The velocity of large size bubbles exceeds the corresponding single-bubble limit at the end of the interaction with their neighboring bubbles.

The interactions between the bubbles can be approximated by a superposition hypothesis [2]. The superposition hypothesis treats each individual bubble and the outer envelope formed by the tips of the bubbles as a set of independent bubbles and spikes. At the tip of the more advanced bubble, the envelope and the bubble are in phase. Therefore the portion of the envelope near the tip of a more advanced bubble is approximated as a long wavelength bubble. Similarly, the portion of the envelope near the tip of a less advanced bubble is approximated as a long wavelength spike, since the bubble and the envelope are out of phase at that point. The bubbles and spikes formed by the outer envelope represent the coupling between neighboring bubbles. To the lowest order of approximation in the interactions between bubbles, the theory assumes that the velocity at the tip of each bubble in the multi-bubble system is the sum of the velocity of the corresponding bubble in a single-bubble system plus the velocity of the (single) spike or the (single) bubble formed by the outer envelope. The velocity of the individual bubble and the velocity of the envelope can be evaluated from the single-bubble theory given in section III. An example of bubble merger, obtained from the front tracking method, is shown in Figure 4.

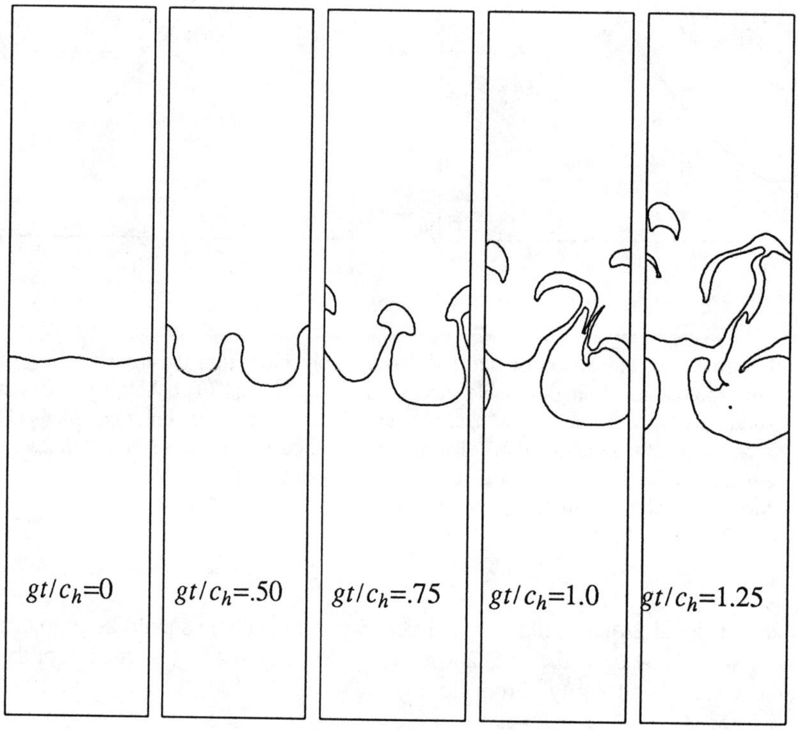

Figure 4. Successive times in a two bubble merger process. The compressibility and density ratio for this case are $M^2 = 0.1$ and Atwood number $A = 2/3$. It can be seen that the large bubble overtakes the smaller one at $gt/c_h = 1.0$. The velocity of the large bubble is accelerated during the merger while the velocity of the small bubble is reversed, see Figure 5.

The velocities at the tips of the bubbles in Figure 4 are recorded during the numerical

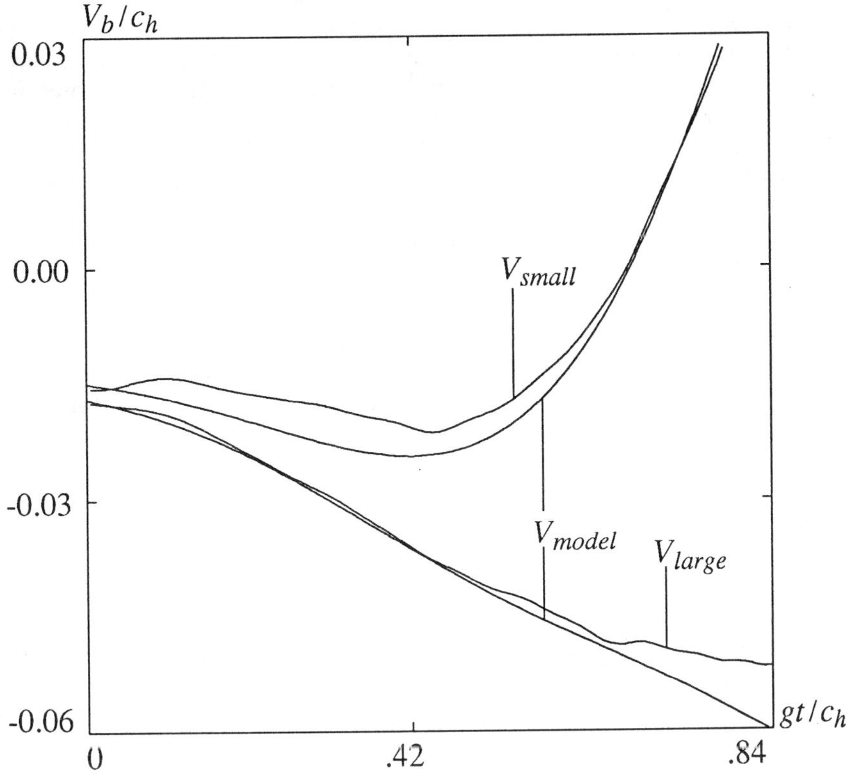

Figure 5. The plot of bubble velocities vs. time for the two bubble merger simulation. The result shows that the small bubble is accelerated at the beginning and is then decelerated after about $gt/c_h = .42$. The small bubble is washed downstream after its velocity is reversed. The large bubble is under constant acceleration. The smooth curves represent the bubble motion as predicted by the superposition hypothesis.

simulation. The comparison of the numerical data with the results of the superposition theory is given in Figure 5. The computations of Figures 4 and 5 were done by X. L. Li. The superposition theory explained the phenomena that the velocity of a more advanced bubble in a multi-bubble system exceeds the velocity of a bubble of same size in single-bubble system and that a less advanced bubble in a multi-bubble system changes its direction at the end of its interaction with its more advanced neighboring bubble. The validity of the superposition approximation is shown more clearly in the behavior of the small bubble in Figure 5. Since a small wavelength bubble has a larger growth rate than the large wavelength bubble in the linear regime, the velocity of the small bubble is dominated by the single bubble velocity initially. As the amplitude of the envelope increases, it reaches a point where the accelerations of the single bubble and the envelope have the same magnitude but different signs. Then the small bubble acceleration reverses. After that point, the contribution of the envelope dominates and the small bubble is washed downstream quickly. The experimental data of Read has also been analyzed [2]. The analysis shows that the agreement between the experiments and the superposition theory is in the range of 1.2-13.5%, which is within the experimental accuracy. The numerical verification of the superposition theory is obtained by comparison of this theory with direct numerical simulation of the full two-fluid Euler equations [2]. We emphasize that the superposition hypothesis contains no free parameters. The study shows that there are three phenomena to which the superposition theory does not apply: (1) the multiple connectivity of the multi-phase flow behind the bubble interface at late times, (2) the occurrence of tip splitting and secondary instability of bubbles, and (3) the exponential density stratification of the initial fluids for isothermal systems with high compressibility. Where applicable, the analysis indicates an agreement of the superposition theory with numerical simulation within the range of 1.4-16.7% [2].

V. The Chaotic Mixing Layer. When one considers an ensemble of bubbles, the successive processes of bubble merger lead to a chaotic mixing layer. During the processes of bubble merger, the large bubbles expand and the length scale of the mixing layer changes dynamically. The change of length scale places the chaotic mixing layer in a state of constant acceleration. Experiments show that the acceleration constant can be expressed as αAg, where α is a dimensionless constant insensitive to the fluid media used in the experiments. The value of α is about 0.06 [19].

Recently a statistical model, based on the renormalization group method, was constructed to describe the chaotic mixing process of Rayleigh-Taylor unstable interfaces [4]. The statistical model results from a simplification of the dynamics defined by the two-fluid Euler equations (Eqs. (1)). The state space of the model is the set of piecewise constant functions on the line, each constant piece representing a single bubble. The theory considers an ensemble of bubbles distributed uniformly over an interval $[-a(t)+\bar{h}(t), a(t)+\bar{h}(t)]$. Here $\bar{h}$ is the average position of the bubbles. Uniform distribution on the interval is not preserved dynamically, and so the dynamics is modified (approximated) to preserve this assumption. For simplicity the theory assumes that the bubbles have a common radius $r(t)$ at time t. In the molecular chaos approximation of the model, the nearest neighbor pairs have uncorrelated heights h_1 and h_2. Therefore the pair interactions are evaluated by drawing two adjacent bubbles randomly from the ensemble. Possible limitations to this approximation have been discussed in reference [17].

Since the radius of the bubbles changes dynamically with time as a consequence of bubble merger, we scale all variables accordingly. We define the following scaled variables: $t' = (Ag/r)^{1/2} t$ is the scaled time, $h' = (h-\bar{h})/r$ is the scaled height, $a' = a/r$ is the scaled width. h_m' is the scaled height separation of adjacent bubbles when merger is completed and $t_m' = t_m'(h_2' - h_1')$ is the scaled time to merger for adjacent bubbles with initial height separation $h_2' - h_1'$, where $h_2' \geq h_1'$. $dt' = (Ag/r)^{1/2} dt$ is the scaled differential time and $v' = (Agr)^{-1/2} v$ is a scaled velocity. Then $t_m'(h_m') = 0$. The statistical average of a function $F(h')$ over the mixing layer is defined as $\langle F(h') \rangle = \int_{-a'}^{a'} F(h')dh' / 2a'$, and the average of the two variable function $F(h_1', h_2')$ is defined as $\langle F(h_1', h_2') \rangle = \int_{-a'}^{a'} \int_{-a'}^{a'} F(h_1', h_2')\, dh_1' dh_2' / (2a')^2$.

The dynamics of the statistical model is that at merger height separation, h'_m, the higher bubble doubles in size (merges with its neighbor) while the lower bubble is removed from the ensemble. Before merger, each bubble moves with a scaled velocity $v' = v_b' + v_e'$ which is a sum of a scaled single bubble velocity v_b' and an envelope velocity v_e' (the superposition approximation) [2]. In the spirit of the molecular chaos approximation, one ignores here the fact that each bubble has two neighbors. Here v_b' is an absolute number, the single bubble terminal velocity, while $v_e' = v_e'(h_2' - h_1')$ depends on $h_2' - h_1'$ [1,17].

The growth of the bubble radius, in the continuous time approximation, is modeled by the equation [4]

$$\frac{dr}{dt} = (Ag)^{1/2} < \frac{1}{t_m'(h_2' - h_1')} > r^{1/2}, \tag{7}$$

which expresses the hypothesis that the more advanced bubble doubles in size at the end of its merger interval. This scaling is based on a picture of the merger process in which the radii of the two neighboring bubbles are frozen until merger is completed. Thus the time dependent radius $r(t)$ applies to the ensemble, but not to the pair of bubbles selected from the ensemble for merger. The same picture is used below to determine the bubble velocities.

The foregoing discussion, which specifies the statistical model and the approximations, leads to the fundamental dynamical equation for dh' / dt'. This equation includes a contribution from v_b' and a statistical contribution from v_e'. The dynamics for h' can be formulated either through its action on the stochastic variable h' or, by duality, on the probability measure which defines the distribution of h'. It is convenient to use both of these formulations, for distinct terms in the equations. Let G be a function of h'. The result is:

$$< \frac{dG}{dt'} > = v_b' < \frac{dG(h')}{dh'} > + < \frac{h_m' - h_2' + h_1'}{2\, t_m'(h_2' - h_1')} \frac{dG(h_2')}{dh'} >$$

$$- < G(h_1') \int \frac{dh_2'}{2a'\, t_m'(h_2' - h_1')} > + < \frac{1}{t_m(h_2' - h_1')'} > < G(h_1') > .$$

The first term is the uniform drift in h' from the single bubble velocity v_b'. The remaining terms derive from the envelope velocity v_e'. The second term is the contribution from merger for a bubble that is higher than its neighbor. It is expressed in terms of the action of the dynamics on the stochastic variable h'. Merger, per se, has no effect on the height of the higher (surviving) bubble. However, the instantaneous envelope velocity v_e', applied to the higher bubble, time averaged from the present to the time of merger, can be computed as $(1 / 2)\, \Delta$ (height separation) $/\ \Delta$ (time). This time averaged instantaneous scaled envelope velocity is the fraction $(h_m' - h_2' + h_1') / 2\, t_m'(h_2' - h_1')$ in the second term.

The third and fourth terms are the contributions to merger for a bubble below its neighbor. In these terms the dynamics is expressed in terms of its action on the measure. The third term corresponds to removal of the bubble from the ensemble and the fourth to its replacement with a bubble drawn at random from the ensemble. Consider the third term. The integral over h_2' is the probability per unit time for a bubble of height h_1' to be removed from the ensemble by merger. This is then integrated over all bubbles in the ensemble, i.e. over h_1'. The fourth term adds back a bubble in a probabilistic sense. Since $<1 / t_m(h_2' - h_1')'>$ is the probability per unit (scaled) time for removing a bubble of height h_1', integrated over h_1', i.e. the total probability per unit time of removing a bubble, the fourth term establishes conservation of probability. For all terms, the envelope dynamics is not given by the instantaneous velocity v_e' combined with merger dynamics, but instead this velocity is time averaged up to and including merger. This is the continuous time approximation, used also in (7).

By setting $G = \frac{h}{r}, \frac{\bar{h}'^2}{2}$, one has [4]

$$\frac{d(\bar{h} / r)}{dt'} = v_b' + \frac{1}{2} < \frac{h_m' - h_2' + h_1'}{t_m'(h_2' - h_1')} > - < \frac{h_1'}{t_m'(h_2' - h_1')} > \tag{8}$$

and

$$\frac{dw'^2}{dt'} = 2 < h_2' \frac{h_m' - h_2' + h_1'}{t_m'(h_2' - h_1')} > - < \frac{h_1'^2}{t_m'(h_2' - h_1')} > + w'^2 < \frac{1}{t_m'(h_2' - h_1')} >, \tag{9}$$

where w' is the variance of the scaled height, defined by the equation $w^2 = \int (h - \bar{h})^2\, dh / 2a$.

For a uniform distribution, one has $w' = a' / \sqrt{3}$. The natural interval for w' is given by $0 \le a'(t) \le h_m' / 2$. The lower limit is the trivial fixed point corresponding to an unstable interface consisting of bubbles of identical height. At the upper endpoint, defined by instantaneous merger for bubble pairs with extreme separation, the renormalization group equation directs the flow into the interval. One can show that the right hand side of Eq.(9) is positive at the lower end point, discontinuous at the upper endpoint, and negatively infinite for $a'(t) > h_m'/2$, i.e. above the upper endpoint. The existence of a nontrivial renormalization group fixed point follows.

Let $< \cdot >_{w*}$ denote the expectation evaluated at the renormalization group fixed point $w*$. Solving the equations (7)-(8) with $w' = w*$ yields $r(t) = \frac{Ag}{4} <1 / t_m'>_{w*}^2 \, t^2$, and

$$\bar{h}(t) = \frac{Ag}{2}\left[v_b' + < \frac{h_m' - h_2' - h_1'}{2\, t_m'(h_2' - h_1')} >_{w*} \right] <1 / t_m'>_{w*} t^2 \,. \tag{10}$$

By comparison with the expression $\bar{h} = \alpha A g t^2$, we have

$$\alpha = \frac{1}{2}\left[v_b' + < \frac{h_m' - h_2' - h_1'}{2\, t_m'(h_2' - h_1')} >_{w*} \right] <1 / t_m'>_{w*} \,. \tag{11}$$

Since the two variable function t_m' depends on the difference $h_2' - h_1'$ only, it is convenient to introduce the variables $z = h_2' - h_1'$ and $Z = (h_2' + h_1') / 2$. Then Eq.(11) become

$$\alpha = (v_b' + \frac{h_m'}{2T}) \frac{1}{2T} \,, \tag{12}$$

where $\frac{1}{T} = \int_0^{2a'} (2a' - z) / (4a'^2 t_m(z)) \, dz$ and the integration is evaluated at the fixed point. Here we have integrated over Z in Eq. (12). The variable z is associated with the envelope formed by the bubbles at the heights h_1' and h_2'. Using Eqs (3) and (4), we can evaluate α in Eq. (12) numerically for a given value of h_m'. The value for v_b' can be obtained from Eq. (5), and h_m' can be determined either from the study of bubble pair-interaction [2] or from experimental pictures [19].

The results of Read's experiments [19] on two dimensional nearly incompressible fluids show that for $D = 3.00$, the value of h_m' is in the range of 0.5-1.0. The value of α for nearly incompressible fluids with $D = 1.60$, 1.74 and 2.74 is also available from reference [19], although the experimental pictures of the mixing interfaces for such systems are not presented in that reference. We assume that h_m' is in the same range 0.5-1.0 for these systems.

In Figure 6, we show $\frac{dw'}{dt'}$ as a function of $2a'/h_m'$ for nearly incompressible systems, $M^2 = 0.1$, $D = 1.60$, 1.74, 2.74 and 3.00. Here the dimensionless compressibility M depends on the wavelength, which is measured from the envelope. The solid curves correspond to $h_m' = 0.5$ and the dashed curves correspond to $h_m' = 1.0$. The results show that $\frac{dw'}{dt'}$ is positive when $0 < a' < h_m'/2$. $\frac{dw'}{dt'}$ becomes negative and infinite when $a' > h_m'/2$ (see Eq. (9)). Therefore the fixed point appears at the upper limit, i.e. $a' = h_m'/2$. The dependence of α on h_m' appears explicitly in Eq. (12) and implicitly through $\frac{1}{T}$.

In Figure 7 we plot α as a function of h_m' for the systems described above. Figure 7 shows that the value of α is not sensitive to the value of h_m' for the parameter range studied in the experiments. The range of experimental values of α is also shown in Figure 7 for comparison.

In Table 1, we list the comparison between the theoretical values of α for h_m' in the range 0.5 - 1.0, and the experimental values for systems with density ratio $D = 1.60$, 1.74, 2.74 and 3.00. The last column of Table 1 is the average relative error, defined as $\frac{\Delta\alpha}{\alpha} = (<\alpha_{theory}> -$

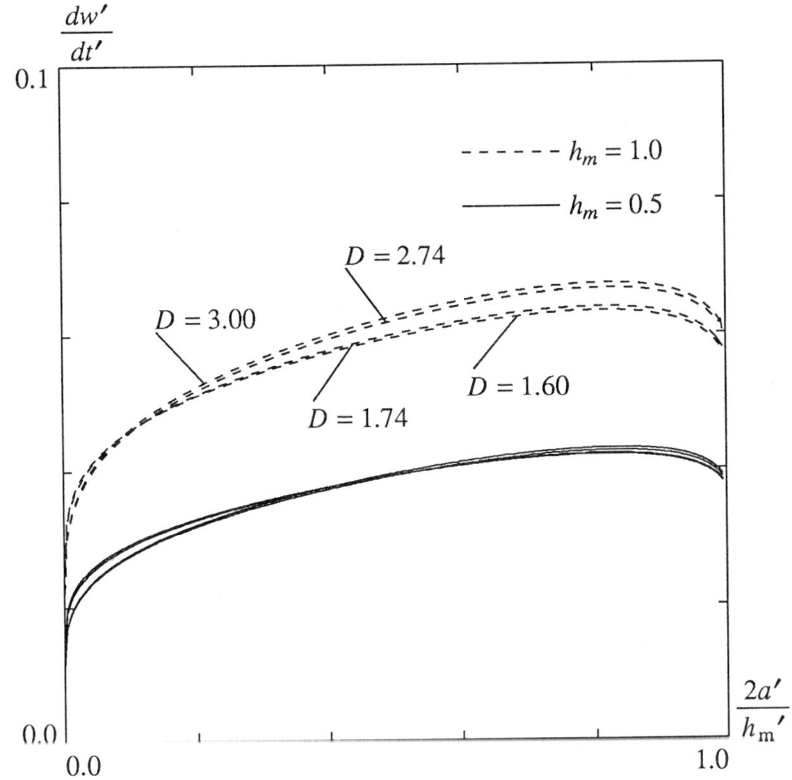

Figure 6. Plot of $\frac{dw'}{dt'}$ as a function of $2a'/h_m'$ for nearly incompressible systems, $M^2 = 0.1$, D = 1.60, 1.74, 2.74 and 3.00. The solid curves correspond to $h_m' = 0.5$ and the dashed curves correspond to $h_m' = 1.0$. The results show that $\frac{dw'}{dt'}$ is positive when $0 < a' < h_m'/2$. $\frac{dw'}{dt'}$ becomes negative and infinite when $a' > h_m'/2$ (see Eq. (9)). Therefore the fixed point appears at the upper limit, i.e. $a' = h_m'/2$.

$\langle\alpha_{experiment}\rangle) / \langle\alpha_{experiment}\rangle$. Figure 7 and Table 1 show that the theoretical values of α obtained from the renormalization group model are in good agreement with the experimental data. The experimental values of α for systems with very high density ratio ($D = 600$) are also available [19]. However, such nearly incompressible, very high density ratio systems have not been studied in detail in our studies of single-mode fingering growth and the pair-interaction between bubbles [1,2]. We refer to reference [14] for the study of the incompressible, infinity density ratio ($D = \infty$) system, where a value for α in the range of 0.05 - 0.06 is reported. The dimensionless acceleration constant α has also been studied in the numerical simulations of Youngs [16]. Youngs reported a value for α from numerical simulation as about 0.04 - 0.05 in reference [16]. In numerical simulations using the front tracking method, the value of α is about 0.055 - 0.065 until the interface becomes highly multiply connected. At late time, α decreases to 0.038 -0.044, due to a strictly two dimensional effect of change of multiphase connectivity of the front [2].

The constant acceleration of the mixing layer is the consequence of the dynamical change of the length scales. When the size of the bubble reaches the size of the physical domain, the length scale is fixed. Then the dimensional argument of a constant velocity holds. Such behavior

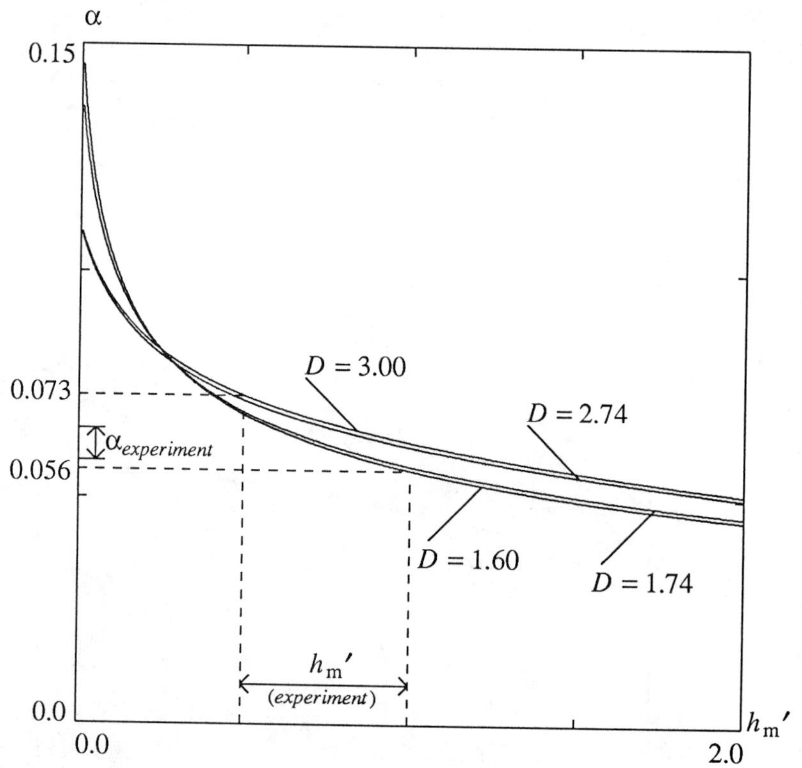

Figure 7. The dependence of α on the scaled merger height h_m' for nearly incompressible systems, $M^2 = 0.1$, $D = 1.60$, 1.74, 2.74 and 3.00. The range of theoretical values of α obtained from the renormalization group fixed point model with $h_m' = 0.5 - 1.0$, is marked by two horizontal dashed lines. The range of experimental value of α is also presented here for comparison. It shows that the results of the renormalization group fixed point model are in good agreement with the experimental data and that the theoretical values of α are not sensitive to the value of h_m' for the parameter range studied in the experiments.

has been observed in the numerical simulation of two bubble system (see the behavior of large bubble in Figures 4 and 5). However, there are about 20 to 40 bubbles initially both in experiments and in numerical simulations of the random interface. It will take about 5 generations of bubble merger to reach the stage of fixed length scale; while both experiments and numerical simulations are only carried up to about 2 generations of bubble merger.

VI. Conclusions. The single-bubble theory captures the dominant behavior of Rayleigh-Taylor unstable interfaces with a single bubble and spike. It covers the entire process of the development of the instability from the initial growth to the asymptotic regime at late times. The results of numerical simulations using the front tracking method show a good agreement with the theory. The dependence of the physical parameters which characterize the dynamic behavior of a single bubble or spike on the Atwood number A and the compressibility M are established for the parameter range being studied. It would be interesting to test the validity of the dependence for parameters outside the range being studied or to modify the dependence to cover a wider parameter range if more data were available.

D	α_{theory}	$\alpha_{experiment}$	$\frac{\Delta\alpha}{\alpha}$
1.60	0.056 – 0.068	0.058	7%
1.74	0.057 – 0.069	0.065	3%
2.74	0.061 – 0.071	0.060 – 0.064	9%
3.00	0.062 – 0.073	0.058 – 0.065	9%

Table 1. Comparison between the theoretical values of α for h_m' in the range 0.5 - 1.0, and the experimental data for systems with density ratio D = 1.60, 1.74, 2.74 and 3.00. The last column is the average relative error, defined as $\frac{\Delta\alpha}{\alpha} = (\langle\alpha_{theory}\rangle - \langle\alpha_{experiment}\rangle) / \langle\alpha_{experiment}\rangle$.

The study of the interaction between bubbles shows that the real bubble velocity can be approximated as the sum of the velocity of a single bubble and the velocity of the long wavelength envelope up to $1\frac{1}{2}$ generations of bubble merger. This simple addition of bubble velocity is not valid in several situations: for strongly gravitationally stratified fluids in isothermal systems; when secondary instability occurs; in the strongly nonlinear late time regime in which the light fluid becomes multiply connected. The multi-connectivity of the fluid is due to the fact that the numerical simulations are in two dimensions. The experiments are approximately two-dimensional. The superposition model has been supported by both the results of our numerical simulation and the analysis of the experiments. Beyond $1\frac{1}{2}$ generations of bubble merger, the dominant discrepancy between our simulations and experiment appears to be due to three-dimensional effects in the small scale structure in the mixing region in the experiments. Therefore for further comparison between the superposition theory and the experimental results, three-dimensional computations are required.

The results of the numerical simulation of multi-bubble systems for nearly incompressible flow using the front tracking method show a good agreement with the experimental results in the acceleration rate of the overall bubble envelope.

For isothermal systems with high compressibility, the exponentially stratified initial density distribution has very strong influences on the dynamic behavior of the bubbles. It would be interesting to study the behavior of Rayleigh-Taylor unstable interfaces under different initializations, e.g. the solutions of the linearized Euler equations for adiabatic systems.

The renormalization group fixed point model captures the dominant statistical behavior of bubbles in the chaotic mixing region. The model shows that the constant acceleration of the chaotic mixing layer is a direct consequence of scaling and the existence of a renormalization group fixed point. The model provides a purely theoretical explanation of constant acceleration and the first direct evidence in support of a renormalization group fixed point to describe the outer envelope of the mixing region for a Rayleigh-Taylor unstable interface. The value of α from a quantitative study of the renormalization group fixed point model agrees with the experimental data as well as with the data from numerical simulations [2,13,14,16,19,27].

The single-bubble theory, the superposition hypothesis and the renormalization group fixed point model can easily be extended to three dimensions. A three-dimensional front tracking code is currently under development. It would be very interesting to study these theories in three dimensions and to compare these theories with the three-dimensional experimental data.

REFERENCE

1. Q. Zhang, *The Motion of Single Bubble or Spike in Rayleigh-Taylor Unstable Interfaces*, preprint.

2. J. Glimm, R. Menikoff, X. L. Li, D.H.Sharp and Q. Zhang, *A Numerical Study of Bubble Interactions in Rayleigh-Taylor Instability for Compressible Fluids*, accepted by Physics of Fluids A.

3. D. H. Sharp and J. A. Wheeler, See AIP Document No. PAPS PFLDA-31-447-50 for 50 pages of Institute for Defense Analysis Log No. U-1997 (1961) by David H. Sharp and John A. Wheeler. Order by PAPS number and journal reference from American Institute of Physics, Physics Auxiliary Publication Service, 335 East 45th Street, New York, NY 10017, The price is $1.50 for each microfiche (98 pages) or $5.00 for photocopies of up to 30 pages, and $0.15 for each additional page over 30 pages, airmail additional. Make checks payable to the American Institute of Physics.

4. J. Glimm and D. H. Sharp, *Chaotic Mixing as a Renormalization Group Fixed Point*, Physical Review Letters, 64 (1990), pp. 2137-2139.

5. Q. Zhang, *Validation of the Chaotic Mixing Renormalization Group Fixed Point*, accepted by Physics Letters A.

6. Lord Rayleigh, *Investigation of the Character of the Equilibrium of an Incompressible Heavy Fluid of Variable Density*, Scientific Papers, Vol. II (Cambridge Univ. Press, Cambridge, England, 1900), pp. 200.

7. G. I. Taylor, *The instability of liquid surfaces when accelerated in a direction perpendicular to their planes. I*, Proc. R. Soc. London Ser. A 201 (1950), pp. 192-196.

8. G. Birkhoff and D. Carter, *Rising Plane Bubbles*, J. Math. Mech., 6 (1957), pp. 769.

9. P. R. Garabedian, *On steady-state bubbles generated by Taylor instability*, Proc. R. Soc. London A, 241 (1957), pp. 423-431.

10. G. R. Baker, D. I. Meiron and S. A. Orszag, *Vortex Simulation of the Rayleigh-Taylor Instability*, Phys. Fluids, 23 (1980), pp. 1485-1490.

11. C. R. Verdon, R. L. McCrory, R. L. Morse, G. R. Baker D. I. Meiron and S. A. Orszag, *Nonlinear Effects of Multifrequency Hydrodynamic Instabilities on Ablatively Accelerated Thin Shells*, Phys. Fluids, 25 (1982), 1653-1673.

12. R. Menikoff and C. Zemach, *Rayleigh-Taylor Instability and the Use of Conformal Maps for Ideal Fluid Flow*, Journal of Computational Physics, 51 (1983), 28-64.

13. J. Glimm and X. L. Li, *Validation of the Sharp-Wheeler Bubble Merge Model from Experimental and Computational Data*, Phys. of Fluids, 31 (1988), pp. 2077-2085.

14. J. A. Zufiria, *Vortex-in-Cell Simulation of Bubble Competition in Rayleigh-Taylor Instability*, (1988).

15. G. Tryggvason, *Numerical Simulation of the Rayleigh-Taylor Instability*, Journal of Computational Physics, 75 (1988), pp. 253-282.

16. D. L. Youngs, *Numerical Simulation of Turbulent Mixing by Rayleigh-Taylor Instability*, Physica D, 12 (1984), pp. 32-44.

17. C. L. Gardner, J. Glimm, O. McBryan, R. Menikoff, D. H. Sharp and Q. Zhang, *The dynamics of bubble growth for Rayleigh-Taylor unstable interfaces*, Phys. Fluids, 31 (1988), pp. 447-465.

18. J. Glimm, O. McBryan, R. Menikoff and D. H. Sharp, *Front Tracking Applied to Rayleigh-Taylor Instability*, SIAM J. Sci. Stat. Comput., 7 (1986), pp. 230-251.

19. K. I. Read, *Experimental Investigation of Turbulent Mixing by Rayleigh-Taylor Instability*, Physica D, 12 (1984), pp. 45-48.

20. D. J. Lewis, *The instability of liquid surfaces when accelerated in a direction perpendicular to their planes. II*, Proc. R. Soc. London A, 202 (1950), pp. 81-96.

21. H. W. Emmons, C. T. Chang and B. C. Watson, *Taylor instability of finite surface waves*, J. Fluid Mech., 7 (1960), pp. 177-193.

22. D. H. Sharp, *An Overview of Rayleigh-Taylor Instability*, Physica D, 12 (1984), pp. 32.

23. J. Glimm and O. McBryan, *Computational Model for Interfaces*, J. Adv. Appl. Math., 6 (1985), pp. 422-435.

24. I.-L. Chern, J. Glimm, O. McBryan, B. Plohr and S. Yaniv *Front Tracking for Gas Dynamics*, J. Comp Phys., 62 (1986), pp. 83-110.

25. J. Glimm, J. Grove, B. Lindquist, O. McBryan and G. Tryggvason, *The Bifurcation of Tracked Scalar Waves*, SIAM J. Sci. Comp., 9 (1988), pp. 61-79.

26. I. B. Bernstein and D. L. Book, *Effect of compressibility on the Rayleigh-Taylor instability*, Phys. Fluids, 26 (1983), pp. 453-458.

CHAPTER 6

Remarks on the Stability of Viscous Shock Waves*

Jonathan Goodman†

As was particularly evident at this conference, nonlinear travelling waves arise in many physical contexts and their stability is studied in a variety of ways. The most general way is what I call Sattinger's method because it was systematized in [9]. An important problem that could not be handeled by Sattinger's method is the stability problem for viscous shock profiles [4], [5], [7]. My purpose here is to suggest that the stability of viscous shock waves can be proven by techniques that closely resemble Sattinger's method. This would make the viscous profile proof simpler and somewhat sharper.

Let me quickly outline some parts of Sattinger's method. A one dimensional travelling wave is a solution of the form $\phi(x-st)$ where $\phi(x) \to u_l$ as $x \to -\infty$ and $\phi(x) \to u_r$ as $x \to \infty$. Here it is x and not ϕ that is one dimensional. We want to know whether travelling waves are asymptotically stable, that is, whether perturbations of ϕ decay to zero as $t \to \infty$. This cannot hold in general because of translation invariance: $\phi(x-\delta)$ will evolve as $\phi(x-\delta-st)$ and will not approach $\phi(x-st)$ as $t \to \infty$, no matter how small δ is. Moreover, a general perturbation of $\phi(x)$ will converge to $\phi(x-\delta-st)$ for some δ that depends on the perturbation and may not be known in advance. Thus we decompose a general perturbation of ϕ, $u(x,t)$, as

$$u(x,t) = \phi(x-\delta(t)-st) + v(x,t)\,,$$

and try to show for suitable $\delta(t)$ that

$$\|v(x,t)\| \to 0 \quad as \quad t \to \infty\,, \tag{1}$$

$$\int_0^\infty \left|\dot{\delta}(t)\right| dt < \infty\ . \tag{2}$$

In the Sattinger stability proof these estimates are interrelated; (1) cannot hold unless $\dot{\delta}$ satisfies some kind of orthogonality condition and (2) cannot hold unless (1) has a sufficient rate. In most cases where the theory applies (see elsewhere in this volume for examples

*I thank Anders Szepessy and Zhouping Xin for many helpful conversations on this material. The work was supported by an NSF PYI award and a Sloan Foundation Fellowship.

†Courant Institute of Mathematical Sciences, New York University.

and counterexamples), the estimates (1) and (2) follow from semigroup theory and a basic estimate for the linearized problem (with positive μ, C)

$$\|v(\cdot,T)\| \le C\left\{e^{-\mu T}\|v(\cdot,0)\| + \int_0^T e^{-\mu(T-t)}\left|\dot\delta(t)\right| dt\right\} , \tag{3}$$

if $\dot\delta(t)$ is chosen properly.

The current stability proof for viscous shock profiles ([4], [5], [7], which I call the old proof) is completely different. There is no estimate like (3) and its substitute requires that δ be determined in advance from the initial perturbation. While that is possible, it requires that the perturbation be localized and leads to proofs that are technical and not very sharp. Here I will suggest ways in which these difficulties might be overcome, but I emphasize that I cannot handle all the nonlinear details. These suggestions come in the form of three observations.

1 The Use of Flux Variables

We can see the need for some trick by considering the simple of a scalar conservation law, $u_t + f(u)_x = u_{xx}$. Write $u(x,t) = \phi(x-st) + \overline{u}(x,t)$; I use $\overline{u}$ instead of v to emphasize that there is no δ. Work in coordinates that move with the wave ($x \leftarrow x - st$), and linearize to find that $\overline{u}$ approximately satisfies (with $a(x) = f'(\phi(x))$)

$$\overline{u}_t + (a(x)\overline{u})_x = \overline{u}_{xx} \ . \tag{4}$$

Multiplication by $\overline{u}$ and integration by parts leads to

$$\frac{1}{2}\frac{d}{dt}\int \overline{u}(x,t)^2 dx = -\frac{1}{2}\int a_x\overline{u}^2 - \int \overline{u}_x^2 \ .$$

However, for a convex flux, f, and a compressive shock, we know that $a_x < 0$, so this estimate does not by itself lead to stability. In fact, we should not expect an L_2 estimate for (4) because *(i)* we do not expect that $\overline{u} \to 0$ as $t \to \infty$ (the shift problem), and *(ii)* the only natural norm is L_1 since (4) conserves mass and compresses the support of $\overline{u}$ so the amplitude must grow.

The old cure for this [4] was to integrate. If δ is chosen so that

$$\int (u(x,0) - \phi(x-\delta))\, dx = 0 \ ,$$

then, it follows formally from the evolution equation that $\int (u(x,t) - \phi(x-\delta))\, dx = 0$ for all t. Thus we may write $u(x,t) - \phi(x-\delta) = \partial_x U(x,t)$ with the hope that $U(\cdot,t) \in L_2(R)$. In any case, a calculation with (2) leads to the "integrated" evolution equation

$$U_t + a(x)U_x = U_{xx} \tag{5}$$

and to the estimate

$$\frac{1}{2}\frac{d}{dt}\int U(x,t)^2 dx = \frac{1}{2}\int a_x U^2 - \int U_x^2 \ .$$

In this estimate $a_x < 0$ is good and implies that $\|U(\cdot,t)\|_{L_2} \to 0$ as $t \to \infty$.

The weakness of this approach is that $\delta(t)$ is constant and must be chosen in advance. This requires the perturbation $\overline{u}(x,0) = u(x,0) - \phi(x)$ to be localized; otherwise $U = \int^x (u-$

ϕ) will not be square integrable with small L_2 norm. For example, if $\overline{u}(x,0) = g(x - x_0)$ for some small localized g then $\|\overline{u}\|$ remains small but still no matter how δ is chosen $\|U\|_{L_2} \to \infty$ as $x_0 \to \infty$.

The first of my three observations in that there is a different way to cure the $a_x < 0$ problem. This is to use the *flux variable*

$$w = \mathcal{F}\overline{u} = f'(\phi)\overline{u} - \overline{u}_x \ .$$

This variable also satisfies the *integrated equation* (5). Moreover the transformation $\overline{u} \to w$ does not require integration so for example $\|w\|$ is uniformly small when $\overline{u}(x,0) = g(x - x_0)$ and $x_0 \to \infty$ as above.

The flux transformation also treats δ in a more natural way. The operator $\mathcal{F}$ has ϕ_x in its kernel, as follows from the equation for ϕ: $f(\phi)_x = \phi_{xx}$. That this is the whole kernel is a consequence of the Poincaré type inequality:

Proposition 1 *There is a C so that for any p and $u \in L_p(R)$ there is a δ so that*

$$\|u - \delta\phi_x\|_{L_p} \leq C\,\|\mathcal{F}u\|_{L_p} \ .$$

In particular,

$$\delta = \frac{\int \phi_x(x)u(x)dx}{\int \phi_x(x)^2 dx} \tag{6}$$

works.

Proof: Integrate and use Young's inequality. See [3] for more details.

Remark 1: The formula (6) arises from minimizing

$$\|u - \delta\phi_x\|_{L_2} \ ,$$

so it might be surprising that it works for all L_p. Furthermore, although ϕ_x is an eigenfunction of $Lu = u_{xx} - (a(x)u)_x$, (6) gives a δ that evolves in time since L is not self adjoint. In general when expanding a function in terms of eigenfunctions, the expansion coefficients are gotten by taking inner products with left eigenfunctions (i.e. eigenfunctions of the adjoint operator). The right and left eigenfunctions will generally be different if the operator is not normal. We could expect the expansion coefficient corresponding to a zero eigenvalue to be time independent. In our case, the null eigenfunction for the adjoint operator ($L^* = \partial_{xx} + a(x)\partial_x$) is $\psi(x) \equiv 1$ and the expansion coefficient, $\int_{-\infty}^{\infty} \psi(x)u(x,t)dx = \int_{-\infty}^{\infty} u(x,t)dx$, is indeed constant in time. However, since ϕ_x is not a left eigenfunction we should not expect that $\frac{d}{dt}\int \phi_x(x)u(x,t)dx = 0$. Substitution into (4) shows in fact that it is not.

Remark 2: The flux variable transformation also works for systems and without linearization. Suppose $u_t + f(u)_x = u_{xx}$ represents a nonlinear system. Then the time derivative, $v = u_t$ satisfies $v_t + (f(u)'v)_x = v_{xx}$. However, since $v = \partial_x(-w)$ where $w = f(u) - u_x$ is the flux variable, the v equation integrates to $w_t + f(u)'w_x = w_{xx}$. Proposition 1 also holds for the vector valued case that arises from linearizing the nonlinear flux variable transformation for systems.

Remark 3: The motivation for finding $\mathcal{F}$ came from the linearized evolution equation $u_t = Lu$ where $Lu = u_{xx} - (f'(\phi)u)_x$. Since L has a kernel, $L\phi_x = 0$, it is natural to seek an operator, $\mathcal{G}$, that removes the kernel of L. Then it might be possible to prove that $\mathcal{G}u \to 0$ as $t \to \infty$. Clearly, $\mathcal{G} = L$ is one choice but that leads back to the same equation for Lu. However, Lu is better than u in that Lu can be integrated while u cannot in general.

2 Weighted Norm Estimates

One way or another (by integrating or by using flux variables) one arrives at the *integrated equation*

$$U_t + A(x)U_x = U_{xx} \ .$$

I write $A(x)$ in place of $a(x)$ to indicate that this is a system of equations rather than a scalar equation. This is diagonalized (see [4]) by using $L(x)$ and $R(x)$ (matrices of left and right eigenvectors of $A(x)$) to give roughly

$$V_t + \Lambda(x)V_x + M(x)V = V_{xx}$$

where $\Lambda(x) = diag(\lambda_1(x), \cdots, \lambda_n(x))$ and $M(x) \leq C \cdot |\phi_x(x)|$. The old thing to do [4] is to multiply by V^t (the transpose of V) and integrate, which gives

$$\frac{d}{dt}\int |V|^2 \leq \int \lambda_{lx}V_l^2 + C\int |\phi_x|\,|V'|^2 - 2\int |V_x|^2$$

where V' represents all components of V except V_l. Here l is the family of the shock. The analogue of $a_x < 0$ for scalars is $\lambda_{lx} < 0$. Unfortunately, there is nothing to say about the sign of other derivatives. This does not give a simple stability estimate $\frac{d}{dt}\int |V|^2 \leq 0$ but it can be nursed along with help of the *vertical estimate* [4]

$$\int_0^\infty |V'(x,t)|^2\,dt \leq \int |V(x',t)|^2\,dx' + \epsilon \int\int |\phi(x')|\,|V(x',t)|^2\,dx'dt$$

where ϵ can be made small by taking weak shocks.

A much simpler approach (my second observation) uses weighted norms. This is the same as using weighted variables $\tilde{V}_k$ defined from the V_k using weights α_k by $\tilde{V}_k = \alpha_k(x)V_k$. Then proper choice of the α_k leads to essentially

$$\tilde{V}_t + \Lambda(x)\tilde{V}_x + \tilde{M}(x) = \tilde{V}_{xx}$$

where $\tilde{M}$ has positive diagonals and is diagonally dominant. Solutions of such an equation satisfy

$$\frac{d}{dt}\left\|\tilde{V}(\cdot,t)\right\|_{L_p} \leq 0$$

for all p. To do this, it suffices to take the weights to satisfy, for $k \neq l$,

$$\frac{\alpha_{kx}(x)}{\alpha_k(x)} = C \cdot \frac{|\phi_x(x)|}{\lambda_k(x)}$$

with suitable C. The α_k will be bounded since $|\phi_x(x)| \to 0$ exponentially as $|x| \to \infty$ and λ_k never vanishes if $k \neq l$ [4]. We make take $\alpha_l = 1$.

Sattinger and Pego [8] used weighed norms to prove estimates like (3) for the linearized problem. That does not seem to work for nonlinear systems, as an example of Pego [8] shows clearly.

There is an anology between the weights used here and the functional that Glimm [2] used in his existence theorem for inviscid conservation laws. The Glimm functional is roughly

$$F = \sum_{i,j} |\beta_i(j,k)| + \sum_{\substack{i,j,i',j' \\ app}} |\beta_i(j,k)\beta_{i'}(j',k)| \ .$$

Here $\beta_i(j,k)$ is the strength of the wave of family i at the grid point (x_j, t_k) and *app* means the all pairs i,j and i',j' that are approaching (in the sense of Glimm). This can be rewritten as

$$F = \sum_{i,j} \alpha_i(j,k)\, |\beta_i(j,k)|$$

where

$$\alpha_i(j,k) = 1 + C \cdot \sum_{\substack{i',j' \\ app(i,j)}} C \cdot |\beta_{i'}(j',k)|$$

where $app(i,j)$ is the set of waves i',j' that are approaching the wave i,j. Glimm's proof that F is nonincreasing in time is roughly that when wave $\beta_i(j,k)$ interacts with wave $\beta_{i'}(j',k)$ the wave strengths $|\beta_i(j,k)|$ and $|\beta_{i'}(j',k)|$ can increase, but this increase is compensated by a decrease in the weights $\alpha_i(j,k)$ and $\alpha_i'(j',k)$, since the waves are no longer approaching after the interaction. This is precisely the argument used above for the interaction of the perturbation components V_k with the profile ϕ; the interaction can produce more waves, but the potential to interact is decreasing at the same time so the total interaction strength is finite.

3 Rates of Decay

Global existence theorems for nonlinear problems can hinge on decay rates for solutions of linearized problems. In the conservation law context, this can be seen in the existence proof of Tartar [10]. Also, rates of decay have been of interest for their own sakes [1]. The third and final observation is that some of the above ideas lead to direct energy method proofs of rates of decay.

Proposition 2 *If* $u_t + f'(\phi)u_x = u_{xx}$ *(scalar) then*

$$\|u(\cdot,t)\|_{L_\infty} \le \frac{C}{\sqrt{t}} \|u(\cdot,0)\|_{L_1}$$

Proof Sketch: For the heat equation the proof is a combination of energy estimates, the Sobolev inequality, and duality:

Step 1, $\|u(\cdot,t)\|_{L_\infty} \le \frac{1}{t^{1/4}} \|u(\cdot,0)\|_{L_2}$**:** This can be found in [4] and uses the inequalities

$$\|u(\cdot,t)\|_{L_\infty} \le \sqrt{\|u(\cdot,t)\|_{L_2} \cdot \|u_x(\cdot,t)\|_{L_2}} \ ,$$

$$\|u(\cdot,t)\|_{L_2} \le \|u(\cdot,s)\|_{L_2} \quad if \quad s \le t \ , \tag{7}$$

$$\|u_x(\cdot,t)\|_{L_2} \le \frac{1}{\sqrt{t}} \|u(\cdot,0)\|_{L_2} \ .$$

The last inequality uses the fact that u_x also satisfies the heat equation so (7) also applies to u_x.

Step 2, Duality: The backwards heat equation is dual to the forwards heat equation: if $v_t = -v_{xx}$ and $u_t = u_{xx}$ then $\frac{d}{dt}\int u(x,t)v(x,t)dx = 0$. Thus, the backward heat equation satisfies a dual estimate to Step 1. Reversing time and replacing v by u then gives, for solutions of the forwards heat equation,

$$\|u(\cdot,t)\|_{L_2} \leq \frac{1}{t^{1/4}} \|u(\cdot,0)\|_{L_1}$$

Step 3, Composition of estimates: Putting together the estimates from Step 1 and Step 2 gives us

$$\|u(\cdot,t)\|_{L_\infty} \leq \frac{2^{1/4}}{t^{1/4}} \|u(\cdot,t/2)\|_{L_2} \leq \frac{2^{1/2}}{t^{1/2}} \|u(\cdot,0)\|_{L_1} \quad ,$$

as desired.

All these steps can be carried out for $u_t + a(x)u_x = u_{xx}$. The analogue of Step 1 is proved by differentiating and then taking the flux transformation: $w = \mathcal{F}u_x$ satisfies $\|w(\cdot,t)\|_{L_2} \leq \|w(\cdot,0)\|_{L_2}$. The dual equation, $v_t - (a(x)v)_x = -v_{xx}$, also has energy estimates. See [3] for details.

Remarks: *(i)* For Burgers' equation, Proposition 2 can be verified by linearizing the Hopf-Cole transformation [6]. *(ii)* If you do not localize, the decay rate cannot be faster than for constant coefficient parabolic equations. Thus, at least the power $t^{-1/2}$ is optimal. *(iii)* For constant coefficient problems, the decay rates follow quickly from obvious properties of the explicit solution kernel. *(iv)* Together, Proposition 2 and Proposition 1 show that solutions of the unintegrated perturbation equation (4) have the decay property that there is a $\delta(t)$ so that $\|\bar{u} - \delta\phi_x\|_{L_\infty} \leq \frac{C}{\sqrt{t}} \|\bar{u}(\cdot,0)\|_{L_1}$

References

[1] I. L. Chern and T. P. Liu, *Convergence to diffusion waves of solutions for viscous conservation laws*, Comm. Math. Phys., 110 (1987), pp. 503–517.

[2] J. Glimm, *Solutions in the large for nonlinear hyperbolic systems of equations*, Comm. Pure Appl. Math., 18 (1965), pp. 697–715.

[3] J. Goodman, *Remarks on the stability of viscous shock waves.* in preparation.

[4] ——, *Nonlinear asymptotic stability of viscous shock profiles for conservation laws*, Arch. Rat. Mech. Anal., 95 (1986), pp. 325–344.

[5] S. Kawashima and A. Matsumura, *Asymptotic stability of travelling wave solutions of systems for one dimensional gas motion*, Comm. Math. Phys,, 101 (1985), pp. 97–127.

[6] P. D. Lax, *Hyperbolic Systems of Conservation Laws and the Mathematical Theory of Scock Waves*, Soc. Ind. Appl. Math., Philadelphia, 1973.

[7] T.-P. Liu, *Nonlinear Stability of Scock Waves for Viscous Conservation Laws*, vol. 328 of AMS Memoirs, Amer. Math. Soc., 1986.

[8] R. Pego, *Remarks on the stability of shock profiles for conservatoin laws with dissipation*, Trans. Amer. Math. Soc., 291 (1985), pp. 353–361.

[9] D. SATTINGER, *On the stability of waves of nonlinear parabolic systems*, Adv. Math., 22 (1976), pp. 312–355.

[10] L. TARTAR, *Conservation laws.* unpublished.

CHAPTER 7

A Phase Transition Problem—A Dynamical Systems Approach

Harumi Hattori*
Konstantin Mischaikow†

Abstract. In this note we discuss a dynamical systems approach to a phase transition problem based on the Korteweg theory of capillarity. We first mention the motivations for this work in the Introduction. Then, we discuss the existence of a global solution to show that we have a dynamical system. We discuss the stability and bifurcation analysis of stationary solutions and then we study the connecting orbit problems in the semiflow. The connection matrix is a useful tool to discuss qualitative aspects of the dynamical behavior of solutions.

1. Introduction. In this note we study the global dynamics for the following system of parabolic equations

$$\begin{aligned} p_t &= \nu p_{xx} - \eta q_{xx} + \sigma(q) - P, \\ q_t &= p_{xx}, \end{aligned} \tag{1.1}$$

where $x \in [0,1]$ and the boundary conditions are given by

$$\begin{aligned} p_x(0,t) &= 0, \quad p(1,t) = 0, \\ q_x(0,t) &= 0, \quad q_x(1,t) = 0. \end{aligned} \tag{1.2}$$

*Supported in part by Army Grant DAAL 03-89-G-0088. Department of Mathematics, West Virginia University, Morgantown, WV 26506
†Supported in part by NSF Grant DMS-8901273. School of Mathematics, Center for Dynamical Systems and Nonlinear Studies, Georgia Institute of Technology, Atlanta, GA 30332

The above system is derived from an equation

$$u_{tt} = \sigma(u_x)_x + \nu u_{xxt} - \eta u_{xxxx} \tag{1.3}$$

with boundary conditions

$$u(0,t) = 0, \quad \sigma(u_x(1,t)) + \nu u_{xt}(1,t) - \eta u_{xxx}(1,t) = P, \tag{1.4}$$

$$u_{xx}(0,t) = 0, \quad u_{xx}(1,t) = 0, \tag{1.5}$$

by setting $p = \int_1^x u_t \, dx$ and $q = u_x$. Equation (1.3) models a bar which goes through a phase transition. The boundary conditions (1.4) show that the bar is under a soft loading device. The boundary conditions (1.5) are the natural boundary conditions for the corresponding variational problem. The terms with the coefficients ν and η are called viscosity and capillarity terms, respectively. In what follows, we assume that σ is given by Fig. 1.1. In this figure $(0, \alpha^*]$ and $[\beta^*, \infty)$ are called the α-phase and the β-phase, respectively. They correspond to the different phases of the material.

There are two motivations for this work. One motivation comes from the following viscous equation. Andrews & Ball [**1,2**], Dafermos [**3**], and Pego[**4**] considered the equation

$$u_{tt} = \sigma(u_x)_x + \nu u_{xxt}. \tag{1.6}$$

Pego has shown for (1.6) with the soft loading device that some discontinuous stationary solutions are dynamically stable. Andrews & Ball [**2**] and Slemrod [**5**] considered equation (1.3).

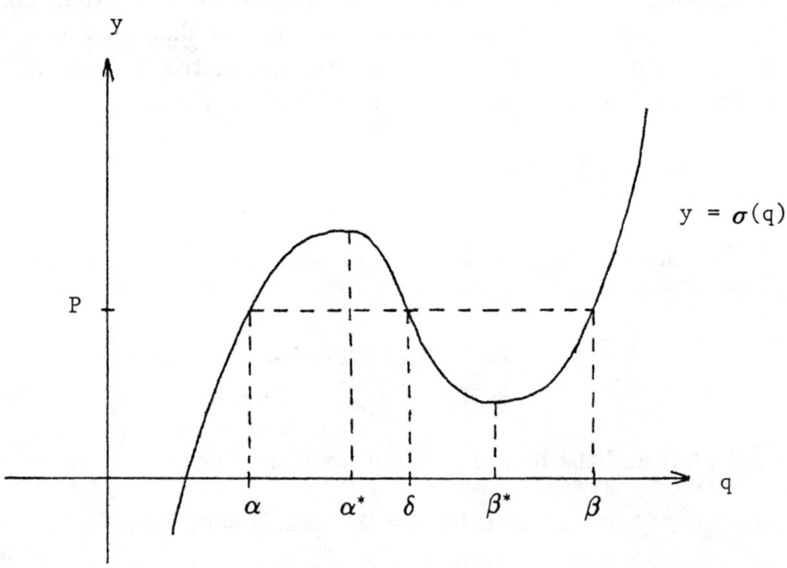

Figure 1.1

Another motivation comes from the corresponding variational problem. Carr, Gurtin, and Slemrod [6] considered the following problem arising from (1.3), (1.4), and (1.5).

Problem S_η: Minimize $E(v) = \int_0^1 \{W(v) - Pv + \frac{\eta}{2}(v')^2\}(x,t)\,dx$.

For this problem they have shown that $v = \alpha, \beta$ are stable (minimizers) and $v = \delta$ and all nonconstant solutions are unstable (not minimizers). (See [6] for the precise statements.) Therefore, it is interesting to see the dynamical consequence of the capillarity term.

In what follows, we discuss first the existence of global solutions, then stability of stationary solutions and the bifurcation diagram, and finally the connecting orbit problems in the semiflow. We omit most of the proofs, as they will appear in [7].

2. Existence of a global solution.

We state the theorem establishing the existence of a global solution to (1.1) and (1.2).

Theorem 2.1 *Suppose σ is Lipshitz continuous and that $(p_o, q_o) \in L^2(0,1)$. Then there exists a unique global solution $(p,q) \in H^2(0,1)$.*

We define the operator A by

$$A\begin{pmatrix} p \\ q \end{pmatrix} = \begin{pmatrix} \nu p_{xx} - \eta q_{xx} \\ p_{xx} \end{pmatrix}$$

and show that A with the boundary conditions (1.2) is an infinitesimal generator of a compact analytic semigroup. It should be mentioned that since the boundary conditions are not typical, they cause some difficulty in estimating the eigenvalues and the resolvent. The compact analytic semigroup gives local existence. Now, we use the following identity

$$E(p,q)(t) + \int_0^t \int_0^1 \nu p_{xx}^2(x,s)\,dx ds = E(p,q)(0), \tag{2.1}$$

where

$$E(p,q)(t) = \int_0^1 \{\frac{1}{2}p_x^2 + W(q) - Pq + \frac{\eta}{2}q_x^2\}(x,t)\,dx, \tag{2.2}$$

as the *a priori* estimate for the L^2 norm of (p,q) so that the continuation argument is possible, and hence we can show the existence of a global solution.

3. Stability and bifurcation analysis.

Here, we discuss the stability of stationary solutions and their bifurcations.

Lemma 3.1 *The constant solutions $(0,\alpha)$, $(0,\beta)$, and $(0,\delta)$ are stationary solutions for all values of $\eta > 0$. Furthermore, their indices are $\mathbf{h}((0,\alpha)) = \mathbf{h}((0,\beta)) = \Sigma^0$. Namely, they are dynamically stable.*

Next, consider the eigenvalue problem corresponding (1.1) and (1.2):

$$\begin{aligned} &\eta q_{xx} - (\nu\lambda + \sigma'(\delta))q = -\lambda p, \\ &p_{xx} = \lambda q. \end{aligned} \tag{3.1}$$

Lemma 3.2 *The eigenvalues of (3.1) cross the origin from left to right of the imaginary axis at $\eta = -\sigma'(\delta)/(n\pi)^2$ as we decrease η. Furthermore zero eigenvalues are simple.*

Lemma 3.3 *If $\sigma \in C^3$, $\sigma'(\delta) < 0$, and $\sigma'''(\delta) > 0$, then there is a supercritical pitchfork bifurcation at $\eta = -\sigma'(\delta)/(n\pi)^2$. Furthermore, if $\sigma(u+\delta)/u < \sigma'(\delta)$ for $\alpha - \delta \le u \le \beta - \delta$ except at $u = 0$, then there is no secondary bifurcation along the non-constant stationary solutions.*

Theorem 3.2 *There exists a global compact attractor $\mathbf{A}$ for (1.1) and (1.2).*

The above lemmas and the theorem imply

Lemma 3.4 *If $\eta > -\sigma'(\delta)/\pi^2$, the stationary solution $(0,\delta)$ has one dimensional unstable manifold or equivalentely the index is $\mathbf{h}((0,\delta)) = \Sigma^1$.*

Combining the above lemmas and the theorem we have

Theorem 3.3 *If $\sigma \in C^3$, $\sigma'(\delta) < 0$, and $\sigma'''(\delta) > 0$, then the following holds:*

(i) For $-\sigma'(\delta)/(n\pi)^2 < \eta < -\sigma'(\delta)/((n-1)\pi)^2$, $(0,\delta)$ is a nondegenerate stationary solution and has an n-dimensional unstable manifold.

(ii) If $M(k^{\pm})$ denote the non-constant stationary solutions which arise from the bifurcation point $\eta = -\sigma'(\delta)/(k\pi)^2$, then $M(k^{\pm})$ are non-degenerate and have k-dimensional unstable manifolds.

4. Conley theory. In this section we briefly explain the connection matrix which will be used in the following section. This is an extension of Conley's index and applies to a Morse decomposition of isolated invariant sets which induces a partial ordering P. The connection matrix is a linear map on this partial ordering satisfying the following properties.

(1) The connection matrix Δ is a linear mapping (denoted by Δ) from the index of a Morse set to the index of a Morse set. In our case it is a mapping from the index of a stationary solution to the index of a stationary solution, because the system is a gradient system.

(2) The elements of the matrix are zero or one ($\mathbf{Z}_2$ field) and the entry with the index Σ^k to the one with the index Σ^{k-1} is possibly one (degree -1 map). If it is one there is a connecting orbit from Σ^k to Σ^{k-1}.

(3) It is strictly upper triangular.

(4) $\Delta^2 = 0$.

(5) $\mathrm{Ker}(\Delta)/\mathrm{Im}(\Delta) = H_*(h(\mathbf{A}))$.

The above properties are valid for the submatrix of Δ if it satisfies additional properties. See [7] for more details.

5. Connecting orbit problems. We now discuss the connecting orbit problem in the semiflow. In the semiflow the connecting orbit means the solutions connecting two stationary solutions, namely,

$$\lim_{t\to-\infty}(p,q)(t) = \text{a stationary solution},$$

$$\lim_{t\to\infty}(p,q)(t) = \text{another stationary solution}.$$

To simplify the notation let $M(0^+) = (0,\alpha)$, $M(0^-) = (0,\beta)$, and $M(n) = (0,\delta)$. Then, we have

Theorem 5.4 *Given a collection* $\{j^*, j+1^*, j+2^*, ..., j+r^* \mid * = +or-\}$ *and* $\epsilon > 0$, *there exists a solution* $(p(t), q(t))$ *of (1.1) and (1.2) and a sequence* $t_1 > t_2 > ... > t_{r-1}$ *such that*

$$\lim_{t\to-\infty}(p(t), q(t)) = M(j+r^*), \quad \lim_{t\to\infty}(p(t), q(t)) = M(j^*), \tag{5.1}$$

and

$$d(M(j+i^*), (p(t_i), q(t_i))) < \epsilon. \tag{5.2}$$

Furthermore,

$$cl(C(M(j+r^*), M(j^*))) \cap cl(C(M(j+i^*), M(j+s^*))) \neq \emptyset, \text{ for } 0 \le s \le r.$$

and $cl(C(M(j+r^*), M(j^*)))$ *contains an* $(r-1)$ *parameter family of orbits.*

This theorem establishes that there is always a connecting orbit from a stationary solution with higher dimensional unstable manifold to a stationary solution with lower one, and if the number of unstable manifolds differs by r, then there is an $(r-1)$ parameter family of connecting orbits. To prove this we apply the connection matrix to the global compact attractor $\mathbf{A}$ whose homology index is Σ^0. When there are $(2n+1)$ stationary solutions, we can show that the connection matrix is given by

$$\begin{array}{c} \\ M(0^*) \\ M(1^*) \\ M(2^*) \\ \vdots \\ M(n) \end{array} \begin{array}{c} \begin{array}{ccccc} M(0^*) & M(1^*) & M(2^*)\cdots & & M(n) \end{array} \\ \left(\begin{array}{ccccc} 0 & B_1 & 0 & \cdots & 0 \\ & 0 & B_2 & \cdots & \vdots \\ & & 0 & & \\ & & & & B_n \\ & & & & 0 \end{array}\right) \end{array},$$

where

$$B_k = \begin{pmatrix} 1 & 1 \\ 1 & 1 \end{pmatrix}, \quad k = 1, \cdots, (n-1)$$

and

$$B_n = \begin{pmatrix} 1 \\ 1 \end{pmatrix}.$$

It is not difficult to verify the properties (2) - (4) of the connection matrix. For (5) we should note that

$$H_*(h(\mathbf{A})) = (Z_2, 0, 0, \cdots).$$

Hence, $H_*(h(\mathbf{A})) = 1$. This implies that $Ker(\Delta) - Im(\Delta) = 1$.

References

[1] Andrews, G., *On the existence of solutions to the equation* $u_{tt} = u_{xxt} + \sigma(u_x)_x$, J. Diff. Eqns. 35 (1980), 200-231.

[2] Andrews, G. and J.M. Ball, *Asymptotic behaviour and change of phase in one-dimensional nonlinear viscoelasticity*, J. Diff. Eqns. 44 (1982), 306-341.

[3] Dafermos, C.M., *The mixed initial-boundary value problem for the equation of nonlinear one-dimensional viscoelasticity*, J. Diff. Eqns. 6 (1969), 71-86.

[4] Pego, R., *Phase transitions: Stability and admissibility in one-dimensional viscoelasticity*, IMA Report # 180 (1985).

[5] Carr, J., M.E. Gurtin, and M. Slemrod, *One dimensional structured phase transformations under prescribed loads*, J. Elasticity 15 (1985), 133-142.

[6] Slemrod, M., *Admissiblity criteria for propagating phase boundaries in a van der Waals fluid*, Arch. Rat. Mech. Anal. 81 (1983), 301-315.

[7] Hattori, H. and K. Mischaikow, *A dynamical systems approach to a phase transition problem*, to appear in J. Diff. Eqns.

CHAPTER 8

Nonphysical Limits of Solutions of the Navier-Stokes Equations for Compressible Flow

David Hoff†
Denis Serre‡

Abstract. We show that physical solutions of the Navier-Stokes equations for one-dimensional, compressible flow need not depend continuously on their initial data, at least when vacuum states are allowed. Specifically, we consider two fluid regions initally separated by a third region of very low density δ. We show that, as $\delta \to 0$, the physical solutions corresponding to $\delta > 0$ do not in fact converge to a physical solution, but rather to a nonphysical weak solution in which the two fluids cannot collide, independent of their initial velocities, and whose separate momenta are not conserved.

We show that physical solutions of the Navier-Stokes equations for one dimensional compressible flow need not depend continuously on their initial data, at least when vacuum states are allowed. Specifically, we consider a gas containing an internal region of very low density δ expanding into a vacuum. When $\delta = 0$ the correct, physical solution thus contains two fluid regions, each of whose momenta is separately conserved, and which may collide at some positive time, depending on their initial velocities. We show that, as $\delta \to 0$, the solutions corresponding to $\delta > 0$ do not in fact converge to this physical solution, but rather to a nonphysical weak solution in which the two fluids cannot collide, and whose separate momenta are not conserved. In the present note we describe these solutions in a fairly general fashion; complete technical details are worked out in Hoff and Serre [3].

The flows which we consider are solutions of the Navier-Stokes equations for viscous, compressible, isentropic flow in one space dimension:

$$\begin{aligned} &\rho_t + (\rho u)_x = 0 \\ &(\rho u)_t + (\rho u^2 + P(\rho))_x = \epsilon u_{xx} \ . \end{aligned} \tag{1}$$

Here ρ, u, and P represent respectively the density, velocity, and pressure, and ϵ is a positive viscosity parameter. P is assumed to be in $C^1(\{\rho \geq 0\})$, and to satisfy $P, P' > 0$ for $\rho > 0$, $P(0) = 0$, and the bound

$$\int_0^1 s^{-2} P(s)\, ds < \infty \ .$$

† Department of Mathematics, Indiana University, Bloomington, IN 47405. Research of the first author was supported in part by the NSF under Grant No. DMS 9001606.

‡ Ecole Normale Superieure de Lyon, 46, alee d'Italie, Lyon Cedex 07, France.

These conditions are satisfied, for example, by the pressure in a polytropic fluid, $P = C\rho^\gamma, \gamma > 1$.

First we describe a particular solution (ρ, u) containing two initially disjoint fluids expanding into a vacuum which approach each other rapidly. Thus fix points $a_0 < b_0 < c_0 < d_0$ and smooth initial data (ρ_0, u_0) defined on the intervals $[a_0, b_0]$ and $[c_0, d_0]$, with ρ_0 strictly positive. We require that equations (1) hold on the sets $\{(x,t) : a(t) \le x \le b(t)\}$ and $\{(x,t) : c(t) \le x \le d(t)\}$, where the curves $a(t), \ldots$ are particle trajectories starting from $a_0, \ldots$; i.e.,

$$\dot{a}(t) = u(a(t), t) \ , \ a(0) = a_0 \ , \ \text{etc.} \tag{2}$$

We also impose the boundary conditions

$$P(\rho) = \epsilon u_x \tag{3}$$

at $x = a+, b-, c+, d-$. A short computation based on the second equation in (1) then shows that, at least formally, the separate momenta

$$\int_a^b \rho u \, dx \text{ and } \int_c^d \rho u \, dx$$

are constant in time. Observe that, since translates

$$\begin{aligned} \bar{\rho}(x,t) &= \rho(x - \sigma t, t) \\ \bar{u}(x,t) &= u(x - \sigma t, t) + \sigma \end{aligned}$$

are also solutions of (1) when (ρ, u) is a solution, we can take u_0 to be large and positive on $[a_0, b_0]$ and large and negative on $[c_0, d_0]$ so as to force a collision $b(t) = c(t)$ in finite time. The global existence of the solution (ρ, u) follows from the results of Serre [4], who considers a single gas expanding into a vacuum. We therefore construct separate solutions for the fluids on $[a, b]$ and on $[c, d]$ and simply take their superposition in x-t space, at least up until the collision time. At the collision time we then solve a new initial value problem for the single fluid on $[a, d]$ expanding into a vacuum. (There is no mathematical difficulty in treating solutions even of the full Navier-Stokes equations in one space dimension with discontinuous initial data; see Hoff [1] and [2].)

To reiterate, we have constructed a solution (ρ, u) which solves the equations (1) on its support (that is, the support of the density) with the following properties: the boundary of the support consists of particle trajectories on which (3) holds; a collision $b(t) = c(t)$ occurs in finite time; and the separate momenta of the two fluids are constant up until the collision time, and the total momentum is constant for all time.

Although we do not expect the Navier-Stokes equations (1) to be a realistic physical model at or near vacuum states, we nevertheless attempt to extend the above solution (ρ, u) to the entire upper half plane $\{t > 0\}$. First observe that, when $\rho = 0$, the equations (1) reduce to the simple requirement that $u_{xx} = 0$, i.e., that $u(\cdot, t)$ be linear in x for each fixed t. (The Navier-Stokes equations thus constrain the velocity even in regions devoid of fluid particles!) Next, by applying the Rankine-Hugoniot condition to the equations (1), we find that, if (ρ, u) is a smooth solution except on a curve of discontinuity $(x(t), t)$, then (ρ, u) will still be a distribution solution provided that

$$[u] = 0 \ , \ \dot{x}(t) = u(x,t) \ , \ \text{and } [P] = [\epsilon u_x] \ , \tag{4}$$

where $[\cdot]$ denotes the jump across the discontinuity curve. The particular solution (ρ, u) constructed above can therefore be extended to the left of the curve $(a(t), t)$ by taking $\rho = 0$ and $u(x,t) = u(a(t), t)$, and similarly to the right of the curve $(d(t), t)$. This construction fails, however, in the cavity region $\{b(t) < x < c(t)\}$ prior to the collision time. Indeed, the boundary condition (3) and the jump relations (4) would require that

$$0 = (P - \epsilon u_x)|_{(b(t)-,t)} = (P - \epsilon u_x)|_{(b(t)+,t)} = -\epsilon u_x|_{(b(t)+,t)} \ .$$

Since $u(\cdot,t)$ is linear in x in the cavity region, it then follows that $u(\cdot,t)$ must in fact be constant there. In particular, $u(b(t),t) = u(c(t),t)$ for all t, contradicting our stipulation that a collision must occur in finite time.

Of course, we do not expect the above solution to have physical significance off its support, so the failure of the above construction is of no physical consequence. On the other hand, it leads us to consider the modified problem in which the initial vacuum on $[b_0, c_0]$ is replaced by a fluid of very low density $\delta > 0$. We obtain as above the physical solution (ρ^δ, u^δ) for this single fluid, initially supported on $[a_0, d_0]$, expanding into the surrounding vacuum, and we inquire about its limit as $\delta \to 0$. Specifically, we define

$$\rho_0^\delta = \begin{cases} 0 & x < a_0 \text{ or } x > d_0 \\ \rho_0 & x \in [a_0, b_0] \cup [c_0, d_0] \\ \delta & x \in (b_0, c_0) \quad , \end{cases}$$

and we define u_0^δ to agree with u_0 on $[a_0, b_0] \cup [c_0, d_0]$, to be constant on $(-\infty, a_0]$ and on $[d_0, \infty)$, and to be linear on $[b_0, c_0]$. Again, the results of Serre [3] apply to establish the global existence of a solution (ρ^δ, u^δ) in the sense of (1) - (3), and the considerations indicated above in (4) enable us to extend this solution to all of $\{t > 0\}$. The surprizing result is that the family $\{(\rho^\delta, u^\delta)\}$ not only fails to converge as $\delta \to 0$ to the physical solution (ρ, u) constructed above, but also that its limit points are solutions with strikingly different, nonphysical properties:

Theorem. a. *Given a sequence $\delta_j \to 0$, there is a subsequence $\delta = \delta_{j_k}$ such that:*

i. u^δ converges uniformly on compact sets in $\{t > 0\}$ to a Holder continuous function $\tilde{u}$.

ii. The particle trajectories $a^\delta, \dots$ emanating from the initial points $a_0, \dots$ in the solutions (ρ^δ, u^δ) converge uniformly on bounded time intervals to C^1 curves $\tilde{a}, \dots$.

iii. ρ^δ converges uniformly on compact sets away from the curves $\tilde{a}, \dots$ to a function $\tilde{\rho}$.

b. *$(\tilde{\rho}, \tilde{u})$ is a weak solution of (1) in $\{t > 0\}$ with initial data (ρ_0, u_0) (extended to all of $\mathbf{R}$ as described above) with the following properties: $\operatorname{supp} \tilde{\rho}(\cdot, t) = [\tilde{a}(t), \tilde{b}(t)] \cup [\tilde{c}(t), \tilde{d}(t)]$; the momentum conservation boundary conditions (3) hold at $\tilde{a}$ and $\tilde{d}$, and the jump conditions (4) hold at $\tilde{b}$ and $\tilde{c}$; and $\tilde{\rho}$ is zero and $\tilde{u}$ is linear in x in the cavity region $[\tilde{b}, \tilde{c}]$.*

c. *The separate momenta*

$$\int_{\tilde{a}}^{\tilde{b}} \tilde{\rho}\tilde{u}\, dx \text{ and } \int_{\tilde{c}}^{\tilde{d}} \tilde{\rho}\tilde{u}\, dx$$

are not constant in t, and no collision between the two fluids in the solution $(\tilde{\rho}, \tilde{u})$ can occur; that is, there is a constant C such that, for all t,

$$\tilde{c}(t) - \tilde{b}(t) \geq 1/C \,. \tag{5}$$

(ρ, u) and $(\tilde{\rho}, \tilde{u})$ are therefore solutions of (1) with the same initial data (ρ_0, u_0), but with distinctly different physical properties. Of course, they are solutions in slightly different senses: (ρ, u) satisfies (1) only on its support and satisfies the boundary conditions (3) everywhere on the boundary of its support; $(\tilde{\rho}, \tilde{u})$, on the other hand, satisfies (1) everywhere, but fails to satisfy (3) on the cavity boundaries $\tilde{b}$ and $\tilde{c}$. Still, the fact remains that (ρ, u) and (ρ^δ, u^δ) are solutions *in exactly the same sense;* the data for (ρ^δ, u^δ) tends to that for (ρ, u), but the corresponding solutions do not.

The proof of the above theorem requires a fairly technical analysis of the precise regularity and compactness properties of the solutions (ρ^δ, u^δ). Complete details may be found in Hoff and Serre [3]. Two estimates lie at the heart of the matter, however, and these can be readily derived in Lagrangean coordinates. We therefore define $v = 1/\rho$ and $p(v) = P(\rho)$, and we introduce the Lagrangean variable y defined by $dy = \rho dx$. (Thus the lines y = constant correspond to particle trajectories.) The equations (1) then become

$$\begin{aligned} &v_t - u_y = 0 \\ &u_t + p(v)_y = \left(\frac{\epsilon u_y}{v}\right)_y \ , \quad 0 < y < M \ , \ t > 0 \ , \end{aligned} \tag{6}$$

where M is the total mass. The boundary condition (3) transforms to

$$p = \frac{\epsilon u_y}{v}\,, \quad y = 0+, M- \;. \tag{7}$$

We multiply the first equation in (6) by $-p$ and the second by u and integrate to obtain the energy balance relation

$$\int_0^M \left[\frac{u^2}{2} + \psi(v)\right](y,t)\,dy\,dt + \int_0^M \int_0^T \frac{\epsilon u_y^2}{v}\,dy\,dt = C_0 \tag{8}$$

where $\psi(v) = \int_v^\infty p(s)\,ds$ is the potential energy and C_0 is the total initial energy. C_0 depends on an upper bound for $\rho(x,0)$, but is independent of a lower bound for $\rho(x,0)$. Next we define the function

$$H(y,t) = \int_0^y u(z,t)\,dz - \epsilon \log v(y,t)\;.$$

A short computation based on (6) and (7) shows that $H_t = -p < 0$, so that H is decreasing in t. Since

$$\left|\int_0^y u\,dz\right| \le (2MC_0)^{1/2}$$

by (8), we thus obtain the bound

$$\log v(y,t) \ge \log v(y,0) - 2(2MC_0)^{1/2}\;.$$

Applying this result to the solution (ρ^δ, u^δ) at points within the cavity region $b^\delta(t) \le x \le c^\delta(t)$, and using the fact that $\rho^\delta(x,0) = \delta$ for $b_0 \le x \le c_0$, we then find that

$$\rho^\delta(x,t) \le C\delta\;,\; b^\delta(t) \le x \le c^\delta(t)\;, \tag{9}$$

where the constant C depends only on M and C_0. Since the total mass on the interval $[b^\delta(t), c^\delta(t)]$ is independent of t, we then obtain that

$$\begin{aligned}(c_0 - b_0)\delta &= \int_{b^\delta(t)}^{c^\delta(t)} \rho^\delta(x,t)\,dx \\ &\le C[c^\delta(t) - b^\delta(t)]\delta\;.\end{aligned}$$

Dividing by δ and then letting $\delta \to 0$, we conclude finally that

$$0 < \frac{c_0 - b_0}{C} \le \tilde{c}(t) - \tilde{b}(t)$$

for all t. This proves the first assertion in (c) that no collision occurs in the solution $(\tilde{\rho}, \tilde{u})$. To prove the other, we apply the conclusions of (b) to obtain

$$\begin{aligned}\frac{d}{dt}\int_{\tilde{a}}^{\tilde{b}} (\tilde{\rho}\tilde{u})(x,t)\,dx &= (\epsilon \tilde{u}_x - P(\tilde{\rho}))(\tilde{b}-,t) \\ &= \epsilon \tilde{u}_x(\tilde{b}+,t)\;.\end{aligned}$$

However, $\tilde{u}_x(\cdot,t)$ is constant on the interval $(\tilde{b}(t), \tilde{c}(t))$, and is nonzero, at least for t near zero, because $\tilde{u}(\tilde{c}(t),t) - \tilde{u}(\tilde{b}(t),t)$ is continuous in t and negative at $t = 0$ by construction. This proves that the momentum of the fluid on the left is decreasing for t small and positive.

Recall that the original solution (ρ, u) was constructed to include two particular physical features, namely, the collision of two initially disjoint fluids, and the conservation in time of their separate momenta before the collision time. The failure of the limiting solution $(\tilde{\rho}, \tilde{u})$ to include these two physical features may be understood as follows. When $\delta > 0$ the two fluids on $[a^\delta, b^\delta]$ and $[c^\delta, d^\delta]$ are physically linked by the intervening fluid in $[b^\delta, c^\delta]$. The presence of this intervening fluid thus prevents a collision, for otherwise the average density in $[b^\delta, c^\delta]$ would have to become infinite; this is precluded, however, by the upper bound (9) for ρ^δ, which therefore is the key estimate in the entire analysis. Since the constant C in (9) is independent of δ, the physical coupling is thus "remembered" in the limit as $\delta \to 0$, and no collision of the two fluids can o

REFERENCES

[1] David Hoff, *Discontinuous solutions of the Navier-Stokes equations for compressible flow,* to appear in Archive Rational Mech. Anal.

[2] David Hoff, *Global well-posedness of the Cauchy problem for the Navier-Stokes equations of nonisentropic flow with discontinuous initial data,* to appear in J. Diff. Eqns.

[3] David Hoff and Denis Serre *The failure of continuous dependence on initial data for the Navier-Stokes equations of compressible flow,* submitted to SIAM J. Appl. Math.

[4] Denis Serre *Solutions faibles globales des equations de Navier-Stokes pour un fluide compressible,* C.R. Acad. Sc. Paris t. 303, Serie I, 13, 1986.

CHAPTER 9

Change of Type in Simple Models for Two-Phase Flow*

Barbara Lee Keyfitz†

Abstract. A standard model for two-phase compressible non-reacting flow in one space dimension is studied. This system of equations is known to exhibit change of type (hyperbolic/nonhyperbolic). The boundary between hyperbolic and nonhyperbolic regions is examined here. It is shown that at this boundary the slow wave speeds coalesce and become complex. If attention is restricted to a cross-section of state space in which the slow waves dominate, then a two-equation model is obtained: it is in conservation form and is identical to a model for two-way traffic flow which was discovered about thirty years ago.

1. Introduction. The aim of this paper is to show some connections between models in wide use in the study of two-phase flows and results on change of type in systems of conservation laws. The focus of the comparison is the mathematical phenomenon of change of type. The theory of conservation laws suggests how the linear ill-posedness of the equations may appear as an instability which has a

*Research supported by the AFOSR, Grant Number AFOSR-86-0088, and by the AFOSR and NSF, Grant Number DMS-89-03768. Support from the Texas Advanced Research Program, Project Number 003652-077, is also acknowledged.

†Department of Mathematics, University of Houston, Houston, Texas 77204-3476.

reasonable physical interpretation, when nonlinear effects are present. It is possible that these results may have implications for modelling two-phase flows: mathematical predictions from conservation law theory can be compared with experimental and numerical results. To do this, the mathematical theory must be further developed, and some gaps bridged. This paper outlines a way that two-phase flow models may fit into the theory being developed.

We rely extensively on the review article by Stewart and Wendroff [14], where the model we use is carefully derived by averaging. This model is also mentioned in several other papers, mainly as a starting point for refinements whose purpose is to remove the change of type (Ramshaw and Trapp [12], Ransom and Hicks [13]). We consider the simplest case: one-dimensional motion of isentropic, compressible, non-reacting fluids. (We could include an energy equation; this increases the number of equations but does not otherwise change the comparison.) Following Stewart and Wendroff, we assume the fluids are inviscid and barotropic. Labelling them with subscripts 1 and 2, we write the basic equations for conservation of mass and transport of momentum as

$$\left.\begin{aligned} &\partial_t(\alpha_i\rho_i) + \partial_x(\alpha_i\rho_i u_i) = 0 \\ &\partial_t(\alpha_i\rho_i u_i) + \partial_x(\alpha_i\rho_i u_i^2) + \alpha_i\partial_x p = 0 \end{aligned}\right\} \quad i = 1,\ 2 \tag{1.1}$$

where

$$\alpha_1 + \alpha_2 = 1, \qquad \rho_i = \rho_i(p)\ , \quad i = 1,\ 2. \tag{1.2}$$

Here α_i is the volume fraction of the *i*th fluid, $\rho_i(p)$ its density and u_i its velocity. Using (1.2), we see that (1.1) is a system of four equations in the four unknowns $n \equiv \alpha_1$, p, u_1 and u_2. It is not in conservation form: indeed, the momentum equations express transfer of momentum between fluids. The total momentum is conserved, but a fourth conserved quantity can be found only by considering a nonlinear function of the variables. Such a substitution produces weak solutions which will not be equivalent to the weak solutions of (1.1); however, there is no canonical way to define weak solutions of a system that is not in conservation form. This is presently being considered by several researchers (LeFloch [9]; see also related work of Cauret, Colombeau and LeRoux [2]) and we leave it aside, with two comments: first, the form of the equation does not determine the type -- the characteristic speeds of (1.1) are the same if it is rewritten in conservation form; second, the nonlinear considerations we use here will have to appear, in some form, if the primitive nonconservative equations (1.1) are used.

A conservative form of (1.1), the one we shall use, can be found by introducing as state variable

$$u = (m_1, u_1, m_2, u_2) \tag{1.3}$$

where $m_i = \alpha_i \rho_i(p)$, $i = 1$ and 2, and (m_1, m_2) replace n and p. One finds p as a function of m_1 and m_2 from the relation

$$\alpha_1 + \alpha_2 = \frac{m_1}{\rho_1(p)} + \frac{m_2}{\rho_2(p)} = 1. \tag{1.4}$$

Since

$$\frac{dm_1}{\rho_1} - \frac{m_1\rho_1'}{\rho_1^2}dp + \frac{dm_2}{\rho_2} - \frac{m_2\rho_2'}{\rho_2^2}dp = 0 ,$$

one obtains

$$\frac{\partial p}{\partial m_i} = \frac{1/\rho_i}{m_1\rho_1'/\rho_1^2 + m_2\rho_2'/\rho_2^2} , \quad i = 1, 2 . \tag{1.5}$$

The denominator is positive if $m_1 + m_2 > 0$; we note that since α_1 is a dimensionless quantity, m_i has the dimensions of density, and

$$\frac{d\rho_i}{dp} = \frac{1}{dp/d\rho_i} = \frac{1}{c_i^2}$$

where c_i is the speed of sound in the *i*th fluid and may be assumed positive, finite, and a function of m_1 and m_2 in the two-fluid flow.

Manipulation of (1.1) results in the conservation law system

$$\partial_t \begin{pmatrix} m_1 \\ u_1 \\ m_2 \\ u_2 \end{pmatrix} + \partial_x \begin{pmatrix} m_1 u_1 \\ \frac{1}{2} u_1^2 + q_1 \\ m_2 u_2 \\ \frac{1}{2} u_2^2 + q_2 \end{pmatrix} = 0 . \tag{1.6}$$

Here $q_i(p) = \int dp/\rho_i$ can be expressed as a function of m_1 and m_2 by (1.4); the first and third equations express conservation of mass, the second and fourth conservation of enthalpy in each fluid. We shall write (1.6) frequently as a conservation law system

$$u_t + f(u)_x = 0 \tag{1.7}$$

and we will also use the quasilinear form

$$u_t + A(u)u_x = 0 \tag{1.8}$$

where $A = df$ is explicitly given by

$$A = \begin{pmatrix} u_1 & m_1 & 0 & 0 \\ a_{11} & u_1 & a_{12} & 0 \\ 0 & 0 & u_2 & m_2 \\ a_{21} & 0 & a_{22} & u_2 \end{pmatrix} . \tag{1.9}$$

Here

$$a_{ij} = \frac{\partial q_i}{\partial m_j} = \frac{1}{\rho_i \rho_j} \cdot S \tag{1.10}$$

where

$$(1.11)\qquad \frac{1}{s} = \frac{m_1\rho_1'}{\rho_1^2} + \frac{m_2\rho_2'}{\rho_2^2} = \frac{m_1}{c_1^2\rho_1^2} + \frac{m_2}{c_2^2\rho_2^2}$$

is a symmetric function of m_1 and m_2. From (1.10) it is clear that $a_{12} = a_{21}$ and $a_{12}a_{21} = a_{11}a_{22}$.

The characteristic speeds of (1.9) or (1.1) are the zeroes of

$$(1.12)\qquad P(\lambda) \equiv (\lambda-u_1)^2(\lambda-u_2)^2-a_{11}m_1(\lambda-u_2)^2-a_{22}m_2(\lambda-u_1)^2 .$$

Introducing the notation

$$\bar{u} = \frac{1}{2}(u_1+u_2)\ ,\quad \delta = \frac{1}{2}(u_1-u_2)\ ,\quad \bar{\lambda} = \lambda - \bar{u},$$

we have

$$(1.13)\qquad P(\lambda) = \bar{P}(\bar{\lambda}) = (\bar{\lambda}^2-\delta^2)^2- a_{11}m_1(\bar{\lambda}-\delta)^2- a_{22}m_2(\bar{\lambda}+\delta)^2 .$$

Finally, we simplify by letting $\mu = \bar{\lambda}/\delta$ and dividing by δ^4: $P(\lambda) = 0$ is equivalent to $Q(\mu) = 0$ where

$$(1.14)\qquad Q(\mu) = \mu^4 - (2 + r)\mu^2 + 2(r_1 - r_2)\mu + (1 - r)$$

and

$$(1.15)\qquad r = r_1 + r_2\ ,\quad r_i = a_{ii}m_i/\delta^2\ ,\quad r = 1,\ 2\ .$$

The root structure of Q was elucidated by Stewart and Wendroff. By examining $Q(0)$ and $Q(\pm 1)$, one sees that Q always has at least two distinct real roots, and that Q has four real roots if $r \le 1$. Solving

$$Q(\mu) = Q'(\mu) = 0$$

for r gives the locus of coincident roots; this is a lengthy calculation which, Stewart and Wendroff realized, results in the simple equation

$$(1.16)\qquad r_1^{1/3}+ r_2^{1/3} = 4^{1/3}\ .$$

Since $r_i \ge 0$, the first quadrant is physically

interesting; it is divided by the curve (1.16) into a bounded set

$$H = \{(r_1,r_2) \mid r_1^{1/3} + r_2^{1/3} < 4^{1/3}\}$$

in which the corresponding equation (1.6) is hyperbolic, and the complementary unbounded set

$$NH = \{(r_1,r_2) \mid r_1^{1/3} + r_2^{1/3} > 4^{1/3}\}$$

in which it is not. See Figure 1.1.

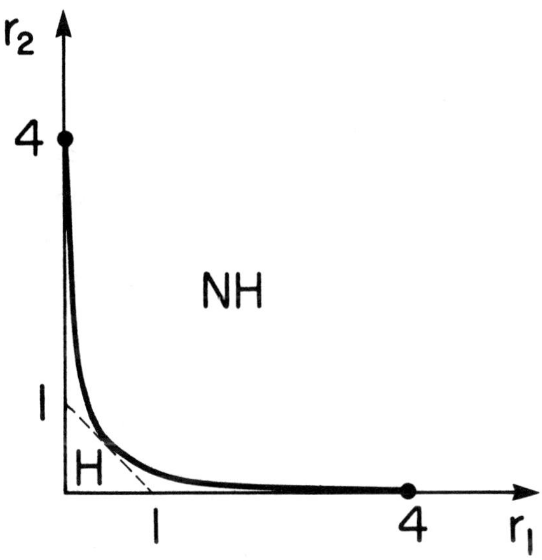

Figure 1.1. Hyperbolic and non-hyperbolic regions in (r_1,r_2) space.

It is quite interesting to describe the hyperbolic and nonhyperbolic regions in state space. Let $\mathcal{D} = \mathbb{R}^+ \times \mathbb{R} \times \mathbb{R}^+ \times \mathbb{R}$ represent the physically feasible region of state space and let

$$(1.17) \qquad \mathcal{B} = \{u \in \mathcal{D} \mid (a_{11}m_1)^{1/3} + (a_{22}m_2)^{1/3} = (u_2 - u_1)^{2/3}\} .$$

Then $\mathcal{B}$ is the boundary between the set of hyperbolic points

$$(1.18)\qquad \mathcal{H} = \{u \in \mathcal{D} \mid (a_{11}m_1)^{1/3} + (a_{22}m_2)^{1/3} < (u_2 - u_1)^{2/3}\}$$

and the complementary nonhyperbolic set, $\mathcal{NH}$. We note that the velocity variables appear only once in (1.17) or (1.18), and only in the combination $u_2 - u_1$; thus it is relatively simple to describe the projections of $\mathcal{H}$ and $\mathcal{NH}$ onto a three-dimensional subspace $\{(m_1, m_2, \delta)\}$ of $\mathcal{D}$, or, in terms of the primitive variables, $\{(n, p, \delta)\}$. Using (1.10) and (1.11),

$$(a_{11}m_1)^{1/3} + (a_{22}m_2)^{1/3} = \frac{\left(\frac{\alpha_1}{\rho_1}\right)^{1/3} + \left(\frac{\alpha_2}{\rho_2}\right)^{1/3}}{\frac{\alpha_1}{c_1^2\rho_1} + \frac{\alpha_2}{c_2^2\rho_2}} .$$

For each fixed p, this is a concave function of $n = \alpha_1$ on $[0, 1]$; for a fixed n, the expression is an "effective sound speed" and increases with p. (For γ-law gases, $\rho_i c_i^2 = \gamma_i p$, and the calculation simplifies further.) Some cross-sections are illustrated in Figures 1.2 - 1.6. As functions of m_1 and m_2,

$$(1.19)\qquad (a_{11}m_1)^{1/3} + (a_{22}m_2)^{1/3} = \frac{\left(\frac{m_1}{\rho_1^2}\right)^{1/3} + \left(\frac{m_2}{\rho_2^2}\right)^{1/3}}{\left(\frac{m_1}{c_1^2\rho_1^2} + \frac{m_2}{c_2^2\rho_2^2}\right)} .$$

with $\rho_i = \rho_i(p)$, $c_i = c_i(p)$ and $p = p(m_1, m_2)$. If one considers a curve in (m_1, m_2) given by $p(m_1, m_2) =$ constant, this is concave; if the ratio m_1/m_2 is fixed, this is an increasing function of m_1 (or of m_2 — i.e. of p); the level curves of (1.19) are shown in Figure 1.5, and another typical cross-section of $\mathcal{D}$ in Figure 1.6.

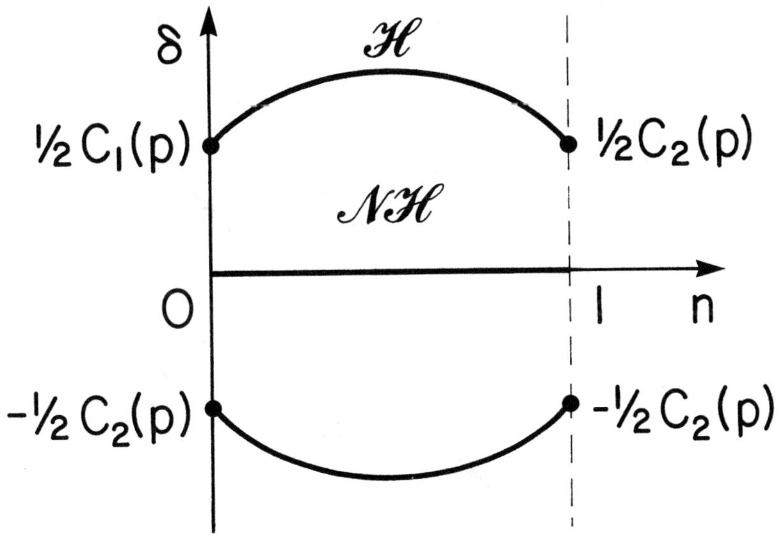

Figure 1.2. Boundary $\mathscr{B}$ for fixed p

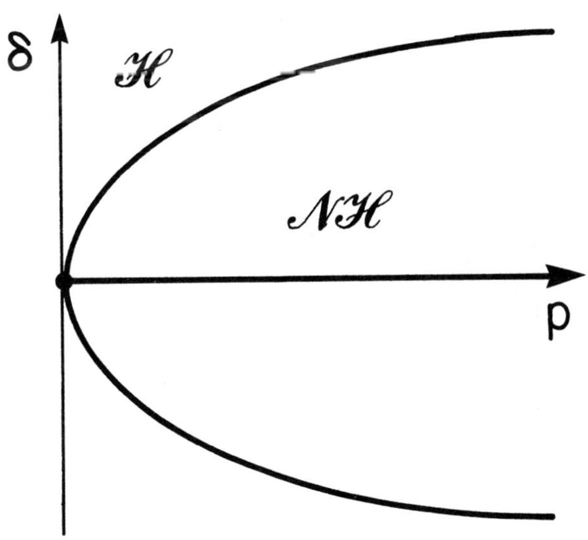

Figure 1.3. Boundary $\mathscr{B}$ for fixed n

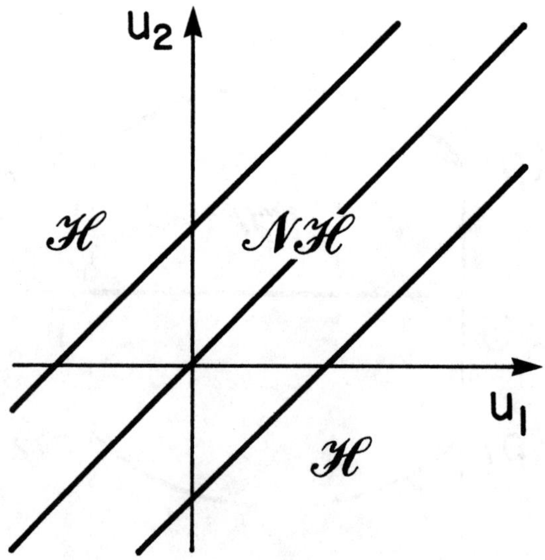

Figure 1.4. Boundary for fixed p and n (or m_1, m_2)

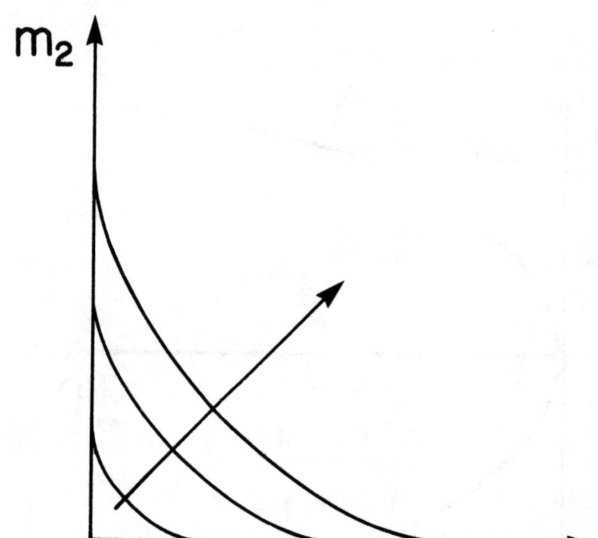

Figure 1.5 : Level curves of $(a_{11}m_1)^{1/3} + (a_{22}m_2)^{1/3}$

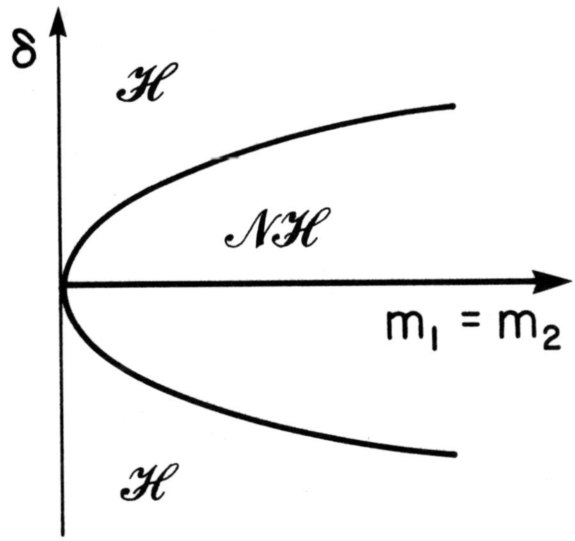

Figure 1.6. Cross-section of boundary with $m_1 = m_2$

The subspace $u_1 = u_2$ must be considered separately, since $\delta = 0$ and (1.14) cannot be used. Instead, from (1.13),

$$\bar{P}(\bar{\lambda}) = \bar{\lambda}^2(\bar{\lambda}^2 - a_{11}m_1 - a_{22}m_2)$$

and the equation has coincident roots. A small perturbation, to $\delta \neq 0$, always puts u into $\mathcal{NH}$. This type of singular behavior could occur with a pair of equations, but it has not been studied (among other things, the local picture is of a very ill-posed equation); on the other hand, the behavior near $\mathcal{B}$ (where $\delta \neq 0$) appears to be that of familiar two-equation models, as we shall now show.

2. Reduction to a 2x2 system. John Hunter (unpublished lecture; see [5] for the general framework) has described the use of weakly nonlinear geometric optics to reduce (1.6) to the transonic small disturbance equation near a point on $\mathcal{B}$. The normal form is

$$(2.1) \qquad \begin{aligned} u_t + v_x &= 0 \\ v_t + uu_x &= 0 \end{aligned}, \qquad A = \begin{pmatrix} 0 & 1 \\ u & 0 \end{pmatrix}.$$

The eigenvalues, $\lambda^2 - u = 0$, clearly change from real to complex along $\mathcal{B}$: $u = 0$. Some further discussion of local normal forms and their uses is found in [6].

In this section we will use a different reduction to show that (1.6) has a structure found in some two-equation models, including some quadratic models. These considerations suggest directions for a more detailed study.

We begin by observing that when (r_1, r_2) crosses B, given by equation (1.16), the largest and smallest roots of Q remain real, and the middle two coalesce and become complex. (This is easily verified at $r_1 = r_2 = 1/2$, and is true, by continuity, up to the end points (1,4) and (4,1) of B, when three roots coincide.) For any point $u^0 \in \mathcal{B}$, we can perform an affine rescaling so that the interior roots are both zero; then, for u in a neighborhood of u^0, the real part of these roots is $\mathcal{O}(|u - u^0|^{1/2})$, while the extreme roots are $\mathcal{O}(1)$. Hence it is reasonable to consider motions $u(x,t)$ in which the waves associated with the extreme speeds have propagated out of the domain of interest, and only the slow disturbances remain.

If there were a set of Riemann invariant coordinates, one could actually diagonalize the system and identify, explicitly, nonlinear functions $w_1(u)$, $w_2(u)$ that would be constant as long as fast waves were not present. (Strictly speaking, they would be only approximately constant for weak solutions with shocks.) Our system does not appear to possess Riemann invariants, but we can do the following: approximate the surfaces $w_i(u)$ = constant by planes,

$$(2.2) \qquad w_i(u) = \ell_i \cdot (u - u^0) = 0\ , \qquad i = 1,\ 2$$

where $\ell_i = \ell_i(u^0)$ is the left eigenvector of $A(u^0)$ corresponding to $\lambda_i(u^0)$; here $i = 1, 2$ correspond to the extreme roots.

This argument derives from weakly nonlinear geometric optics [5], in which a wave propagating with speed λ_i has the form $\sigma(x - \lambda_i t) r_i(u^0)$; thus the linear structure is

$$u - u^0 = \sum_{j=1}^{4} \sigma_j(x - \lambda_j t) r_j(u^0) \; ; \tag{2.3}$$

if $\sigma_1 = \sigma_2 = 0$ then $\ell_i(u^0)\cdot(u - u^0) = 0$, $i = 1, 2$, since $\ell_i \cdot r_j = 0$ if $i \neq j$. (Note: in (2.3), $r_4(u^0)$ must be a generalized eigenvector if $u^0 \in \mathcal{B}$, but the calculation is the same.)

From (A.12) and (A.13) in the appendix, the four roots are

$$-\mu_0 \pm \sqrt{3+(r-1)/3}\,, \quad \mu_0\,, \quad \mu_0\,,$$

where $\mu_0 = \sqrt{(r-1)/3}\,\mathrm{sgn}(r_1-r_2)$; thus, the roots are most widely separated if

$$\mu_0 = 0 \quad \text{or} \quad r_1 = r_2 = \frac{1}{2}\;; \tag{2.4}$$

then they are

$$\pm\sqrt{3}, \; 0, \; 0\,. \tag{2.5}$$

We shall make this choice, for definiteness, but any other value of $r < 4$ will lead to a similar reduction.

Now, (2.4) specifies a two-dimensional subset of u, namely

$$a_{11}m_1 = a_{22}m_2 = \frac{1}{2}\,\delta^2 = \frac{1}{8}\,(u_2 - u_1)^2\,. \tag{2.6}$$

This identifies possible values of $u^0 = (m_1^0, u_1^0, m_2^0, u_2^0)$ and is a subset $\mathcal{B}_0$ of $\mathcal{B}$. If the fluids are identical, the first equation of (2.6) implies $m_1 = m_2$: in general, the first equation of (2.6) yields an equation $g(m_1, m_2) = 0$. In the cross-sections of Figs. 1.2 - 1.6, where $\mathcal{B}$ is a curve, $\mathcal{B}_0$ will appear as isolated points in Fig 1.2; in 1.4, $\mathcal{B}_0$ will be absent or will coincide with the cross-section of $\mathcal{B}$; in 1.3 and 1.6, either may be the case, depending on the constitutive relations. It is again clear that $\mathcal{B}_0$, like $\mathcal{B}$, is a cylinder over the subspace $\{(m_1, m_2, \delta)\}$. (These statements are true for any other choice μ_0 corresponding to $a \equiv r - 1 > 0$ and for the $\mathcal{B}_a$ so determined.)

Fixing $u^0 = (m_1^0, u_1^0, m_2^0, u_2^0) \in \mathcal{B}_0$, we calculate the left eigenvectors of $A - \lambda_i I$, $i = 1, 2$. Note that

$$\lambda_i = \delta \mu_i + \bar{u} = \delta_0 \mu_i + \bar{u}^0 \tag{2.7}$$

where

$$\delta_0 = \frac{1}{2}(u_2^0 - u_1^0), \quad \bar{u}^0 = \frac{1}{2}(u_1^0 + u_2^0), \quad \mu_i = \pm\sqrt{3}.$$

Now,

$$A(u^0) - \lambda_i I = \begin{pmatrix} -\delta_0(1+\mu_i) & m_1^0 & 0 & 0 \\ a_{11}^0 & -\delta_0(1+\mu_i) & a_{12}^0 & 0 \\ 0 & 0 & \delta_0(1-\mu_i) & m_2^0 \\ a_{21}^0 & 0 & a_{22}^0 & \delta_0(1-\mu_i) \end{pmatrix},$$

and so

$$\ell_i^0 = \Big(\delta_0\sqrt{m_2^0}(-1 + \mu_i),\ m_1^0\sqrt{m_2^0}(2 - \mu_i),\ \delta_0\sqrt{m_1^0}(1 + \mu_i),\ m_2^0\sqrt{m_1^0}(2 + \mu_i)\Big), \tag{2.8}$$

where we have used (2.6) to eliminate a_{ij}^0, and $\mu_i^2 = 3$.

We can now write (2.2) in detail, using (2.8):

$$\begin{aligned} \frac{1}{2}\delta_0(m_2\sqrt{m_1^0} - m_1\sqrt{m_2^0}) + \sqrt{m_1^0 m_2^0}\,(u_1\sqrt{m_1^0} + u_2\sqrt{m_2^0}) &= \gamma_1 \\ \delta_0(m_1\sqrt{m_2^0} + m_2\sqrt{m_1^0}) + \sqrt{m_1^0 m_2^0}\,(u_2\sqrt{m_2^0} - u_1\sqrt{m_1^0}) &= \gamma_2 \end{aligned}, \tag{2.9}$$

where γ_1 and γ_2 are given by the left hand side of (2.9) evaluated at $u^0 = (m_1^0, u_1^0, m_2^0, u_2^0)$. Finally, we can solve (2.9) for

$$(2.10)\qquad \begin{aligned} u_1 &= \frac{1}{2m_1^0\sqrt{m_2^0}}\left(\gamma_1 - \gamma_2 + \frac{1}{2}\,\delta_0(3m_1\sqrt{m_2^0} + m_2\sqrt{m_1^0})\right) \\ u_2 &= \frac{1}{2m_2^0\sqrt{m_1^0}}\left(\gamma_1 + \gamma_2 - \frac{1}{2}\,\delta_0(3m_2\sqrt{m_1^0} + m_1\sqrt{m_2^0})\right) \end{aligned} .$$

Thus $u_i = U_i(m_1, m_2)$, $i = 1, 2$, where U_i is an affine function of m_1 and m_2.

If we seek solutions in which the u_i evolve by (2.10) instead of by the original equation (1.6), we obtain a simplified system for (m_1, m_2):

$$(2.11)\qquad \partial_t\begin{pmatrix} m_1 \\ m_2 \end{pmatrix} + \partial_x\begin{pmatrix} m_1U_1 \\ m_2U_2 \end{pmatrix} = 0 .$$

In (2.11), the flux is a quadratic polynomial in m_1 and m_2; furthermore, it has the form of the "two-directional traffic flow equations" [1]; for example, if we let $m_1^0 = m_2^0 = m^0$ (assuming identical fluids) to reduce the complexity of notation, we have

$$\gamma_1 - \gamma_2 = m^0\sqrt{m^0}(5u_1^0 - u_2^0)/2 , \qquad \gamma_1 + \gamma_2 = m^0\sqrt{m^0}(5u_2^0 - u_1^0)/2 .$$

To obtain a closer resemblance to the two-way traffic model, we must assume, say, $u_1 < 0$, $u_1^0 < 0$, and $u_2 > 0$, $u_2^0 > 0$. Then $\gamma_1 - \gamma_2 < 0 < \gamma_1 + \gamma_2$, and the coefficients in U_1 and U_2 have the correct signs. If we scale the m_i by $m^0 = 1$, and choose $u_1^0 = u_2^0 = -1$ so $\delta_0 = 1$, then

$$(2.12)\qquad U_1 = \frac{1}{4}\,(-6 + 3m_1 + m_2) , \; U_2 = \frac{1}{4}\,(6 - 3m_2 - m_1) ,$$

which is an example of the traffic flow equations, valid on a quadrilateral

$$R:\quad m_i \geq 0 , \qquad 3m_1 + m_2 \leq 6 , \qquad 3m_2 + m_1 \leq 6 ,$$

illustrated in Figure 2.1. We note that the point corresponding to u^0, (1, 1), is in the interior of R, and on $\mathcal{B}_R$, the boundary between the hyperbolic and nonhyperbolic regions of R. Thus, results on the behavior of solutions under change of type in (2.11) may have some interpretation for (1.6). We note, however, that the correspondence between solutions of (1.6) and (2.11) is only between "small amplitude" waves in each, since it is only then that we are justified in restricting to the linear subspace of $u \in \mathcal{D}$ that results in (2.11).

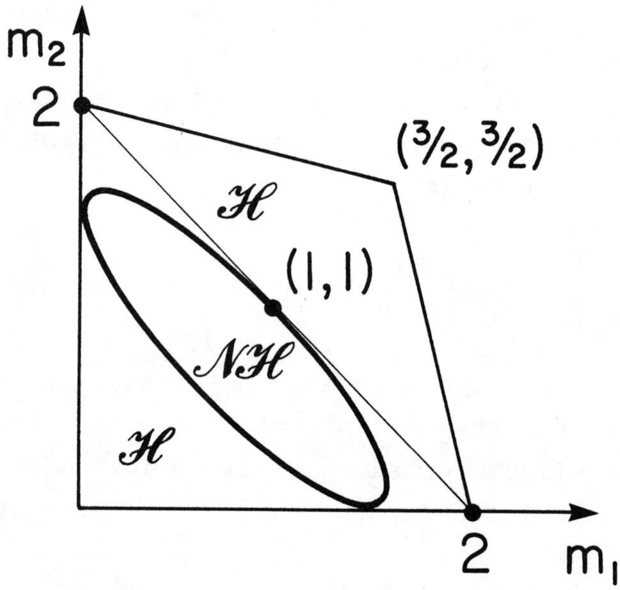

Figure 2.1. Geometry of the kinematic model

3. Applicable results on 2x2 systems and conclusions. In the original paper [1] that derives the two-directional traffic flow model, Bick and Newell work out many details of the solution of the Riemann problem, including determination of rarefaction curves, and of the Hugoniot locus. The system considered by Bick and Newell contains two parameters, but the only invariant, under affine scalings of dependent and independent variables, is the ratio, β, of negative influence of the oncoming traffic to the negative self-interaction of the flow. This ratio is one-third in the symmetric situation described by (2.12). The assumption of identical fluids corresponds to the assumption of identical traffic behavior in Bick and Newell's model. An explicit limit in the traffic model, $\beta \to 0$, is not possible in this reduction.

Holden and Holden [4] have compiled a list of solutions of the Riemann problem for a quadratic model which can be transformed to the example (2.11), (2.12) by an affine change of coordinates; thus there is a correspondence between the Riemann solutions listed in [4] and solutions to (2.11). However, Holden and Holden did not resolve questions of uniqueness of solutions or admissibility of shocks. Some of these questions are addressed in [6] and [8] in which some local results (for small amplitude shocks near $\mathcal{B}$) are given. The use of local theory gives the following qualitative results: some shocks between $\mathcal{H}$ and $\mathcal{NH}$ are admissible under a viscous profile type criterion, and this criterion relates to intrinsic considerations of hyperbolicity [8]. Furthermore, in models such as (2.11) and the traffic flow equations, there are large amplitude shocks which cross the nonhyperbolic region, which satisfy a modified characteristic admissibility criterion (the Lax geometric entropy criterion), and have viscous profiles. These are described in [7].

A general conclusion from the model studies is that waves with well-defined speeds characterize weak solutions for some models that show change of type. We note some recent studies of a different class of models showing change of type, the so-called phase-transition models (or models for propagation of phase boundaries). In these models, a single density function describes the fluid, whose equation of state corresponds to a nonmonotone pV relation: the two monotone decreasing portions delineate the two phases. It was noted in [6] and [7] that these models do not have the same local behavior near $\mathcal{B}$ as the "generic" model given by (2.11). Now we have evidence, from the studies of Pego and Serre [11], that the Glimm scheme does not converge for the phase-transition models. I do not know if this behavior is related to the special local properties of the phase transition equations, or what one could say about the convergence of the Glimm scheme for equations like (2.11).

In fact, it is still not clear what the appropriate

notion of well-posedness for a system such as (1.6) should be; an answer depends on the mathematical properties of (1.6), and on careful physical modelling to understand how (1.1) relates to more complete descriptions of two-phase flows.

We may note one property of the reduction in section 2: the two equations that result in (2.11) are the two which were originally in conservation form in (1.1). This may give some confidence that the shocks found in (2.11) may have some relevance to (1.1).

We should also remark on the fact that, in a broad sense, the system (1.1) or (1.6) contains in it the "kinematic" model, (2.11). Now, although the Richardson-Lighthill model for one-way traffic flow is still used, the two-way traffic model is not regarded as realistic [10]. We do not suggest that its appearance in a reduction of a two-phase flow model provides a justification of the hypotheses of either model. What is suggested by the correspondence is that it may be natural for change of type to occur when simple models are used to describe the evolution of two distinct phases. Experienced researchers have urged a resolution of the paradox in better physical modelling [3]. However, this difficulty keeps reappearing. It may be useful to understand the mathematical behavior of systems such as (1.1) or (2.11) in order to improve the physical modelling.

Appendix: Calculation of B and the roots of Q on B. We summarize the calculations. The boundary set B, in the nonnegative quadrant (r_1, r_2), is found by eliminating μ from $Q = Q' = 0$:

$$Q = (\mu^2 - 1)^2 - (r\mu^2 + 2(r_2 - r_1)\mu + r) = 0 \tag{A.1}$$

$$Q' = 4\mu(\mu^2 - 1) - 2(r\mu + (r_2 - r_1)) = 0 . \tag{A.2}$$

From $-(A.1) + \mu \times (A.2)$ we get

$$(r + 2)\mu^2 + 3(r_2 - r_1)\mu + 2(r - 1) = 0 . \tag{A.3}$$

Now (A.3) $\times\ \mu$ - (A.2) $\times$ (r + 2)/4 gives

(A.4) $$6(r_2 - r_1)\mu^2 + r(r + 8)\mu + (r_2 - r_1)(r + 2) = 0 .$$

From the two quadratics (A.3) and (A.4) we get a linear equation

(A.5) $$A\mu + B = 0$$

where

(A.6) $$A = r(r+2)(r+8) - 18(r_2-r_1)^2 = r(r-4)^2 + 72r_1r_2$$

$$B = (r_2 - r_1)\ (r - 4)^2$$

and we use

(A.7) $$(r_2 - r_1)^2 = r^2 - 4r_1r_2 .$$

Now (A.5) and (A.3) give another linear relation

(A.8) $$[3(r_2 - r_1)A - B(r + 2)]\mu + 2(r - 1)A = 0$$

and finally μ can be eliminated from (A.5) and (A.8):

(A.9) $$(r + 2)B^2 - 3(r_2 - r_1)AB + 2(r - 1)A^2 = 0 .$$

Using (A.6) and (A.7), this reduces to

(A.10) $$(r - 4)^3 + 108r_1r_2 = 0 .$$

Writing this as

$$3(4r_1r_2)^{1/3} = 4 - (r_1 + r_2) ,$$

one can show that the only solution is

(A.11) $$r_1^{1/3} + r_2^{1/3} = 4^{1/3}$$

by letting the difference in (A.11) be γ, substituting in (A.10) and finding $\gamma = 0$ is the only real solution. Finally, we find μ, the double root, to be

$$\mu = - \frac{B}{A} = - \frac{(r_2-r_1)(4-r)^2}{r(4-r)^2 + 72r_1r_2} .$$

Now

$$4(4-r)^2 - r(4-r)^2 = 108r_1r_2 ,$$

so

$$r(4-r)^2 + 72r_1r_2 = 4\left((4-r)^2-9r_1r_2\right) .$$

Hence

$$\mu = -\frac{(r_2 - r_1)}{4\left(1 - \dfrac{9r_1r_2}{(4-r)^2}\right)} ,$$

and

$$(4-r)^2 = (108r_1r_2)^{2/3} = 9(4r_1r_2)^{2/3},$$

so

$$\mu_0 = -\frac{(r_2 - r_1)}{4\left(1 - \dfrac{r_1r_2}{(4r_1r_2)^{2/3}}\right)} = \frac{r_2 - r_1}{(4r_1r_2)^{1/3} - 4}$$

$$= \frac{r_2 - r_1}{\frac{4-r}{3} - 4} = -\frac{3(r_2 - r_1)}{(r + 8)} ;$$

one finds

(A.12) $$\mu_0^2 = \frac{r-1}{3} .$$

Now $Q(\mu)$ factors as

$$(\mu - \mu_0)^2(\mu^2 + a\mu + b) = (\mu^2-2\mu_0\mu+\mu_0^2)(\mu^2+a\mu+b)$$

$$= \mu^4 + (a-2\mu_0)\mu^3 + (\mu_0^2-2a\mu_0+b)\mu^2 + (a\mu_0^2-2\mu_0 b)\mu + \mu_0^2 b ,$$

from which we read off

$$a = 2\mu_0$$

$$b + \mu_0^2 - 2a\mu_0 = -(r + 2)$$

or

$$\mu_0^2 b = 1 - r ,$$

$$b = 3\mu_0^2 - r - 2 = \frac{3\cdot 9(r_2-r_1)^2}{(r + 8)^2} - (r + 2)$$

$$= \frac{27(r^2-4r_1r_2)}{(r+8)^2} - (r+2) = -3 .$$

Hence the other roots satisfy

$$\mu_1, \mu_2 = -\mu_0 \pm \sqrt{\mu_0^2+3} = -\mu_0 \pm \sqrt{3 + \frac{r-1}{3}} . \tag{A.13}$$

REFERENCES

1. J. H. Bick and G. F. Newell, "A continuum model for two directional traffic flow", *Quart. of Appl. Math.*, XVIII (1960), 191-204.

2. J. J. Cauret, J. F. Colombeau and A. Y LeRoux, "Discontinuous generalized solutions of nonlinear nonconservative hyperbolic equations", *Jour. of Math. Anal. Appl.*, 139 (1989), 552-573.

3. D. Gidaspow, in "Suggestions for further research and urgent problems on two-phase flows and heat transfer" by A. E. Bergles. et. al., *Two-Phase Flows and Heat Transfer Vol.III*, (ed. Kakaç & Veziroglu) Hemisphere Publishing Corp., Washington, (1976), 1457-1458.

4. H. Holden and L. Holden, "On the Riemann problem for a prototype of a mixed type conservation law, II", in *Current Progress in Hyperbolic Systems: Riemann Problems and Computations* (ed. B. Lindquist), American Mathematical Society, Providence, (1989), 331-367.

5. J. K. Hunter and J. B. Keller, "Weakly nonlinear high frequency waves", *Comm. Pure App. Math.*, XXXVI (1983), 547-569.

6. B. L. Keyfitz, "The use of vectorfield dynamics in formulating admissibility conditions for shocks in systems that change type", in *Problems Involving Change of Type* (ed. K. Kirchgassner), Springer, 1990, 141-150.

7. B. L. Keyfitz, "A criterion for certain wave structures in systems that change type", in *Current Progress in Hyperbolic Systems: Riemann Problems and Computations* (ed. B. Lindquist), American Mathematical Society, Providence, (1989), 203-213.

8. B. L. Keyfitz, "Shocks near the sonic line: a comparison between steady and unsteady models for change of type", to appear in Proceedings of IMA workshop on Nonlinear Evolution Equations that Change Type (ed. Keyfitz and Shearer).

9. P. LeFloch, "Entropy weak solutions to nonlinear hyperbolic systems under nonconservative form", *Commun. In Partial Diff. Eqns.*, 13 (1988), 669-727.

10. G. F. Newell, private communication, 1988.

11. R. L. Pego and D. Serre, "Instabilities in Glimm's scheme for two systems of mixed type", *Siam J. Numer. Anal.*, 25 (1988), 965-988.

12. J. D. Ramshaw and J. A. Trapp, "Characteristics, stability, and short-wavelength phenomena in two-phase flow equation systems", *Nuclear Sci. and Eng.*, 66 (1978), 93-102.

13. V. H. Ransom and D. L. Hicks, "Hyperbolic two-pressure models for two-phase flow", *Jour. Comp. Phys.*, 53 (1984), 124-151.

14. H. B. Stewart and B. Wendroff, "Two-phase flow: models and methods", *Jour. Comp. Phys.*, 56 (1984), 363-409.

CHAPTER 10

On the Viscosity Criterion for Hyperbolic Conservation Laws*

Tai-Ping Liu†

Abstract

To obtain a well-posed theory for hyperbolic conservation laws one needs to impose certain admissibility criterion for shock waves.

A commonly used one is the viscosity criterion of admitting only inviscid shock waves, which are zero dissipation limits of viscous solutions. This yields a canonical criterion for strictly hyperbolic systems. However, for nonstrictly hyperbolic systems admitting all inviscid shock waves, which are limits of viscous travelling waves, may not yield a well-posed theory. It is therefore natural to study the stability of viscous travelling waves, especially in the zero dissipation limit. We investigate this for a system arising in the study of magnetohydrodynamics and elasticity. It turns out that a viscous travelling wave may be stable for fixed viscosity but not uniformly in the strength of viscosity. This gives a stronger version of the viscosity criterion which yields a well-posed hyperbolic theory. The stability phenomena exhibited here for overcompressive shock waves would hold for general systems, not just for the particular model considered here. Our analysis is based on the consideration of time invariants for varying values of viscosity.

Introduction

Consider hyperbolic conservation laws.

$$u_t + f(u)_x = 0,\ u \in \mathbf{R}^n, -\infty < x < \infty,\ t \geq 0$$

* Research supported in part by NSF grant DMS 90-0226, Army grant DAAL03-87-K-0063, AF0SR-89-0203, and Energy Dept. grant DEFG 02-88-ER 25053.

†Department of Mathematics, Stanford University

It is well known that certain admissibility criterion for the shock waves are needed for a well-posed theory. A commonly used one is the viscosity criterion of admitting only those shock waves which are limits of solutions of the viscous conservation laws

$$u_t + f(x)_x = (b(u)u_x)_x,$$

as the viscosity matrix $b(u)$ tends to zero. Not only does it contain more physics, the viscous system is also well posed because it is essentially parabolic and possesses smooth solutions. For strictly hyperbolic systems the viscosity criteria is equivalent to other canonical criteria Lax [6], Liu [7], Conley-Smoller [2], Conlon [3] Pego [9] etc. For nonstrictly hyperbolic systems the issue is controversial.

If we adopt a weak version of viscosity criterion of admitting shock waves which are limits of some viscous travelling waves, then the hyperbolic model may not be well-posed. It is natural to consider stronger versions. One would be to consider only viscous travelling waves which are nonlinearly stable in time. An even stronger version is to consider those which are stable in time as well as in the zero dissipation limit. We exhibit by a physical model in magnetohydrodynamics and elasticity that the above three versions may not be equivalent. For the particular model the strongest version of viscosity criteria offer a well-posed hyperbolic theory.

2. Inviscid Theory

The following system has been derived from the study of magnetohydrodynamics and elasticity (cf. [5], [4], [1])

$$\begin{aligned} &u_t + (u\phi)_x = 0 \\ &v_t + (v\phi)_x = 0; u \in \mathbf{R}^1, v \in \mathbf{R}^1. \end{aligned} \tag{2.1}$$

Here ϕ is a given function of u and v. For simplicity we take

$$\phi(u,v) = u^2 + v^2$$

The characteristic values λ_1, λ_2 and characteristic directions r_1, r_2 are

$$\lambda_1 = \phi = u^2 + v^2, \quad \lambda_2 = \phi + u\phi_u + v\phi_v = 3(u^2 + v^2)$$

$$r_1 = \begin{pmatrix} v \\ -u \end{pmatrix}, \quad r_2 = \begin{pmatrix} u \\ v \end{pmatrix}$$

The first characteristic field is linearly degenerate and second one is genuinely nonlinear except at the origin in the sense of Lax [6],

$$\nabla\lambda_1 \cdot r_1 \equiv 0, \quad \nabla\lambda_2 \cdot r_2 = 6(u^2 + v^2)$$

Consequently two states (u_1, v_1) and (u_2, v_2) on the circle $\phi =$ constant

$$u_1^2 + v_1^2 = u_2^2 + v_2^2$$

are connected by a contact discontinuity with speed $\lambda = \phi$. It is easy to see that for the characteristic field λ_2 the characteristic curve R_2, the integral curve of r_2, and the Hugoniot curve coincide, cf.[10]. Thus shock and rarefaction waves take values along the radial directions. For simplicity we take the direction to be the u axis. It is clear that along the u-axis (1.1) is reduced to the scalar conservation law

$$u_t + (u^3)_x = 0 \tag{2.2}$$

From the theory of scalar conservation law a shock wave (u_-, u_+), $u_- > 0$, is admissible if it satisfies Oleinik's entropy condition, which in the present case becomes

$$u_- > u_+ \geq u_* \tag{2.3}$$

where u_* is the negative state with, Figure 2.1

$$\frac{(u_-)^3 - (u_*)^3}{u_- - u_*} = 3(u_*)^2$$

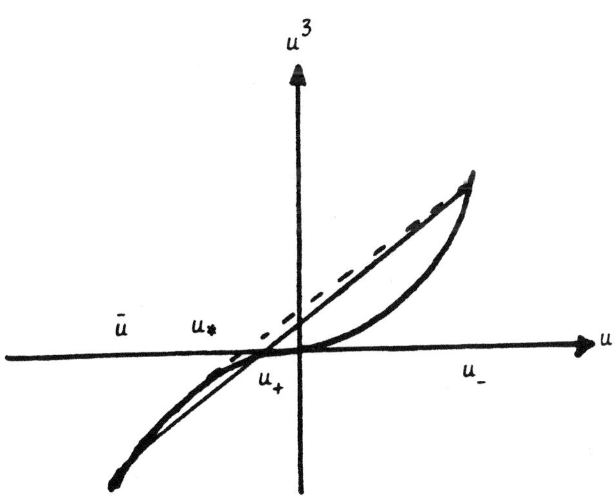

Figure 2.1: Scalar Admissibility Criterion

A rarefaction wave takes values on the radial direction and on the same side of the origin with $\phi_+ > \phi_-$. If all the above waves are admitted then the hyperbolic theory is not well posed. This is easy to see for the Riemann problem

$$(u,v)(x,0) = \begin{cases} (u_-,0) & \text{for } x < 0 \\ (u_+,0) & \text{for } x > 0 \end{cases}$$

and $u_- > 0, 0 > u_+ \geq u_*$. Then the problem can be solved either by a simple shock wave or by the contact discontinuity $(u_-,0;-u_-,0)$ followed by a shock wave $(-u_-,0;u_+,0)$.

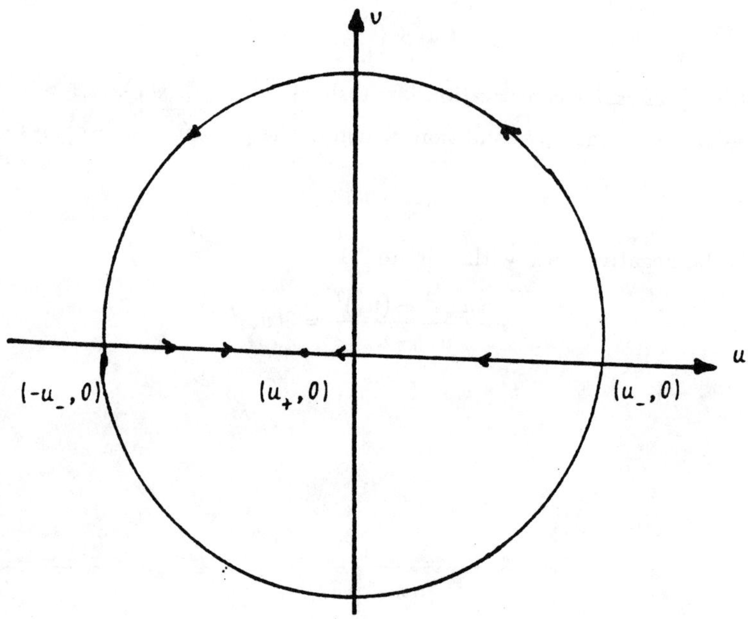

Figure 2.2: Two Solutions to Riemann Problem

The ill-posedness vanishes if we impose a criterion stronger than (2.3).

$$u_- > u+ \geq 0 \tag{2.4}$$

so that only the second solution is allowed.

The question is then on what physical ground we should disqualify those shock waves with $u_* \leq u_+ < 0$. This we investigate in the next section through stability analysis for viscous systems.

3. Stability of Viscous Waves.

Consider the viscous system associated with (1.1)

$$u + (u\phi)_x = \varepsilon u_{xx}$$

(3.1) $$v_t + (v\phi)_x = \varepsilon v_{xx}; \varepsilon > 0.$$

As in the last section, the system can be restricted to the u-axis and is reduced to

(3.2) $$u_t + (u^3)_x = \varepsilon u_{xx}$$

The admissibility condition (2.3) can be obtained from (3.2): the shock is the limit of travelling wave solutions of (3.2) as $\varepsilon \to 0$ if and only if it satisfies (2.3). Clearly a travelling wave solution of (3.2) is also a travelling wave solutions of the system (3.1) by setting $v \equiv 0$. On the other hand when $u_* \leq u_+ \leq 0$ there are infinitely many travelling waves for(3.2) connecting $(u_-, 0)$ to $(u_+, 0)$. This is so because the shock wave is then overcompressive. For $u_* < u_+ < 0$ we have

(3.3) $$\lambda_1(u_+, v_+) < \lambda_2(u_+, v_+) < s < \lambda_1(u_-, v_-) < \lambda_2(u_-, v_-)$$

where s is the shock speed

$$s = \frac{(u_+)^3 - (u_-)^3}{u_+ - u_-}.$$

To find travelling waves

$$(u, v)(x, t) = (u, v)(\frac{x - st}{\varepsilon}) \equiv (u, v)(\xi)$$

we have from (1.1) that

$$u' = -s(u - u_\pm) + u\phi - u_\pm\phi_\pm$$

$$v' = -s(v - v_\pm) + v\phi - v_\pm\phi_\pm,$$

$$v_- = v_+ = 0.$$

This system of o.d.e. has critical points $(u_-,0)$, $(u_+,0)$ and $(\bar{u},0)$. Here $\bar{u}$ is the position with the same shock speed s with u_-. cf. Figure 2.1. From the overcompressibility, (3.3), (u_-,v_-) is an unstable node and (u_+,v_+) a stable node and therefore there are infinitely many connecting orbits, Figure 3.1.

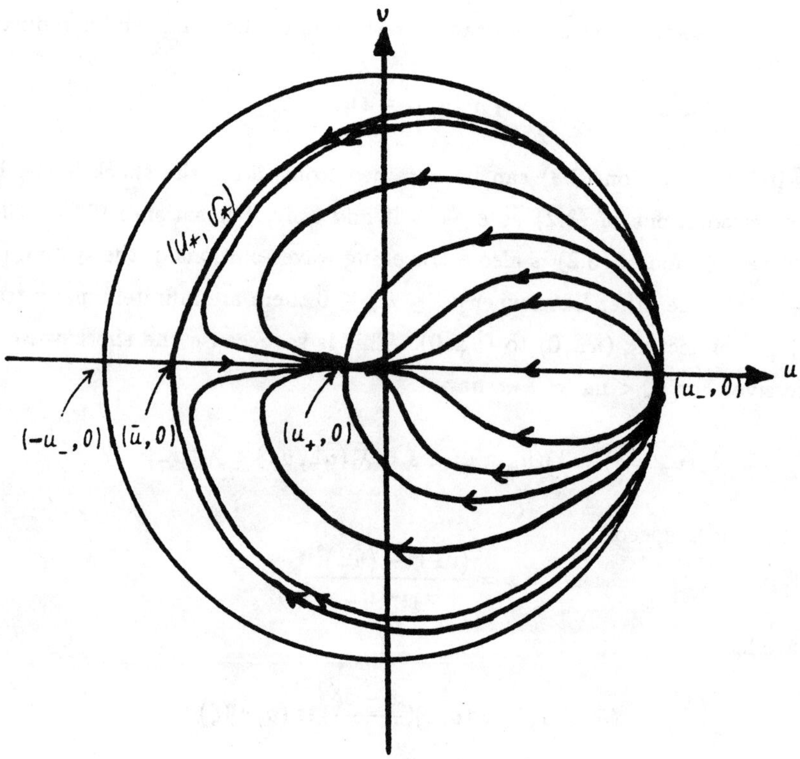

Figure 3.1: Overcompressive Travelling Waves.

Now let us investigate the stability of any of these travelling waves. Given a pertrubation $(\bar{u},\bar{v})$ of a travelling wave $(u_1,v_1)(\xi)$:

$$(u,v)(x,t) = (\bar{u},\bar{v})(x,t) + (u_1,v_1)\left(\frac{x-st}{\varepsilon}\right),$$

(3.5)

$$(\bar{u},\bar{v})(x,0) \to 0 \text{ as } |x| \to \infty.$$

Since (3.1) are conservation laws we have two time invariants

$$\int_{-\infty}^{\infty} \bar{u}(x,t)dx = \int_{-\infty}^{\infty} \bar{u}(x,0)dx,$$
$$\int_{\infty}^{\infty} \bar{v}(x,t)dx = \int_{-\infty}^{\infty} \bar{v}(x,0)dx, \quad t \geq 0$$

To study the stability of a travelling wave (u_1, v_1), we investigate first possible effects of a perturbation. It may create acoustic waves travelling with characteristic speeds λ_1, λ_2. Since (u_1, v_1) is overcompressive, (3.3), acoustic waves move toward the travelling wave. Consequently dissipation and nonlinearity work in the finite region in x around $\xi = \frac{x - st}{\varepsilon}$ = constant. Thus if travelling waves are stable then the stability must be in any $L_p(x)$ – norm, $p \geq 1$, provided that the perturbation decays sufficiently fast at $|x| = \infty$, e.g. $|(\bar{u}, \bar{v})(x,0)| = O(1)|x|^{-\alpha}$ for some $\alpha > 1$ as $|x| \to \infty$. In view of the fact that travelling waves can only be orbitally stable and that there are infinitely many travelling waves with the same end states $(u_{\pm}, 0)$, we see that the pertrubation would produce a different travelling wave, properly translated. Taking the time-invariants (3.6) into account we then conclude that the solution $(u,v)(x,t)$ would converge to some travelling wave (u_2, v_2) translated by certain amount x_0, i.e.

$$(u,v)(x,t) \to (u_2, v_2)(\frac{x + x_0 - st}{\varepsilon}) \text{ as } t \to \infty$$

provided that the corresponding time invariants are zero:

$$\int_{-\infty}^{\infty} \bar{\bar{u}}(x,t)dx = \int_{-\infty}^{\infty} \bar{\bar{u}}(x,0)dx = 0,$$
$$\int_{-\infty}^{\infty} \bar{\bar{v}}(x,t)dx = \int_{-\infty}^{\infty} \bar{\bar{v}}(x,0)dx = 0, t \geq 0,$$

$$(u,v)(x,t) = (\bar{\bar{u}}, \bar{\bar{v}})(x,t) + (u_2, v_2)\left(\frac{x + x_0 - st}{\varepsilon}\right), \tag{3.7}$$

$$(\bar{\bar{u}}, \bar{\bar{v}})(x,t) = (\bar{u}, \bar{v})(x,t) + (u_1, v_1)(\frac{x - st}{\varepsilon}) - (u_2, v_2)(\frac{x + x_0 - st}{\varepsilon}) \tag{3.8}$$

The central question is then the following: Given a perturbation $(\bar{u}, \bar{v})(x,0)$ of a travelling wave (u_1, v_1) can one always find another travelling wave (u_2, v_2) and a translation x_0 so that $(\bar{\bar{u}}, \bar{\bar{v}})(x,0)$ defined by (3.8) carry zero integral (3.7)? There is a one–parameter family of travelling waves, Figure 3.1, and the translation x_0 can be any real number. Thus we have two parameters to vary. With suitable choice of these two parameters we should be able to satisfy the two time–invariants (3.6). This is so if ε is fixed and the perturbation $(\bar{u}, \bar{v})(t, 0)$ is sufficiently small. In other words, all of these travelling waves are nonlinearly stable for fixed viscosity strength ε in the sense that a small perturbation of a travelling wave would produce another nearby travelling wave with a small translation.

For the purpose of establishing a hyperbolic theory for (1.1) we need to study a stronger stability property than the one just mentioned. We need to know whether given a fixed perturbation, though small, of a travelling wave the solution would converge to another travelling wave even if the viscosity strength is very small. In other words, we are interested in the stability uniformly in the viscosity strength. It turns out that the travelling waves here are not stable in this stronger sense. For this we need to show that for a given perturbation $(\bar{u}, \bar{v})(x, 0), (\bar{u}, \bar{v})(x, 0)$ of (3.8) can not satisfy the zero time–invariants of (3.7) for any choices of travelling wave (u_2, v_2) and the translation x_0. This can be shown, as we mentioned before, only when ε becomes small. We choose the perturbation $(\bar{u}, \bar{v})$ with

$$M \equiv \int_{-\infty}^{\infty} \bar{v}(x, 0)dx \neq 0.$$

From (3.7) and (3.8) we have

$$M = \int_{-\infty}^{\infty} \left[v_2(\frac{x + x_0}{\varepsilon}) - v_1 \left(\frac{x}{\varepsilon}\right) \right] dx.$$

At this moment we recall that, although travelling waves sharpen up to become shock waves as ε tends to zero, the functions $(u_1, v_1), (u_2, v_2)$ etc. of $\dfrac{x - st}{\varepsilon}$ are independent of ε because the o.d.e. (3.4) they satisfy is independent of ε. Thus the travelling waves, Figure 3.1, are parameterized independent of ε. From the last identity we have

$$M = \varepsilon \int_{-\infty}^{\infty} [v_2(y + y_0) - v_1(y)]dy, \quad y_0 \equiv \frac{x_0}{\varepsilon}. \tag{3.9}$$

To show that the above identity fails for ε sufficiently small, we need only to show that the integral

$$\int_{-\infty}^{\infty} [v_2(y + y_0) - v_1(y)]dy \tag{3.10}$$

is uniformly bounded for any choices of v_2 and y_0. Denote by v_* the v–component of the upper limit of all travelling waves, that is (u_*, v_*), $v_* > 0$, is the travelling wave connecting

$(u_-,0)$ to $(\bar{u},0)$, Figure 3.1. It can be shown easily that for any given travelling wave (u_2,v_2),

$$-\int_{-\infty}^{\infty} v_*(y)dy - \left|\int_{-\infty}^{\infty} v_1(y)dy\right| < \int_{-\infty}^{\infty}[v_2(y+y_0)-v_1(y)]dy < \int_{-\infty}^{\infty} v_*(y)dy + \left|\int_{-\infty}^{\infty} v_1(y)dy\right|$$

In other words, (3.10) is uniformly bounded and (3.9) fails to be satisfied for ε sufficiently small.

Not all overcompressive travelling waves are not stable uniformly in the viscosity strength. When they are stable in this strong sense they usually split into two waves as the viscosity tends to zero.

This does not occur, strictly speaking, for the model (2.1). However when $u_+ = 0$ it does happen in some limiting sense. This is , for a general fixed perturbation $(\bar{u},\bar{v})$ of any of the above travelling wave (u_1,v_1) connecting $(u_-,0$ and $(u_+,0)$ $u_* < u_+ < 0$, and for any given $\varepsilon > 0$, the corresponding asymptotic state (u_2,v_2) exists provided that $u_+ = u_+(\varepsilon)$ is sufficiently small. Moreover, $u_+(\varepsilon)$ tends to zero in the zero dissipation limit $\varepsilon \to 0$, and the width of each asymptotic wave (u_2,v_2) is uniformly bound in ε. In the limit, (u_2,v_2) tends to the combination of the wave (u_*,v_*) and the horizontal wave connecting $(\bar{u},0)$ and $(u_+,0)$.

These waves, however, converge to the contact discontinuity $(u_-,-u_-,0)$ and the shock wave $(-u_-;0,0)$. In other words, the two waves combination $(u_-,0;-u_-,0)$ and $(-u_-,0;0,0)$ can be realized as the limit of travelling waves connecting $(u_-,0)$ and $(u_+,0)$ in the zero dissipation limit $\varepsilon \to 0$ and as $u_+ \to 0$ in a stable manner.

In conclusion, overcompressive travelling shock waves are either unstable in the zero dissipation limit or split into two waves. Our analysis is carried out for the specific model (2.1),(3.1), but the above conclusion should hold for general systems. Thus the strong viscosity criterion of admitting only waves which are stable in the zero dissipation limit preclude shock waves for (2.1) whose end states lie on the opposite sides of the origin. Therefore only shock waves whose end states are on the same side of the origin are allowed, (2.4). For such a shock wave, the corresponding travelling wave is stable. The stability analysis is again based on the consideration of time–invariants. the major difference is that the corresponding travelling wave is a saddle-node connection of the o.d.e. (3.4) and is therefore unique. On the other hand, the wave is not overcompressive and a perturbation, in addition to cause a translation of the wave, produces an acoustic wave moving away from the travelling wave, [8]. It can be seen easily that the Riemann problem can be solved uniquely with such an admissibility criterion, and so we obtain a well-posed hyperbolic theory for (2.1) through the viscosity consideration. A mostly hyperbolic consideration was put forward in [4] which achieves the same selection.

References

1. Brio, M. and Hunter, J., Rotationally invariant waves (preprint)

2. Conley, C. and Smoller, J., Shock waves as limits of progressive wave solutions of higher order equations, Comm Pure Appl. Math. 24 (1971), 459–472.

3. Conlon, J., A theorem in ordinary differential equations with an application to hyperbolic conservation laws, Adv. Math. 35 (1980), 1- 18.

4. Freistühler, H., Dynamical stability and vanishing viscosity, a case study of non-strictly hyperbolic system. (preprint)

5. Keyfitz, B. and Kranzer, H., A system of non-strictly hyperbolic conservation laws arising in elasticity theory, Arch. Rat. Mech. Anal.72 (1980).

6. Lax, P.D., Hyperbolic system of conservation laws, II, Comm. Pure Appl. Math. 10(1957), 537–566.

7. Liu, T.-P, The Riemann problem for general systems of conservation laws, J.D.E., 18(1975), 218–234

8. Liu, T.-P., Nonlinear stability of shock waves for viscous conservation laws, Memoirs, Amer. Math. Soc. 56 No. 328 (1985)

9. Pego, R., Stable viscosities and shock profiles for systems of conservation laws, Trans. Amer. Math. Soc. 282(1984), 749–763.

10. Temple, B., Global solutions of the Cauchy problem for a class of 2×2 nonstrictly hyperbolic conservation laws, Adv. Appl.Math. 3(1982), 335–375.

CHAPTER 11

Fluid Flow in a Supersonic Peristaltic Pump*

Ralph Menikoff†
Klaus S. Lackner†

Abstract. An axially phased implosion of a channel with a phase velocity exceeding the sound speed in the channel acts as a peristaltic pump which drives a planar shock wave along the channel axis. The flow behind the axial shock is well described by the equations for nozzle flow together with an additional dynamical degree of freedom for the shape of the wall. A unique and stable steady-state solution exists which is similar in its structure to the profile of a CJ detonation wave. The two different physical phenomena are modeled by structurally similar hyperbolic PDEs. The solution in both cases is a shock followed by a smooth decrease in pressure ending in a sonic point. The supersonic peristaltic pump offers a tractable example of a model with an unusual wave structure generated by an additional dynamical degree of freedom coupled to the conservation equations by source terms.

1. Introduction. Quasi-linear hyperbolic partial differential equations (PDEs) frequently arise in physics to describe systems of conservation laws. The system of fluid dynamic equations is the classical example. The most important feature of the solution of hyperbolic PDEs is the wave structure, typically, shock waves and rarefaction waves. The effects of the flux function on the wave structure have been studied in great detail; in the context of equilibrium fluid dynamics, see for example the review [20]. In particular, physical effects such as phase transitions can lead to complex wave structures such as split waves and composite waves.

Source terms in the PDEs also affect the wave structure. There are several types of source terms which are physically important. Geometric source terms, which arise in cylindrically or spherically symmetric geometries break the translation invariance of the equations and cause shock waves to be non-steady. More generally, geometric source terms give rise to additional asymptotic states, for example stationary waves in duct flow [15]. Viscosity or heat conduction terms change the PDEs to parabolic type and smear out discontinuities, thus giving shock waves extended smooth profiles. The description of other physical effects requires the system of equations to be enlarged to account for internal degrees of freedom, see for example [22]. New degrees of freedom introduce new dimensional scales into the dynamics of the system and are reflected in the wave structure [11]. Non-

*Received by the editors July, 1990; This work was supported by the U.S. Department of Energy.
†Theoretical Division, Los Alamos National Laboratory, Los Alamos, NM 87545.

equilibrium effects such as dissociation or ionization lead to relaxation phenomena in which the state behind a shock gradually approaches its equilibrium value [18]. Radiation flux may result in radiation precursors in front of a shock, or smooth out the shock discontinuity. Phase separation, for example between air and water vapor, gives rise to condensation shocks. A fast and energetic chemical reaction can lead to the phenomenon of self-sustaining detonation waves.

Another type of source term is an external source. Here, we describe and analyze a physically interesting problem of this type. It is the supersonically phased implosion of a cylindrical shock tube which in effect acts as a peristaltic pump and drives a shock along the tube axis. The applied pressure driving the implosion is a moving external source. The motivation for studying this system, its physical description, and experimental and numerical results have been presented elsewhere [19]. Here, we concentrate on the mathematical aspects of the model equations. The steady-state solution for the model is structurally similar to a self-sustaining, one-dimensional detonation wave. The phased implosion of a tube illustrates, in a simple and tractable example, how a source term in the conservation equations can have an important effect on the qualitative nature of the wave structure.

We begin by presenting the model equations in section 2. As shown in section 3, the steady-state wave solution reduces the PDEs to three algebraic equations and one ODE. The boundary conditions result in an "eigenvalue" like problem to determine the steady-state wave. Furthermore, it can be shown that there is a unique and stable steady-state wave. Its structure is similar to a self-sustaining detonation wave. This analogy is described in section 4. Concluding remarks are in section 5.

2. Model of supersonic peristaltic pump. We consider an axially phased implosion of a cylindrical shock tube. The phase velocity of the implosion is assumed to exceed the sound speed of the ambient fluid inside the shock tube. For a physical realization, a shock tube may be inserted into a cylinder of high explosives which is initiated at one end. As shown in fig. 1, the detonation wave sweeps over the tube and implodes the shock tube wall resulting in a constriction of the tube wall which travels along the tube axis. The constriction acts as peristaltic pump which, however, is driven at supersonic speed. A planar shock wave forms near the entrance of the constriction. Such a supersonic peristaltic pump can be used to generate very strong shock waves and is described in more detail in [19].

Two approximations are needed to obtain a simple one-dimensional model of the phased implosion of a shock tube: The first neglects the axial motion and internal structure of the wall; the second neglects the radial velocity and radial dependence of variables within the working fluid. These approximations are justified when the wall is thin, and the wall velocity is small compared to the phase velocity. For simplicity we assume that the phase velocity is much larger than the ambient sound speed leading to a strong axial shock. We model the system driving the phased implosion with a pressure pulse applied to the outside of the shock tube wall and traveling at a phase velocity v_ϕ

$$P_a(z,t) = \widetilde{P}_a(t - z/v_\phi) \ . \tag{2.1}$$

We assume that $\widetilde{P}_a(t)$ is 0 for $t > 0$, rises abruptly and then falls monotonically to 0 as $t \to -\infty$. The applied pressure pulse travels in the direction of increasing z.

The dynamics of the fluid inside the shock tube is described by the equations for 1-D flow in a duct of variable and time dependent cross-sectional area, $A = \pi R^2$,

$$\partial_t(\rho A) + \partial_z(\rho A v) = 0 \ , \tag{2.2}$$

$$\partial_t(\rho A v) + \partial_z(\rho A v^2 + PA) = P\partial_z A \ , \tag{2.3}$$

$$\partial_t(\rho A \mathcal{E}) + \partial_z(\rho A \mathcal{E} v + PAv) = -P\partial_t A \ , \tag{2.4}$$

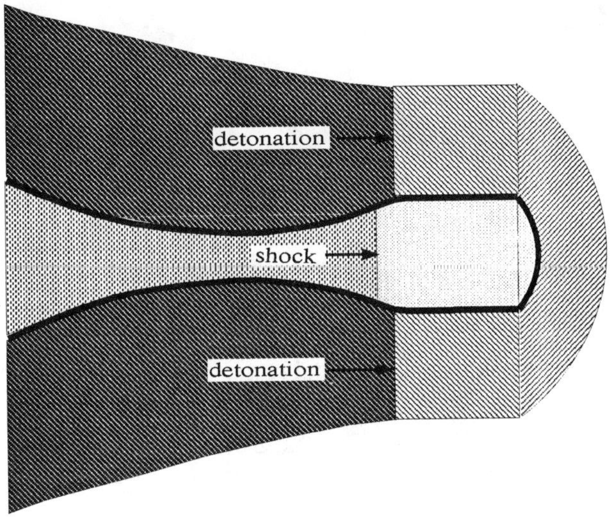

Fig. 1: Schematic cross section of explosively driven phased implosion.

where ρ is the density, v the velocity, and $\mathcal{E} = \frac{1}{2}v^2 + E$ the total specific energy with E the specific internal energy. The pressure is specified by an equation of state (EOS), $P = P(V, E)$, with $V = 1/\rho$ the specific volume. The motion of the wall is governed by Newton's law

$$\sigma \frac{\partial^2}{\partial t^2} R(z,t) = 2\pi R(z,t) \left[P(R,z,t) - P_a(z,t) \right] , \tag{2.5}$$

where σ is the wall mass per unit length.

The model equations (2.2)–(2.5) form a hyperbolic system of PDEs and are written in conservation form. The right hand side of Eqs. (2.3) and (2.4) are source terms for the 1-D axial fluid flow. Equation (2.5) allows for a discontinuity in the radial wall acceleration, but the radial wall velocity, and consequently R and A along with their first derivative, are always continuous. Furthermore, from the left hand side of Eqs. (2.2)–(2.4), it can be seen that the mass, momentum and energy densities and their fluxes are all proportional to A. Therefore, A and the source terms do not affect the shock jump relations for the fluid flow. The applied pressure enters Eq. (2.5) as an external source term. Because the source is moving, v_ϕ has the effect of an additional characteristic velocity [17]. The fluid motion is coupled to the wall motion through the cross-sectional area, while the wall motion is coupled to the fluid motion by the fluid pressure.

Qualitatively, the solution to the model equations is easy to understand. The applied pressure pulse implodes the shock tube wall compressing the fluid inside and raising its pressure. The phased nature of the implosion leads to an axial pressure gradient. When the radial compression results in a sound speed exceeding the phase velocity, compressive waves build up and form an axial shock. Asymptotically, the flow approaches a steady-state in which the shock velocity is equal to the phase velocity.

3. Steady-state solution. To calculate the steady-state, it is convenient to use the co-moving reference frame; that is, the frame of reference moving with the phase velocity. In this frame the duct shape is stationary and takes on the form of a converging–diverging nozzle as shown in fig. 2. The incoming flow, with supersonic velocity $-v_\phi$, goes through a stationary shock near the nozzle entrance. We refer to this shock as the axial shock. As shown below, the flow is again transonic at the point of maximum constriction of the nozzle. We call this point the nozzle throat.

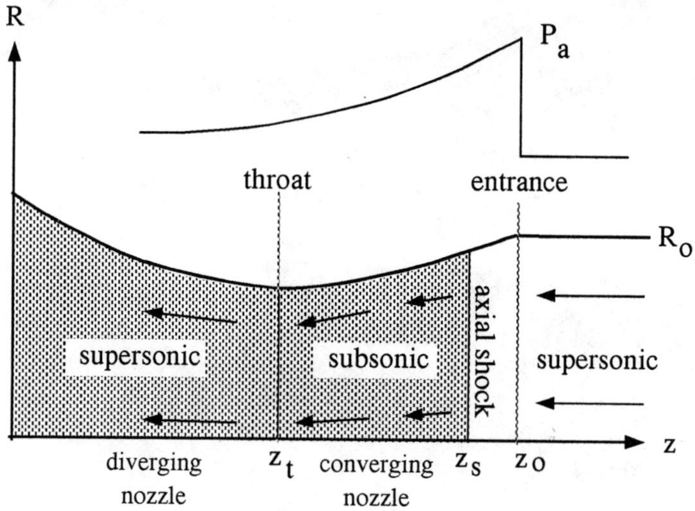

Fig. 2: Sketch of steady-state phased implosion.

Let $q = v_\phi - v$ be the magnitude of the velocity in the co-moving frame. For the steady-state solution, the PDEs reduce to ODEs. Furthermore, the three ODEs for the fluid flow each have a first integral and reduce to the algebraic equations

$$\rho q A = \text{constant}, \tag{3.1}$$

$$\tfrac{1}{2}q^2 + E + PV = \text{constant}, \tag{3.2}$$

$$S = \text{constant}. \tag{3.3}$$

The constants on the right hand side of Eqs. (3.1)–(3.3), corresponding to the flow rate, Bernoulli's constant and entropy, are determined by the state behind the axial shock. The remaining ODE for the nozzle shape is

$$\sigma v_\phi^2 \frac{d^2R}{dz^2} = 2\pi R[P(z) - \widetilde{P}_a(-z/v_\phi)] \, . \tag{3.4}$$

The nozzle shape is determined relative to the onset of the applied pressure pulse which is chosen to occur at $z = 0$. The solution to the algebraic equations defines the fluid pressure in terms of the cross-sectional area normalized to the cross-sectional area at the shock front which is denoted by A_s. Thus, behind the axial shock $P(z) = P_N(A(z)/A_s)$, where P_N is the pressure for steady-state nozzle flow.

It is well known from de Laval nozzle theory [5], that above a critical area ratio, there are two distinct solutions corresponding to a subsonic and a supersonic branch of the flow. The two branches meet at the critical cross-sectional area below which there is no steady-state solution for the nozzle flow. Because the applied pressure pulse falls to zero, the fluid pressure behind the nozzle is unsupported and the wall is driven inward until the pressure behind the shock builds up and limits the inward wall motion to the critical area ratio. As a result the nozzle flow is transonic with the transition occurring at the nozzle throat. The flow is subsonic behind the axial shock and supersonic behind the nozzle throat.

In the case that the axial shock rides behind the nozzle entrance, the model equations (2.2)–(2.4) break down ahead of the axial shock. Because the sound velocity in the ambient fluid is low compared to the velocity of the imploding wall, the time is too short for the fluid to equilibrate with the implosion before being overtaken by the axial shock. Instead oblique radial shocks emanate from the wall and join the axial shock in a Mach-stem configuration. When the axial shock rides near the nozzle entrance, only a negligible volume of fluid is affected by the radial shock waves which are omitted in the 1-D model. Instead of applying the nozzle flow equations in front of the axial shock, we neglect this small flow through the boundary layer, and assume for the bulk of the fluid the ambient state in front of the shock. Thus, we are led to a one-dimensional model, which does not utilize the duct flow equations in front of the axial shock, by the qualitative wave structure of the solution to the physical two-dimensional problem.

In the usual case of nozzle flow with a fixed wall shape, a stationary shock in a converging region of the nozzle is unstable [16, 6]. Consequently, the proper treatment of the flow in front of such a shock has not attracted attention. For the phased implosion, the wall shape is a dynamical degree of freedom which stabilizes the axial shock near the nozzle entrance. The state behind a stationary shock is determined from the state in front of the shock by the Hugoniot jump conditions. A physically correct approximation of the ambient state is important to determine the right hand side of Eqs. (3.1)–(3.3) and hence the fluid flow in the nozzle. The use of the model equations ahead of the axial shock would lead to a precompression in front of the shock and result in a significantly larger and incorrect pressure for the nozzle flow behind the axial shock.

From the above discussion, it follows that the fluid pressure for the ODE (3.4) is given by

$$P(z) = \begin{cases} P_0, & \text{ahead of shock;} \\ P_N^{sub}(A(z)/A_s), & \text{converging nozzle;} \\ P_N^{super}(A(z)/A_s), & \text{diverging nozzle.} \end{cases} \tag{3.5}$$

The function $P_N(A/A_s)$, determined by the state ahead of the shock and the phase velocity, is independent of the nozzle shape. Thus, the right hand side of the ODE (3.4) is well defined. The boundary conditions are the initial wall radius, $R = R_0$, and the initial wall slope, $dR/dz = 0$. In addition, there is a consistency constraint; the minimum radius must occur at the sonic point of the fluid. Because Eq. (3.4) is of second order, only two boundary conditions can be satisfied simultaneously. The axial shock position, relative to the onset of the applied pressure pulse, is the additional degree of freedom needed to satisfy the sonic condition at the nozzle throat. Thus, the shape of the wall is determined by an "eigenvalue" like problem. Based on energy considerations, it was proven in [19] that a unique solution exists to the eigenvalue problem.

The sonic condition at the nozzle throat has an important consequence; the flow in the converging section of the nozzle decouples from the flow behind the throat. Furthermore, energy considerations, experimental data and numerical results [19] strongly indicate that the steady-state solution is the asymptotically stable fixed point of the PDEs. These qualitative aspects of the model imply for applications that the phased implosion is insensitive to the drive conditions and hence a robust means of generating a strong shock wave.

It is frequently important for applications to know how long it takes to reach the steady-state. An approximation to the approach to equilibrium can be obtained by assuming that the solution proceeds through a series of "quasi" steady-states which satisfy the steady-state ODE but have shock velocities different from the phase velocity. The constants for the nozzle flow in Eqs. (3.1)–(3.3) become functions of the shock velocity. Consequently, the nozzle pressure can be parametrized by the shock velocity; $P_N(A/A_s; v_s)$. For a fixed shock position one can solve the ODE with the sonic constraint by varying the shock velocity. Generalizing the energy arguments one can show that v_s increases as the shock lags behind relative to the applied pressure pulse. In the co-moving frame z_s satisfies the dynamical equation

$$\frac{dz_s}{dt} = v_s(z_s) - v_\phi \ . \tag{3.6}$$

Because $dv_s/dz_s < 0$, z_s asymptotically approaches its equilibrium state. The time constant of the convergence is $\tau = -(dv_s/dz_s)^{-1}$. Preliminary numerical experiments for the shock tube experiment presented in [19] indicate that this is a good approximation. Physically, this approximation implies that the internal degrees of freedom of the nozzle solution equilibrate on a fast time-scale relative to the change in the shock velocity.

4. Analogy of phased implosion with detonation wave. There is a striking similarity between the steady-state solution of the model equations for the phased implosion and the Zeldovich–von Neumann–Doering (ZND) model equations for a one-dimensional Chapman-Jouguet (CJ) detonation wave, see table 1. In particular, the flow in the converging nozzle and the reaction zone of a detonation wave both exhibit a shock followed by a smooth region of decreasing pressure ending in a state which is sonic relative to the shock front. The propagating wave in both cases decouples from the flow behind the sonic point.

Table 1: Correspondences between models

	phased implosion	**detonation wave**
initial model	duct flow	1-D fluid dynamics
additional variable	wall radius	reaction progress variable
additional equation	wall equation	rate equation
source term	applied pressure	chemical energy
wave start-up	axial shock formation	detonation wave initiation
spatial wave support	converging nozzle	reaction zone
sonic point	nozzle throat	end of reaction zone
wave speed	axial shock velocity	detonation velocity
quasi-steady parameter	relative shock position	radius of curvature

The similarity of the solutions is rooted in the structural similarity between the model equations describing the two phenomena. Both models add an additional dynamical degree of freedom to the hyperbolic system of conservation equations describing the one-dimensional fluid flow. The extra degree of freedom satisfies a differential equation and couples to the hydrodynamic equations through source terms. The enlarged system of PDEs remains hyperbolic but introduces new types of spatially extended waves; for example, the traveling nozzle solution and the detonation wave.

The equations describing the ZND model for a detonation wave are [8]

$$\partial_t(\rho) + \partial_z(\rho v) = 0 \ , \tag{4.1}$$

$$\partial_t(\rho v) + \partial_z(\rho v^2 + P) = 0 \ , \tag{4.2}$$

$$\partial_t(\rho \mathcal{E}) + \partial_z(\rho \mathcal{E} v + P v) = 0 \ , \tag{4.3}$$

$$\partial_t(\rho \lambda) + \partial_z(\rho v \lambda) = \rho \mathcal{R} \ , \tag{4.4}$$

where λ is the reaction progress variable which is the mass fraction of burned explosive; $\mathcal{R}(V, E, \lambda)$ is the reaction rate, for example, a first-order Arrhenius kinetics reaction rate is

$$\mathcal{R} = K(1-\lambda)\, exp(-E_a/kT) \ . \tag{4.5}$$

with T the temperature and constants K, E_a and k the frequency factor, activation energy and Boltzmann constant respectively. The equation of state $P = P(V, E, \lambda)$ includes the energy release Q of the reaction. For example, an ideal reactive EOS is

$$P(V, E, \lambda) = (\gamma - 1)(E + Q\lambda)/V \ . \tag{4.6}$$

The similarity between the reactive fluid equations (4.1)–(4.3) and the duct flow equations (2.2)–(2.4) is even more striking if the chemical energy release is included in the specific internal energy of the fluid. With the substitution $\widehat{E} = E + Q\lambda$ and $\widehat{\mathcal{E}} = \mathcal{E} + Q\lambda$, the ideal reactive EOS becomes an ideal EOS, $\widehat{P}(V, \widehat{E}) = P(V, E, \lambda)$, which in this special case is independent of λ. The conservation laws have the same form as Eqs. (4.1)–(4.3) but with an energy source term $Q\rho\mathcal{R}$ on the right hand side of Eq. (4.3). Thus, the additional degrees of freedom, R for the phased implosion model and λ for the reactive hydrodynamics, in each case have the effect of adding source terms to the one-dimensional fluid dynamics equations.

For steady-state waves the model again reduces to three algebraic equations and one ODE

$$\rho v = \text{constant}, \tag{4.7}$$
$$\tfrac{1}{2}v^2 + E + PV = \text{constant}, \tag{4.8}$$
$$\rho v^2 + P = \text{constant}, \tag{4.9}$$
$$v\partial_z \lambda = \mathcal{R}. \tag{4.10}$$

In both models the algebraic equations, (4.7)–(4.9) or (3.1)–(3.3), express the local conservation of mass, momentum and energy and determine the hydrodynamic variables as a function of the added variable. The ODE for the additional dynamical degree of freedom then determines the wave profile. In both cases the sonic condition leads to an eigenvalue problem which has a unique solution. The eigenvalue for Eq. (4.10) corresponds to the CJ detonation velocity, and for Eq. (3.4) to the relative position of the axial shock with respect to the applied pressure pulse. In the simple case with one reaction the CJ detonation velocity may be determined algebraically as the entropy extremum along the reactive Hugoniot. With more than one reaction, the CJ detonation velocity needs to be determined from the eigenvalue problem for the ODEs [8].

For the hydrodynamic equations the initial and final state of a shock wave are related by the Hugoniot jump conditions. Typically source terms such as viscosity, which gives rise to a shock profile, or a relaxation phenomenon, which adds structure after the shock front, do not change the asymptotic state behind the wave. The additional degree of freedom in our model equations varies continuously even at a shock discontinuity. Across the extended steady waves, there is a non-zero change in the additional variable because of the source terms which have the effect of giving the extended wave a profile. The source terms also modify the jump conditions across the wave. For an explosive, because the initial state ($\lambda = 0$) is metastable, the reaction once initiated runs to completion ($\lambda = 1$). Moreover, the source terms do not occur in the hydrodynamic equations. As a consequence, the modification in the jump conditions simply consists of using the EOS of the reactants for the initial state and the EOS of the reaction products for the final state. In the special case when the ideal reactive EOS is replaced by an ideal EOS and the chemical energy release shows up as a source term in Eq. (4.3), the energy jump across the detonation wave is the heat release Q.

The jump conditions for the wave driven by the phased implosion is determined by integrating the ODEs for the steady-state equations across the spatial extent of the wave. The results are

$$J\left[\rho A(v - v_\phi)\right]_{throat}^{entrance} = 0 \ , \tag{4.11}$$

$$J\left[\rho A\left((v-v_\phi)v+PV\right)\right]_{throat}^{entrance}=\int_{A_t}^{A_0} P dA\,, \tag{4.12}$$

$$J\left[\rho A\left((v-v_\phi)\mathcal{E}+vPV\right)\right]_{throat}^{entrance}=v_\phi\int_{A_t}^{A_0} P dA\,. \tag{4.13}$$

We note that the source terms from the applied pressure pulse add both momentum and energy to the fluid flow. For the self-sustaining wave which moves with the phase velocity, the hydrodynamic variables at the final state are determined by the algebraic equations for nozzle flow and the sonic condition. As a result, the right hand side of the jump equation may be calculated and used to determine the final state of the additional dynamical variable, namely the wall radius at the sonic point.

It can be shown that between the nozzle throat and entrance $\int PdA = \int \widetilde{P}_a dA$. The right hand side of Eq. (4.13) represents an external energy and corresponds to the chemical energy released in a detonation wave. Consequently, Eq. (4.13) expresses the energy balance between the work needed to drive the fluid flow in the converging region of the nozzle and the external work supplied to the nozzle by the applied pressure pulse. This observation plays a key role in the proof of the existence and uniqueness of a steady-state solution. In [19] it was shown that the energy difference is a monotonic function of the shock position and that the axial shock is overdriven when sufficiently far back in the throat and underdriven when sufficiently far ahead of the onset of the pressure pulse. Consequently, there is one and only one value of the shock position for which the energy balances. The existence proof together with the experimental data and numerical simulations [19] indicates that the steady-state axial shock solution is either an asymptotic fixed-point of the model equations or is at the center of an asymptotic limit cycle.

The possibility of an oscillatory asymptotic solution, for some parameter range of the model, is strongly suggested by the analogy with detonation waves. Galloping detonation waves were predicted from a stability analysis of one-dimensional detonations [7, 14, 3], have been observed numerically [9] and seen experimentally [1]. The galloping instability is usually difficult to observe because two-dimensional instabilities giving rise to a cellular wave structure [8] frequently dominate the behavior of the detonation wave. The axial shock driven by a phased implosion may be a better test bed for studying such instabilities because it should lack multi-dimensional instability modes, and because the length of the nozzle is large compared to the width of the reaction zone facilitating diagnostic access. Furthermore, the values of the critical parameters governing the dynamics of the system are better known and can be varied over a wide range.

Additional physically motivated source terms may be included for both models. Geometric source terms for cylindrically or spherically diverging detonation waves are mentioned here because they lead to a further analogy between detonation waves and axially phased implosions. Diverging detonation waves are time dependent but can be approximated with quasi steady-state solutions of an "eigenvalue" like problem from solving the steady-state ODEs together with a sonic condition [13, 4]. In this approximation the detonation velocity is a function of the radius, $D = D(R)$. For an alternative derivation of $D(R)$, see [21]. Thus, for the quasi-steady approximation the radius of the detonation wave corresponds to the shock position relative to the onset of the applied pressure pulse in the phased implosion.

Finally, there is an analogy between the formation of the axial shock in the one-dimensional model of a phased implosion and the initiation of a detonation wave [2]. In both cases compressive waves are generated near a sonic point and build up to form a self-sustaining wave. As we have seen, the extensive analogy between the wave structure for a phased implosion and reactive hydrodynamics is due to the structural similarity of the model PDEs describing the two phenomena.

5. Conclusions. To understand physical phenomena, one frequently starts with highly simplified but tractable models of reality. Refinements of the model may introduce more dynamical degrees of freedom which may be viewed as perturbations to the structure of the original model equations. If the model is stable under such structural refinement, it can be an excellent approximation to physical reality even though it may leave out many details. For example the dissipative mechanism in fluid dynamics, be it viscosity or heat conduction, affects the details of shock profiles but has little effect on the global wave structure, and shock propagation. In other cases new wave types appear and dramatically change the behavior of the system. It is therefore an interesting question how, for a given set of hyperbolic equations, the wave structure is affected by the inclusion of additional, physically motivated, dynamical degrees of freedom.

Nonunique or nonexistent solutions to the Riemann problem for hyperbolic equations are frequently indicative of a structural instability of the model. Too many wave types lead to non-uniqueness and the admissibility of a particular solution may depend on dynamics which is left out of the simple model. A noteworthy example occurs in the non-equilibrium description of phase transitions [10]. Too few waves lead to the non-existence of a solution to the Riemann problem and suggests that important dynamics has been left out of the model. An example of this occurs for two-dimensional steady-state supersonic flow [12].

An alternative question for a model which is structurally unstable is what physical corrections may account for a particular wave structure which may be experimentally observed. Frequently, there are small length scales or additional degrees of freedom which can only be described by enlarging the set of model equations. For example, reactive hydrodynamics introduces the reaction progress variable as an additional degree of freedom to the fluid dynamics equations. Another case is the elastic–plastic flow.

We have given a simple example of this type. To the nozzle flow equations we added the radius as an additional dynamical degree of freedom. When the system is driven by an external source, the applied pressure pulse, the dynamic system exhibits a steady-state self-sustaining solution which corresponds to a new wave type. The nozzle equations alone would suggest that this wave is unstable. However the energy balance of the whole system indicates the stability of the wave structure. The observation of the close similarity to the behavior of detonation fronts suggests to us that a further study of the general relationship between wave structures and additional degrees of freedom is an area relevant to physics.

REFERENCES

[1] R. Alpert and T.-Y. Toong, *Periodicity in exothermic hypersonic flows and about blunt projectiles*, Astronautica Acta., 17 (1972), pp. 539–560.

[2] J. Bdzil and A. Kapila, *Shock-to-detonation transition: model problem*, Los Alamos National Lab. preprint, LA-UR-89-1720, 1989.

[3] A. Bourlioux, A. Majda and V. Roytburd, *Theoretical and numerical structure for unstable one-dimensional detonations*, Princeton Univ. preprint, 1989.

[4] B. Bukiet, *The effect of curvature on detonation speed*, SIAM J. Appl. Math., 49 (1989), pp. 1433–1446.

[5] R. Courant and K. O. Friedrichs, *Supersonic flow and shock waves*, Springer-Verlag, New York, 1976.

[6] P. Embid, J. Goodman and A. Majda, *Multiple steady states for 1-D transonic flow*, SIAM J. Sci. Stat. Comput. 5 (1984), pp. 21–41.

[7] J. Erpenbeck, *Stability of idealized one-reaction detonation*, Phys. Fluids, 7 (1964), pp. 684–696.

[8] W. Fickett and W. Davis, *Detonation*, Univ. of Calif. Press, Berkeley, 1979.

[9] F. Fickett and W. Wood, *Flow calculations for one-dimensional detonations*, Phys. Fluids, 9 (1966), pp. 903–916.

[10] J. Glimm, *The continuous structure of discontinuities*, Proceedings of Nice Conference, 1988.

[11] ———, *Nonlinear waves: overview and problems*, 1989.

[12] J. Grove and R. Menikoff, *Anomalous reflection of a shock wave at a fluid interface*, J. Fluid Mech., 219 (1990), pp. 313–336.

[13] J. Jones, *Asymptotic analysis of an expanding detonation*, NYU Thesis, 1987.

[14] H. Lee and D. Stewart, *Calculation of linear detonation instability: One-dimensional instability of plane detonations*, J. Fluid Mech., 216 (1990), pp. 103–132.

[15] T.-P. Liu, *Quasilinear hyperbolic systems*, Commun. Math. Phys., 68 (1979), pp. 141–172.

[16] T.-P. Liu, *Nonlinear stability and instability of transonic flows through a nozzle*, Commun. Math. Phys., 83 (1982), pp. 243–260.

[17] T.-P. Liu, *Nonlinear resonance for quasilinear hyperbolic equations*, J. Math. Phys., 28 (1987), pp. 2593–2602.

[18] T.-P. Liu, *Hyperbolic conservation laws with relaxation*, Commun. Math. Phys., 108 (1987), pp. 153–175.

[19] R. Menikoff, K. Lackner, N. Johnson, S. Colgate, J. Hyman and G. Miranda, *Shock Wave Driven by a phased implosion*, Phys. Fluids A (1991), pp. 201–218.

[20] R. Menikoff and B. Plohr, *The Riemann problem for fluid flow of real materials*, Rev. Mod. Phys., 61 (1989), pp. 75–130.

[21] D. Stewart and J. Bdzil, *The shock dynamics of stable multidimensional detonations*, Combustion and Flame, 72 (1988), pp. 311–323.

[22] Ya. B. Zel'dovich and Yu. P. Raizer, *Physics of shock waves and high-temperature phenomena*, Academic Press, New York, 1966.

CHAPTER 12

Travelling Waves for a Cooperative and a Competitive-Cooperative System*

Konstantin Mischaikow†

Abstract. This paper is intended to give a clear exposition of a technique for proving the existence of bistable travelling waves. Two simple examples are considered. The first involves two obligate mutualist species and the second, two obligate mutualist species both of which compete with a third.

1. Introduction. This paper addresses the question of existence of travelling wave solutions for systems of reaction diffusion equations. In particular, it is intended to give a clear exposition of a technique for proving the existence of monotone waves for systems which model interacting species (possibly chemical or biological). Of course the hope is that the ideas presented herein will, in a modified form, be applicable to more general systems including those arising from hyperbolic problems and models for phase transitions. The general type of reaction diffusion system considered here is of the form

$$u_{it} = d_i u_{ixx} + u_i f^i(\mathbf{u}) \tag{1.1}$$

where $i = 1, ..., n$, $\mathbf{u} = (u_1, ..., u_n) \in \mathbf{R}^n$, $d_i > 0$, $x, t \in \mathbf{R}$, $f^i \in C^1(\mathbf{R}^n, \mathbf{R})$, and $\frac{\partial f^i}{\partial u_j}(\mathbf{u}) \neq 0$ for $\mathbf{u} \in \mathbf{R}^{n+}$, for $j = 1, ..., n$. Travelling waves are solutions of the form $u(x,t) = u(x - \theta t)$ where θ represents the speed of the wave. A simple substitution shows that finding travelling waves is equivalent to finding solutions to the $2n$-dimensional system of ODEs

$$\begin{aligned} u_i' &= v_i \\ d_i v_i' &= -\theta v_i - u_i f^i(\mathbf{u}) \end{aligned} \tag{1.2}$$

*Supported in part by a grant from the National Science Foundation.
†Center for Dynamical Systems and Non-linear Studies, School of Mathematics, Georgia Institute of Technology, Atlanta, Georgia 30332.

which satisfy the boundary conditions

$$\begin{aligned}\lim_{t\to\infty}(u(t),v(t)) &= B = (b,0)\\ \lim_{t\to-\infty}(u(t),v(t)) &= A = (a,0)\end{aligned} \tag{1.3}$$

where A and B are critical points to (1.2). Such a solution will be denoted by $A \to B$. We shall specialize a bit further and look for "bi-stable" waves, i.e. solutions $A \to B$ where both a and b are hyperbolic attracting fixed points for the reaction system

$$u_i' = u_i f^i(\mathbf{u}). \tag{1.4}$$

This assumption implies that, in general, bistable waves occur only for a discrete set of wave speeds, i.e. the problem becomes one of showing that there exists a wave speed θ^* for which an $A \to B$ connection occurs.

The techniques presented in this paper are based on the Conley index and are not new in the sense that Charles Conley promoted its use in proving existence of connecting orbits in the early seventies. Since then numerous people have successfully used his ideas. Within the context of the types of problems discussed here the most notable applications can be found in the works of Conley [1], Conley & Gardner [2], and Gardner [3]. For the single species problem one can find our approach sketched out in [1] (in particular, see the example IV.2.6). That these ideas could be applied to 2-species problems was first demonstrated in [2] (competing species) and [3] (predator-prey). The contribution of this author and his co-workers (V. Hutson and J. Reineck) is to show that this approach is in fact a very general technique and can be straightforwardly applied to n-species problems of mixed type, i.e. systems involving cooperative, competitive, and predating species simultaneously. It should be remarked that on the level at which the proofs are presented here, our technique is indistinguishable from that in [2] and [3]. However, whereas these two papers use a connection index to find connecting orbits, we use the connected simple system associated with the Conley index to obtain existence. Results of [5] seem to imply that these two approaches are theoretically equivalent *. Nevertheless, it is the author's opinion that the connected simple system is the more basic concept and, thus, more natural to work with. Furthermore, the global bifurcation results of [5] are obtained in the context of connected simple systems, and hence, they can be applied directly to the results obtain by this technique. (For examples of how these theorems can be applied, the reader is referred to [4] and [7]). At any rate, for the purposes of this paper, the reader need not be concerned with this subtle difference as it is hidden in the proof of the abstract existence theorem (Theorem 2.1) which is not presented here.

Since the emphasis of this paper is on clarity, we shall not strive for generality, but rather restrict our attention to two rather simple (but new) examples. More general results of this nature can be found in the forthcoming papers [4], [6], and [7]. The first and simpler example is a 2-species system with the following assumptions on the corresponding reaction system

$$\begin{aligned}u_1' &= u_1 f^1(\mathbf{u})\\ u_2' &= u_2 f^2(\mathbf{u}).\end{aligned} \tag{1.5}$$

A1. *The interaction matrix* $\begin{bmatrix} \frac{\partial f^1}{\partial u_1} & \frac{\partial f^1}{\partial u_2} \\ \frac{\partial f^2}{\partial u_1} & \frac{\partial f^2}{\partial u_2} \end{bmatrix}$ *has the following signs* $\begin{bmatrix} - & + \\ + & - \end{bmatrix}$.

(A1) implies that the reaction dynamics form a cooperative (or mutualist) system with intraspecies competition.

* Robert Gardner has informed me that Conley originally considered using connected simple systems in their work, but then later formulated the connection index approach which first appeared in [2].

A2. *There are exactly 3 hyperbolic critical points as indicated in the following table where $b_i > c_i > 0$ and W^u denotes the unstable manifold of the critical point.*

critical point	dimension W^u
$a = (0, 0)$	0
$b = (b_1, b_2)$	0
$c = (c_1, c_2)$	1

Figure 1 indicates what the level sets $f^i(\mathbf{u}) = 0$ might look like.

Theorem A. *Given assumptions (A1) and (A2) there exists a wave speed θ^* for which a monotone $A \rightarrow B$ travelling wave occurs.*

The proof of this theorem will be given in Section 3.

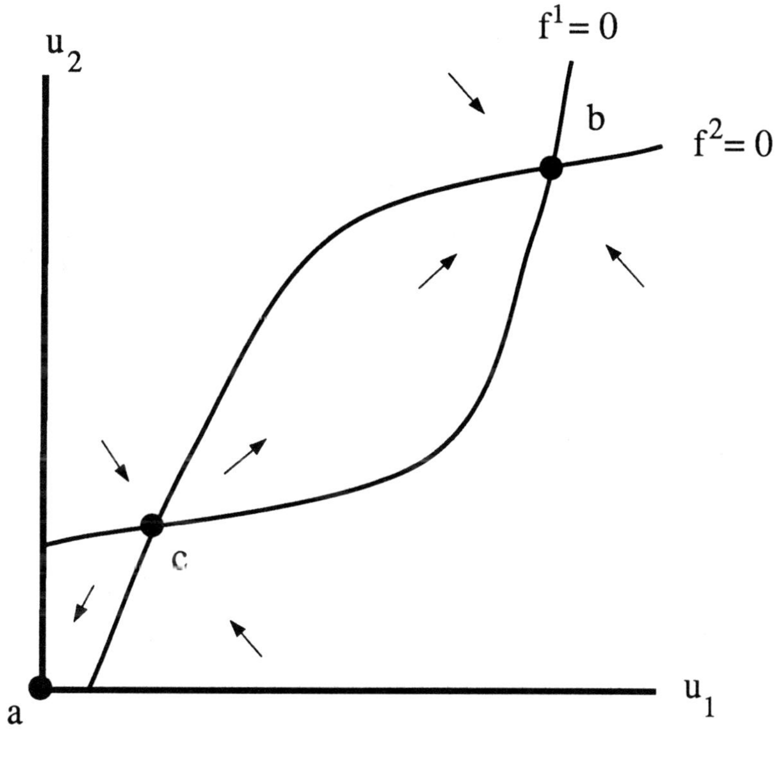

Figure 1

The second example involves adding a third competing species to the first example. Notice that the dynamics on the u_1, u_2-plane is the same as that of the first example. In particular we make the following assumptions.

B1. *The interaction matrix has the following sign structure.*

$$\begin{bmatrix} - & + & - \\ + & - & - \\ - & - & - \end{bmatrix}$$

Therefore, species 1 and 2 are cooperative, and both compete with specie 3.

B2. *As the following table indicates for the reaction system, there are exactly 5 hyperbolic critical points where $b_i > c_i > \bar{c}_i$ for $i = 1, 2$.*

critical point	dimension W^u
$a = (0,0,0)$	1
$b = (1,1,0)$	0
$c = (c_1, c_2, 0)$	2
$\bar{a} = (0,0,\bar{a}_3)$	0
$\bar{c} = (\bar{c}_1, \bar{c}_2, \bar{c}_3)$	1

A remark on the hypothesis that $b = (1,1,0)$ is perhaps in order. This is really not a restriction in the sense that by a rescaling of the $\mathbf{u}$ variables b can be located anywhere in the positive orthant of the u_1, u_2-plane. However, this hypothesis makes the statement of assumption (B3) simpler and avoids carrying constants around in the proof of Lemma 4.8.

Figure 2 indicates what the level sets $f^i(\mathbf{u}) = 0$ might look like.

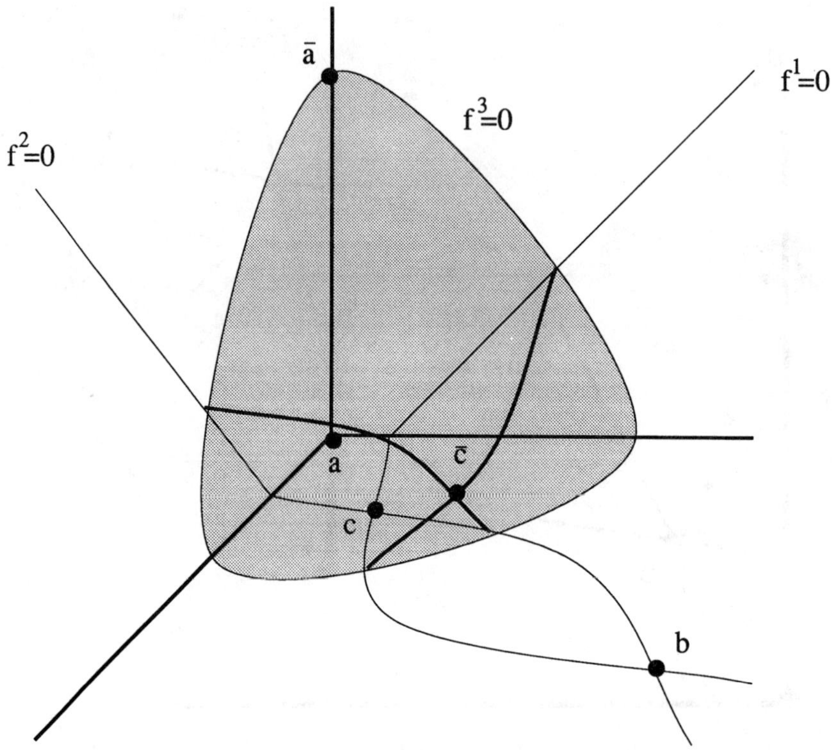

Figure 2

Define

$$V(s) := min\{f^1(s,s,0), f^2(s,s,0)\}.$$

Let $K_i := min_{\mathbf{u}\in[0,1]\times[0,1]\times\{0\}} f^i(\mathbf{u})$ for $i = 1, 2$.

B3. $\int_0^1 sV(s)ds > -(K_1 + K_2)$.

Theorem B. *Given assumptions (B1), (B2) and (B3), there exists a wave speed θ^* for which a monotone $\bar{A} \rightarrow B$ travelling wave occurs.*

Remark: While (B3) is a sufficient condition to guarentee the existence of the $\bar{A} \rightarrow B$ travelling wave it is far from necessary. An important, but probably difficult, open problem is to find necessary conditions.

2. Abstract Existence Theorem. Before stating the basic existence theorem some notation needs to be introduced. Let $\phi : \mathbf{R} \times X \to X$ be a flow on a locally compact space. Given a compact set N let

$$N^T = \{x \in N | \phi([-T,T],x) \subset N\}.$$

N is called an *isolating neighborhood* if there exists $T > 0$ such that $N^T \subset int(N)$. The maximal invariant set in N is given by $N^\infty = \bigcap_{T>0} N^T$. Clearly if N is an isolating neighborhood and $x \in N^\infty$, then $x \notin \partial N^\infty$. An invariant set S is called *isolated* if there exists an isolating neighborhood N of S such that $S = N^\infty$.

A simple, but as we shall see, useful decomposition of an invariant set S is that of an *attractor-repeller* (A-R) pair. Let $\omega(U)$ and $\omega^*(U)$ denote the omega and alpha limit sets of U respectively. $A \subset S$ is called an *attractor* in S if there exists a neighborhood U of A such that $\omega(U \cap S) = A$. The *dual repeller* of A, denoted by A^* is defined by $A^* = \{x \in S | \omega(x) \cap A = \emptyset\}$. The pair (A, A^*) make up an A-R pair. Notice that given an A-R pair decomposition of S, if $x \in S$ then $x \in A \cup A^*$ or $\omega(x) \subset A$ and $\omega^*(x) \subset A^*$, i.e. S is made up of the attractor A, its dual repeller A^*, and connecting orbits from A^* to A.

One of the basic ideas of this technique is to homotope the complicated n-species problem to a simple 1 species problem for which the existence of a travelling wave is well known. For this reason a discussion of what is meant by continuing isolating neighborhoods and isolated invariant sets is necessary. Consider a parameterized family of flows $\phi(t,x,\sigma) = \phi_\sigma(t,x)$ where $\sigma \in [0,1]$ is the parameter value. Define the parameter flow to be $\Phi : \mathbf{R} \times X \times [0,1] \to X \times [0,1]$ by

$$\Phi(t,x,\sigma) = (\phi_\sigma(t,x),\sigma).$$

Let N_σ denote an isolating neighborhood for the flow ϕ_σ. One says that the isolating neighborhood N_1 continues to N_0 if there exists N an isolating neighborhood of Φ, the parameter flow, such that $N|_{X\times\{0\}} = N_0$ and $N|_{X\times\{1\}} = N_1$. Similarly, if S_σ denotes an isolated invariant set under ϕ_σ, then S_0 continues to S_1 if there exist corresponding isolating neighborhoods which continue.

We are now ready to present the hypotheses for the abstract theorem and start with the homotopy discussed in the previous paragraph. With this in mind we write down the following two parameter set of equations

$$\begin{aligned} u_i' &= v_i \\ d_i v_i' &= -\theta v_i - u_i f^i(\mathbf{u},\sigma) \end{aligned} \tag{$2.1_{\theta,\sigma}$}$$

where now $f^i : \mathbf{R}^n \times [0,1] \to \mathbf{R}$. The two parameters are θ, the wave speed, and σ the homotopy parameter. Of course, the corresponding reaction system is given by

$$u_i' = u_i f^i(\mathbf{u},\sigma). \tag{2.2_σ}$$

We shall adopt the convention that at $\sigma = 0$ we are at the original system we wish to study, i.e. $f^i(\mathbf{u},0) = f^i(\mathbf{u})$ and at $\sigma = 1$ we have a simple system for which we can prove the existence of a travelling wave solution. We formalize this latter statement via the following assumption.

Assumption I. *For the reaction system* (2.2_1) *there exists a 1-dimensional attracting invariant subspace* $L \subset \mathbf{R}^n$. *Furthermore, the dynamics on* L *are as in Figure 3, i.e. the set of bounded solutions consist of the hyperbolic critical points* $\{a,b,c\}$ *and the heteroclinic connections* $c \to a$ *and* $c \to b$.

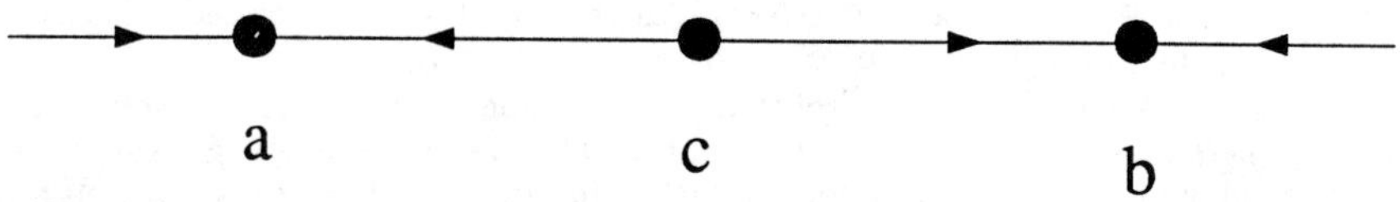

Figure 3

The assumption concerning the homotopy from $\sigma = 1$ to $\sigma = 0$ is as follows. Let $A_1 = (a, 0), B_1 = (b, 0)$, and $C_1 = (c, 0)$.

Assumption II. *A_σ and B_σ continue as critical points for (2.2_σ) and N_σ continues as an isolating neighborhood for $(2.1_{\theta,\sigma})$ such that (B_σ, A_σ) is an attractor-repeller pair for N_σ^∞. Furthermore, for $\sigma = 1$, if there exists an $A_1 \to B_1$ solution to $(2.1_{\theta,1})$ for some value of θ, then the connecting orbit belongs to N_1^∞.*

Theorem 2.1. *Given Assumptions I and II, there exists a travelling wave from A_0 to B_0 with wave speed θ_0^*.*

The proof of this theorem and comments concerning its generalizations can be found in [6]. As will be seen in the following sections for the two example mentioned in the introduction Assumption I is easily satisfied, and it is Assumption II where work is required. However, one of the points to be made in this article is that for a wide class of problems, verifying this second hypothesis is surprisingly straight forward. In the concluding remarks presented in Section 5, two example will be given where it is not clear how to satisfy Assumptions I or II.

In order to demonstrate that the choice of an isolating neighborhood can be made in a natural way we shall briefly indicate how one can choose such a neighborhood in the case of a single species problem, i.e. for the pair of equations

$$\begin{aligned} u' &= v \\ v' &= -\theta v - f(u). \end{aligned} \tag{2.3}$$

See [1] for a more complete discussion.

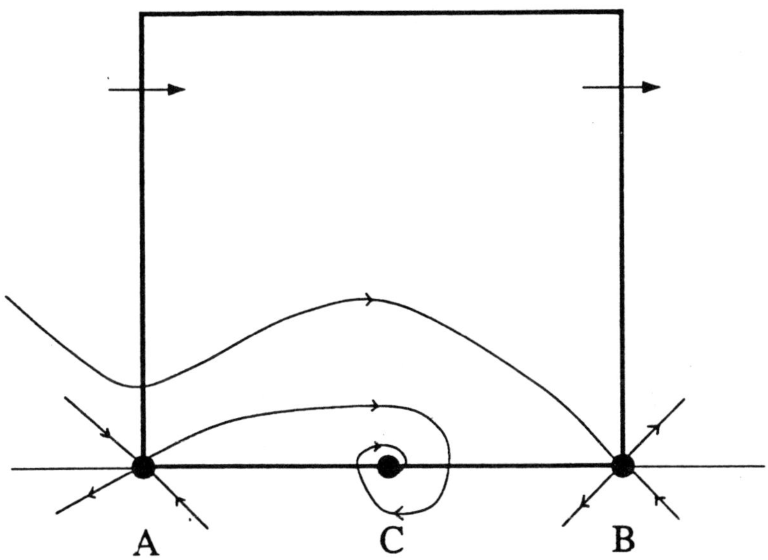

Figure 4

Assume that there are only three critical points $A = (a, 0), B = (b, 0)$, and $C = (c, 0)$ where $a < c < b$. Since we are interested in detecting connecting orbits $A \to B$ we choose a *monotone flow box*

$$\mathcal{R} = [a, b] \times [0, K] := \mathcal{R}_u \times \mathcal{R}_v$$

where K is sufficiently large (See Figure 4 which is drawn for $\theta > 0$). Notice that we are only considering values of $v \geq 0$ and hence u is not decreasing in $\mathcal{R}$. Also notice that $\mathcal{R}$ is written as the product of a "spacial" component $[a, b] = \mathcal{R}_u$, and a "velocity" component $[0, K] = \mathcal{R}_v$. To verify that $\mathcal{R}$ is an isolating neighborhood we only need to show that $\mathcal{R}^\infty \bigcap \partial\mathcal{R} = \emptyset$. Of course

$$\partial\mathcal{R} = \partial[a, b] \times [0, K] \bigcup [a, b] \times \{0\} \bigcup [a, b] \times \{K\}.$$

For simplicity, we shall refer to $\partial[a, b] \times [0, K]$ as the sides of $\mathcal{R}$, $[a, b] \times \{K\}$ as the top of $\mathcal{R}$, and $[a, b] \times \{0\}$ as the bottom of $\mathcal{R}$. Obviously, $\mathcal{R}$ is *not* an isolating neighborhood since the critical points are all elements of the invariant set $\mathcal{R}^\infty$ and elements of the bottom of $\mathcal{R}$. On the other hand, for K large enough any element of the top of the monotone flow box must leave the box eventually in forward or backward time. (The velocity of the orbit is simply too large for it to remain bounded.) Furthermore, on the sides (excluding the critical points) the orbits are entering or leaving. Similarly, on the bottom the orbits are entering or leaving except at the critical points. This suggests modifying the monotone flow box slightly. Since, we wish to make (B, A) into an attractor repellor pair we need to include the points A and B, and exclude the point C. How this is done is indicated in Figure 5. Notice that we add $T_\epsilon(A) := [a - \epsilon, a + \epsilon] \times [-K, K]$ and $T_\epsilon(B) := [b - \epsilon, b + \epsilon] \times [-K, K]$ and delete $B_\epsilon(C) := [c - \epsilon, c + \epsilon] \times [-\epsilon, \epsilon]$. We leave it to the reader to check that this new set

$$N = (\mathcal{R} \cup T_\epsilon(A) \cup T_\epsilon(B)) \backslash B_\epsilon(C)$$

is in fact an isolating neighborhood for all θ. A little more terminology is convenient. A and B are *augmentable* critical points and C is a *removable* critical point because adding a neighborhood about A and B and deleting a neighborhood of C transformed $\mathcal{R}$ into an isolating neighborhood.

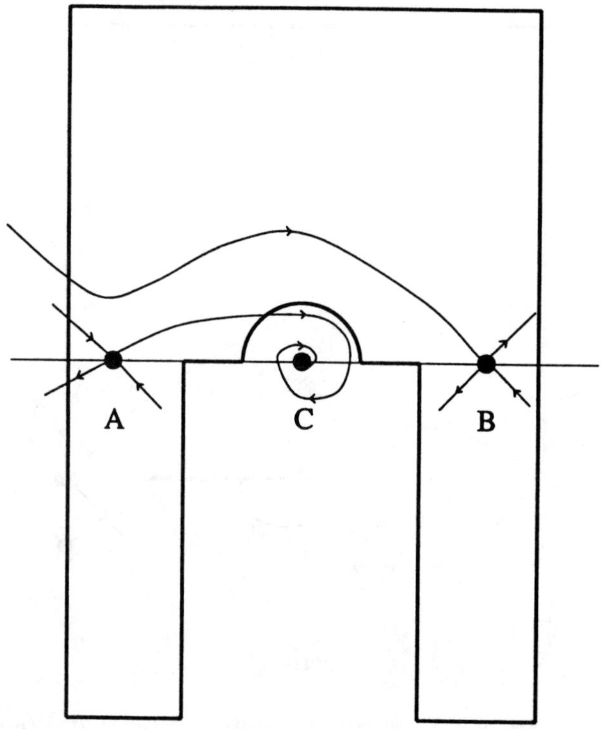

Figure 5

3. Proof of Theorem A. In this section Theorem A is proved. Obviously, by the previous section this involves verifying Assumptions I and II, and then applying Theorem 2.1. This shall be done via a series of lemmas. As will become evident, it is assumption II which requires work.

Lemma 3.1. *Assumption I is satisfied if*

$$f^1(u_1, u_2, 1) = f^2(u_2, u_1, 1).$$

We leave the proof of this lemma to the reader with the comment that the reflective symmetry of the reaction terms makes it relatively easy to check and the remark that a homotopy involving the convex combination of $f^1(\cdot, 1)$ and $f^1(\cdot, 0)$ can be constructed.

The first step towards satisfying Assumption II is the construction of a monotone box. Define

$$\mathcal{R} = [0, \beta_1] \times [0, \beta_2] \times [0, K] \times [0, K]$$

where $\beta_i \geq b_i$ and K is sufficiently large. In line with what was done in section 2, we hope to show that

$$N = (\mathcal{R} \cup T_\epsilon(A) \cup T_\epsilon(B)) \backslash B_\epsilon(C)$$

is an isolating neighborhood independent of the parameter values θ and σ. An important observation is the following: if $(u, v) \in \partial N \cap N^\infty$, then (u, v) is necessarily at a point of internal tangency. Therefore, to show that $\partial N \cap N^\infty = \emptyset$ it only needs to be shown that $(u, v) \in N^\infty$ implies that (u, v) is not a point of internal tangency of N.

Lemma 3.2. $N^\infty \cap ((\partial \mathcal{R}_u \times \mathcal{R}_v) \backslash \{A, B, C\}) = \emptyset$, *i.e.,* N^∞ *does not intersect the sides of* $\mathcal{R}$.

proof: There are four cases to consider.

1. $u = (0, u_2)$. If (u, v) is a point of internal tangency, then $v = (0, v_2)$. In this case equation (1.2) reduces to

$$\begin{aligned} u_2' &= v_2 \\ d_2 v_2' &= -\theta v_2 - u_2 f^2(0, u_2, \sigma). \end{aligned}$$

But the only critical point on the u_2-axis is $(0,0)$, and since $v_2 \geq 0$, the orbit is unbounded in forward time. Therefore, $(u, v) \notin N^\infty$.

2. $u = (u_1, 0)$. An argument similar to that of 1 applies.

3. $(u_1, u_2) = (\beta_1, u_2)$. Given a tangency we know that $u_1' = v_1 = 0$. Now

$$\begin{aligned} u_1'' &= v_1' \\ &= [-\theta v_1 - u_1 f^1(\beta_1, u_2, \sigma)]/d_1 \\ &= [-u_1 f^1(\beta_1, u_2)]/d_1 \\ &> 0. \end{aligned} \tag{3.1}$$

Of course this implies that u_1 is at a local minimum, and hence, we have an external rather than an internal tangency.

4. $(u_1, u_2) = (u_1, \beta_2)$. This follows from an argument similar to that of 3. ∎

Lemma 3.3. *For K sufficiently large, if $(u, v) \in N^\infty$ then $|v_i| < K$.*

proof: This is intuitively clear; for a formal proof see [6]. ∎

Lemma 3.4. *If $(u, v) \in N^\infty \backslash (A \cup B)$ then $v_i \neq 0$.*

proof: To have an internal tangency at $v_i = 0$, it must be that $v_i' = 0$. Since $v_i' = \theta v_i - u_i f^i(\mathbf{u}, \sigma)$, it follows that $u_i f^i(\mathbf{u}, \sigma) = 0$. If $u_i = 0$, then (u, v) is on the side of $\mathcal{R}$ and hence, Lemma 3.2 applies. So assume that $f^i(\mathbf{u}, \sigma) = 0$ and $u_i \neq 0$. Now consider the case $i = 1$ and compute

$$\begin{aligned} v_1'' &= \left[-\theta v_1' - v_1 f^1 - u_1 \frac{\partial f^1}{\partial u_1} v_1 - u_1 \frac{\partial f^1}{\partial u_2} v_2\right] \Big/ d_1 \\ &= \left[-u_1 \frac{\partial f^1}{\partial u_2} v_2\right] \Big/ d_1 . \end{aligned}$$

Thus, if $v_2 > 0$, then $v_1'' < 0$, and hence, v_i is a local maximum, i.e., one is at an external rather than an internal tangency. On the other hand, if $v_2 = 0$, then to be at a point of tangency it must be that $v_2' = 0$. But by the argument above this implies that $f^2(\mathbf{u}, \sigma) = 0 = f^1(\mathbf{u}, \sigma)$, i.e., we are at a critical point. This is a contradiction, therefore there are no internal tangencies for $v_i = 0$. ∎

Lemma 3.5. *C is removable.*

This is a special case of a lemma in [6], alternatively, the reader can trivially modify the proof of Lemma 4.5 of the next section to obtain this result.

Lemma 3.6. *B is augmentable.*

proof: We need to worry about $A \to B$ trajectories which are tangent to $\partial T_\epsilon(B)$ and do not lie in the interior of $\mathcal{R}$. First observe that if $v_i = \pm K$, then Lemma 3.3 rules out it being an element of N^∞. Thus it is only necessary to check orbits which are tangent to the "sides" of $T_\epsilon(B)$. Notice that on such a tragectory the maximum value of u_i, $i = 1, 2$, occurs in the ball $B_\epsilon(B)$. One can now check that this implies that either the maximum of u_1 occurs to the right of the level set $f^1 = 0$ or the maximum of u_2 lies above the level set $f^2 = 0$ or both. In all these cases the inequality corresponding to (3.1) is violated. ∎

By a similar argument we obtain:

Lemma 3.7. *A is augmentable.*

Since N^∞ does not intersect the sides, top, or bottom, of $\mathcal{R}$ (except at the critical points) and since A and B are augmentable and C is removable, N is an isolating neighborhood for all values of $\sigma \in [0,1]$. Thus Assumptions I and II are satisfied, and hence, Theorem 2.1 guarentees the existence of a travelling wave from A to B.

4. Proof of Theorem B. As in the case of the proof of Theorem A, all that needs to be done is to show that Assumptions I and II are satisfied. We begin with Assumption I. As in Lemma 3.1, let $f^1(u_1,u_2,u_3,1) = f^2(u_2,u_1,u_3,1)$. Again it is left to the reader to check that that this assumption on f^1 and f^2 implies that the plane given by $u_1 = u_2$ is invariant under the corresponding reaction system. Furthermore, this defines an attracting invariant 4 dimensional subspace for the travelling wave system. Additional restrictions are obviously necessary to further reduce the dimension of this attracting subspace. It is reasonable to believe that one can construct a reaction system with an attracting invariant line from $\bar{a}$ to b while at the same time preserving the hypothesis **B2**, however, this seems to be rather tricky. To get around this problem, let $w = (u_1+u_2)/2$, and for $u_1 = u_2$ assume that the reaction terms are given by

$$f^3(w,u_3) = 4 - 6w - 4u_3^2$$
$$g^3(w,u_3) = 4 - 6u_3 - 4w^2.$$

This system now contains the attracting line from $\bar{a}$ to b (see [2], section 3.A), but the critical point c no longer exists and a has a 3-dimensional unstable manifold. We will return to this point at the end of this section. For the moment we can summarize this discussion with the following:

Lemma 4.1. *Assumption I can be satisfied.*

As before, to verify Assumption II we begin by constructing a monotone box. Let

$$\mathcal{R} = [0,\beta_1]\times[0,\beta_2]\times[0,\alpha]\times[0,K]\times[0,K]\times[-K,0]$$

where $\beta_i \geq b_i$, and α and K are sufficiently large. In particular, for the proof of Lemma 4.2, it is required that α is large enough that $f^i(\mathbf{u}) = 0$, for $i = 1,2$, implies that $\mathbf{u} \neq (u_1,u_2,\alpha)$. In line with what was done in section 3, the goal is to show that

$$N = (\mathcal{R}\cup T_\epsilon(\bar{A})\cup T_\epsilon(B))\backslash(B_\epsilon(C)\cup B_\epsilon(\bar{C})\cup B_\epsilon(A))$$

is an isolating neighborhood independent of the parameter values θ and σ.

Lemma 4.2. $N^\infty \cap ((\partial\mathcal{R}_u\times\mathcal{R}_v)\backslash\{A,\bar{A},B,C,\bar{C}\}) = \emptyset$, *i.e.* N^∞ *does not intersect the sides of* $\mathcal{R}$.

proof: Clearly,

$$\partial\mathcal{R}_u = ((\partial[0,1]\times[0,1]\times[0,\alpha])\cup([0,1]\times\partial[0,1]\times[0,\alpha])\cup([0,1]\times[0,1]\times\partial[0,\alpha]).$$

For the four cases involving the faces

$$\mathbf{u}\in(\partial[0,1]\times[0,1]\times[0,\alpha])\cup([0,1]\times(\partial[0,1])\times[0,\alpha]),$$

the proof is *exactly* the same as in Lemma 3.2. If

$$\mathbf{u}\in[0,1]\times[0,1]\times\partial[0,\alpha],$$

then a similar argument applies. ∎

The proofs of Lemmas 4.3 and 4.4 are similar to those of the previous section and are left to the reader.

Lemma 4.3. *For K sufficiently large, if $(u,v) \in N^{\infty}$ then $|v_i| < K$.*

Lemma 4.4. *If $(u,v) \in N^{\infty} \backslash (A \cup B)$ then $v_i \neq 0$.*

Lemma 4.5. *$\bar{C}$ is removable.*

proof: The proof follows from examining the linearized equations about the critical point $\bar{C}$. What needs to be shown is that a monotone orbit from A to B cannot pass arbitrarly close to $\bar{C}$. The first step is to notice that complex eigenvalues for the linearized equations induce oscillation. To be more precise, let

$$z' = \Gamma z \tag{4.1}$$

be the system obtained by linearizing the travelling wave system (1.2) about $\bar{C}$. Let $\psi_j = \psi_j^1 + \imath\psi_j^2$, $j = 1, ..., 2n$, denote the eigenvalues of Γ. For $\theta \neq 0$, $\bar{C}$ is hyperbolic therefore the components of the solutions to (4.1) can be written as linear sums of the functions $e^{\psi_j^1 t} \cos \psi_j^2 t$ and $e^{\psi_j^1 t} \sin \psi_j^2 t$.

Now, for ϵ sufficiently small, solutions to (1.2) in the neighborhood $B_\epsilon(\bar{C})$ can be approximated by solutions of (4.1). In particular, choose ϵ small enough such that solutions can be approximated for a time $T > \frac{4\pi}{min\{\psi_j^2 \neq 0\}}$. Now let $(u,v) \in \partial B_\epsilon(\bar{C})$ and without loss of generality assume that it is close enough to the local stable manifold of $\bar{C}$ such that (u,v) stays in $B_\epsilon(\bar{C})$ for time greater than T. Now it is clear that depending on the location of (u,v) it either oscillates about $\bar{C}$, i.e. it is not an element of N^{∞}, or it is well approximated by trajectories on the stable manifold which are asymptotic to the subspaces defined by the real eigenvectors. A similar argument applies to orbits which leave $B_\epsilon(\bar{C})$. Thus the problem has been reduced to understanding the behavior of those trajectories which are asymptotic in forward or backward time to subspaces determined by real eigenvalues. Thus for the remainder of the proof we shall assume that ψ denotes a real eigenvalue of Γ.

Notice that the matrix for the linearized equation at $\bar{C}$ takes the form

$$P = \begin{pmatrix} 0 & I \\ E & -\theta D^{-1} \end{pmatrix} \tag{4.2}$$

where

$$E = \begin{pmatrix} -\frac{\bar{c}_1}{d_1}\frac{\partial f^1}{\partial u_1} & -\frac{\bar{c}_1}{d_1}\frac{\partial f^1}{\partial u_2} & -\frac{\bar{c}_1}{d_1}\frac{\partial f^1}{\partial u_3} \\ -\frac{\bar{c}_2}{d_2}\frac{\partial f^2}{\partial u_1} & -\frac{\bar{c}_2}{d_2}\frac{\partial f^2}{\partial u_2} & -\frac{\bar{c}_2}{d_2}\frac{\partial f^2}{\partial u_3} \\ -\frac{\bar{c}_3}{d_3}\frac{\partial f^3}{\partial u_1} & -\frac{\bar{c}_3}{d_3}\frac{\partial f^3}{\partial u_2} & -\frac{\bar{c}_3}{d_3}\frac{\partial f^3}{\partial u_3} \end{pmatrix}.$$

If we let (μ, ν) denote an eigenvector of P with eigenvalue ψ, then it satisfies the following system of equations

$$\begin{aligned} \nu_i &= \psi\mu_i \\ \sum_{j=1}^{3} -\frac{\bar{c}_i}{d_i}\frac{\partial f^i}{\partial u_j}\mu_j - \frac{\theta\nu_i}{d_i} &= \psi\nu_i . \end{aligned} \tag{4.3}$$

We now claim that the following three sets of equalities and inequalities are not possible:
(1) $\nu_1 > 0$, $\nu_2 > 0$, and $\nu_3 = 0$.
(2) $\nu_1 > 0$, $\nu_2 = 0$, and $\nu_3 < 0$.
(3) $\nu_1 = 0$, $\nu_2 > 0$, and $\nu_3 < 0$.

To see this consider first case (1). Notice that $\nu_3 = 0$ implies that $\mu_3 = 0$. Thus equation (4.3) reduces to

$$\frac{\bar{c}_3}{d_3}\frac{\partial f^3}{\partial u_1}\mu_1 + \frac{\bar{c}_3}{d_3}\frac{\partial f^3}{\partial u_2}\mu_2 = 0,$$

i.e.

$$\frac{\partial f^3}{\partial u_1}\mu_1 + \frac{\partial f^3}{\partial u_2}\mu_2 = 0. \tag{4.4}$$

By assumption, $\frac{\partial f^3}{\partial u_1} < 0$ and $\frac{\partial f^3}{\partial u_2} < 0$ have the same sign. Similarly, by equation (4.3) μ_1 and μ_2 have the same sign. Of course this implies that equation (4.4) cannot hold, i.e. case (1) is impossible. Similar arguments give cases (2) and (3).

Therefore, if one wishes to understand how many eigenvectors of P, "point" in the direction of N, one need only study the eigenvectors of (4.2) for an arbitrary choice of values for $u_i > 0$, $d_i > 0$, and θ. So let $u_i = 1$ and $d_i = 1$ for $i = 1, 2, 3$. Thus the problem has been reduced to understanding the eigenvectors of

$$P = \begin{pmatrix} 0 & I \\ E' & -\theta I \end{pmatrix} \tag{4.5}$$

where

$$E' = \begin{pmatrix} -\frac{\partial f^1}{\partial u_1} & -\frac{\partial f^1}{\partial u_2} & -\frac{\partial f^1}{\partial u_3} \\ -\frac{\partial f^2}{\partial u_1} & -\frac{\partial f^2}{\partial u_2} & -\frac{\partial f^2}{\partial u_3} \\ -\frac{\partial f^3}{\partial u_1} & -\frac{\partial f^3}{\partial u_2} & -\frac{\partial f^3}{\partial u_3} \end{pmatrix}.$$

Finding the eigenvalues of P involves solving

$$\begin{aligned} 0 &= \det\begin{pmatrix} -\psi I & I \\ E' & -(\theta+\psi)I \end{pmatrix} \\ &= \det\begin{pmatrix} -\psi I & 0 \\ E' & -(\theta+\psi)I + \frac{1}{\psi}E' \end{pmatrix} \\ &= \det(-\psi I)\det\left(\frac{-(\theta+\psi)\psi I + E'}{\psi}\right) \\ &= \det(-(\theta+\psi)\psi I + E'). \end{aligned}$$

Let $\zeta = (\theta + \psi)\psi$, then

$$0 = \det(E' - \zeta I).$$

Therefore ζ is an eigenvalue of E'. Let these eigenvalues be denoted by $\zeta_1, \zeta_2, \zeta_3$. Since $\bar{c}$ has a one-dimensional unstable manifold for the reaction system, there exists a unique negative eigenvalue. Assume it is ζ_1.

Of course, the eigenvalues which are of interest are

$$\psi_i^{\pm} = \frac{-\theta \pm \sqrt{\theta^2 + 4\zeta_i}}{2}.$$

Notice that $\psi_i^+ > 0$ and $\psi_i^- < 0$ for $i = 2, 3$, while $\psi_1^{\pm} < 0$, for $\theta > 0$, sufficiently large.

Now an orbit entering $B_{2\epsilon}(\bar{C})$ can be approximated by the stable manifold of $\bar{C}$ and an orbit leaving $B_{2\epsilon}(\bar{C})$ can be approximated by the unstable manifold of $\bar{C}$. The directions of these manifolds are determined by the eigenvectors. In particular, it will be shown that there is no eigenvector (μ, ν) such that $\mu_i \geq 0$ and $\nu_i \geq 0$ for $i = 1, 2$ and $\mu_3 \geq 0$, $\nu_3 \leq 0$. This will imply that any orbit entering $B_{2\epsilon}(\bar{C})$ will leave N.

Assume that there exists an eigenvector (μ, ν) such that $\mu_i \geq 0$ and $\nu_i \geq 0$ for $i = 1, 2$ and $\nu_3 \leq 0$. Since $\mu_i \geq 0$ and $\nu_i \geq 0$, the corresponding eigenvalue must be positive. So without loss of generality, we assume that the eigenvalue is ψ_2^+. Since (μ, ν) is an eigenvector, it must satisfy the equations

$$\begin{aligned} \nu_i &= \psi_2^+ \mu_i \\ -\sum_{j=1}^{3} \frac{\partial f^i}{\partial u_j} \mu_i - \theta \nu_i &= \psi_2^+ \nu_i . \end{aligned} \tag{4.6}$$

Substituting the first equation into the second, we obtain

$$-\sum_{j=1}^{3} \frac{\partial f^i}{\partial u_j} \mu_i = \zeta_2 \mu_i .$$

Therefore, $\zeta_2 > 0$ is an eigenvalue of the reaction system linearized about the critical point $\bar{c}$. However, inspection of Figure 2, or a rather tedious calculation, shows that the unstable eigenvector of the critical point $\bar{c}$ for this system, does not point in the direction of the positive orthant; a contradiction.

The argument for $\theta < 0$ is similar. Finally, by following the stable and unstable center manifolds one can draw the same conclusion for $\theta = 0$. ∎

As in the previous section, the arguments needed to show that B and $\bar{A}$ are augmentable follow from Lemma 4.2 where the tangency conditions for the sides of $\mathcal{R}$ are checked. Thus the following two lemmas are also left for the reader.

Lemma 4.6. *B is augmentable.*

Lemma 4.7. *$\bar{A}$ is augmentable.*

The removal of the critical point A is the first step which differs significantly from that of the previous section.

Lemma 4.8. *Given assumption* **B2**, *A is removable.*

proof: The proof is by contradiction. So assume that A is not removable, then given any homotopy from $(2.1_{\theta,0})$ to $(2.1_{\theta,1})$ there exists a sequence $\{\sigma_n\} \subset [0, 1]$ such that $\lim_{n\to\infty} \sigma_n = \sigma^*$ and a connecting orbit $\bar{A}_{\sigma_n} \to B_{\sigma_n}$ passes within a distance $\frac{1}{n}$ of A. Of course, this implies that at σ^* the connecting orbit $\bar{A}_{\sigma^*} \to B_{\sigma^*}$ is really made up of two orbits $\bar{A}_{\sigma^*} \to A_{\sigma^*}$ and $A_{\sigma^*} \to B_{\sigma^*}$. This leads to a contradiction as follows. As will be shown, the $\bar{A}_{\sigma^*} \to A_{\sigma^*}$ connection can only occur for $\theta > 0$ and the $A_{\sigma^*} \to B_{\sigma^*}$ can only occur when $\theta < 0$, i.e. the two waves cannot occur for the same wave speed and hence cannot arise as the limit of the sequence of $\bar{A}_{\sigma_n} \to B_{\sigma_n}$ orbits.

To see why these conditions on the wave speeds hold, notice that the monotonicity assumption forces the $\bar{A}_{\sigma^*} \to A_{\sigma^*}$ connection to lie on th u_3-axis, and the $A_{\sigma^*} \to B_{\sigma^*}$ to lie in the u_1, u_2-plane. Also, notice that the u_3, v_3-plane is invariant under $(2.1_{\theta,\sigma})$, so to study the $\bar{A}_{\sigma^*} \to A_{\sigma^*}$ orbit one can restrict one's attention to the equations

$$\begin{aligned} u_3' &= v_3 \\ d_3 v_3' &= -\theta v_3 - u_3 f^3(0, 0, u_3). \end{aligned} \tag{4.7}$$

Define $F^3 : \mathbf{R} \to \mathbf{R}$ by $\frac{dF^3}{du_3}(u_3) = u_3 f^3(0, 0, u_3)$. Then, the corresponding reaction system is

$$u_3' = u_3 f^3(0, 0, u_3) = \frac{dF^3}{du_3}(u_3),$$

i.e. it is a gradient system. Since, in the reaction system (2.2_σ) one has a connecting orbit from a to $\bar{a}$, it must be that $F^3(a) < F^3(\bar{a})$. Now let

$$H^3(u_3, v_3) = \frac{v_3^2}{2} + F^3(u_3).$$

It is easy to check that $H^{3\prime} = -\theta v_3^2$, i.e., for $\theta \neq 0$, H^3 is a Lyapunov function for the system (4.7). Obviously, $H^3(\bar{A}) > H^3(A)$, thus an $\bar{A}_{\sigma^*} \to A_{\sigma^*}$ exists if $H^{3\prime} = -\theta v_3^2 < 0$, i.e. if $\theta > 0$.

To insure that $A_{\sigma^*} \to B_{\sigma^*}$ can only occur when $\theta < 0$ is the reason for the hypothesis (B3). Notice that for any trajectory in N^∞

$$\begin{aligned} 0 &= v_i^2|_{-\infty}^{\infty} \\ &= \int_{-\infty}^{\infty} v_i v_i' dt \\ &= \int_{-\infty}^{\infty} -\theta v_i^2 - u_i v_i f^i(u_1, u_2) dt. \end{aligned} \tag{4.8}$$

Since N^∞ only contains monotone orbits, $A_{\sigma^*} \to B_{\sigma^*}$ is monotone and hence, u_1 and u_2 can be parameterized by position, i.e. $s \in (0,1)$. In doing so, (4.8) becomes

$$\int_{-\infty}^{\infty} \theta v_1^2 = -\int_0^1 s f^1(s, u_2(s)) ds \tag{4.9}$$

and

$$\int_{-\infty}^{\infty} \theta v_2^2 = -\int_0^1 s f^2(u_1(s), s) ds. \tag{4.10}$$

Let $Q = \{s | u_2(s) > s\}$ and $Q^c = (0,1) \backslash Q$. Then

$$\int_0^1 s f^1(s, u_2(s)) ds = \int_Q s f^1(s, u_2(s)) ds + \int_{Q^c} s f^1(s, u_2(s)) ds.$$

Since $\frac{\partial f^1}{\partial u_2} > 0$,

$$\int_0^1 s f^1(s, u_2(s)) ds > \int_Q s f^1(s, s) ds + \int_{Q^c} s f^1(s, u_2(s)) ds. \tag{4.11}$$

Similarly,

$$\int_0^1 s f^2(u_1(s), s) ds > \int_Q s f^2(s, u_2(s)) ds + \int_{Q^c} s f^2(s, s) ds. \tag{4.12}$$

Now summing equations (4.9) and (4.10) and using the inequalities (4.11) and (4.12) one has

$$\begin{aligned} -\theta \int_{-\infty}^{\infty} v_1^2 + v_2^2 > \int_Q s f^1(s,s) ds + \int_{Q^c} s f^2(s,s) ds + \int_{Q^c} s f^1(s, u_2(s)) ds \\ + \int_Q s f^2(s, u_2(s)) ds > \int_0^1 s V(s) ds + K_1 + K_2. \end{aligned}$$

Therefore,

$$\theta \int_{-\infty}^{\infty} v_1^2 + v_2^2 < -\left(\int_0^1 s V(s) ds + K_1 + K_2\right) < 0,$$

i.e. $\theta < 0$. ∎

Notice that the proof of Lemma 4.8, presupposes the existence of the critical point c for the system.

Lemma 4.9. *If a is a repeller for the reaction system, then A is removable.*

proof: See [2, 4.D] or simply notice that: if $\theta > 0$, then A is an attractor; if $\theta < 0$, then A is a repeller; and if $\theta = 0$ then A has purely imaginary eigenvalues, in which case orbits spiral "past". In any of these cases it is impossible to have a monotone $\bar{A} \to B$ orbit pass arbitrarily close to A. ∎

Lemma 4.10. *C is removable.*

proof: Again the proof is by contradiction. C not removable implies the existence of a σ^* for which there exist $\bar{A}_{\sigma^*} \to C_{\sigma^*}$ and $C_{\sigma^*} \to B_{\sigma^*}$ connecting orbits. As before, the monotonicity implies that $C_{\sigma^*} \to B_{\sigma^*}$ lies in the u_1, u_2-plane.

Let $W^u(C)$ denote the unstable manifold of C. Then a simple eigenvalue calulation shows that for $\theta > 0$, $dimW^u(C) = 1$ and that $W^u(C)$ is not tangent to the u_1, u_2-plane. Thus, for $\theta > 0$, $C_{\sigma^*} \to B_{\sigma^*}$ does not exist. If $\theta < 0$, then $dimW^u(C) = 5$, and the stable manifold lies on the u_1, u_2-plane. Therefore, a portion of $\bar{A}_{\sigma^*} \to C_{\sigma^*}$ lies on the u_1, u_2-plane which is invariant. Thus, $\bar{A}_{\sigma^*} \to C_{\sigma^*}$ consists of $A_{\sigma^*} \to \bar{A}_{\sigma^*}$ and $A_{\sigma^*} \to C_{\sigma^*}$. But, by the previous proof, $A_{\sigma^*} \to \bar{A}_{\sigma^*}$ only occurs for $\theta > 0$, a contradiction. Finally, notice that if $|\theta|$ is small, then the eigenvalues of the linearized equation at C corresponding to the eigenvectors which are not tangent to the u_1, u_2, v_1, v_2-space are imaginary. Thus the orbits spiral into or out of C off the u_1, u_2-plane. In particular, these are not monotone orbits, and hence, not in N^∞. ∎

5. Concluding Remarks. It is hoped that the reader has not taken offense at the rather cavalier treatment of the proofs of Lemmas 4.2, 4.3, 4.4, 4.6, and 4.7. However, if the reader has taken the effort to fill in the details of the proofs in Section 4, he or she will quickly appreciate the striking similiarities between the calculations for proofs of Theorem A and Theorem B. Furthermore, it is also clear that with a little additional effort these results can be extended to higher dimensions. A proof of "Theorem A", but for n mutualist species, appears in [6] and the same authors are currently developing a general treatment of systems involving competitive and mutualist species. There is, however, a simple 2 competing species model for which it is not clear how to satisfy Assumption I. Consider Figure 6. It is reasonable to expect that there exists an A to $\bar{B}$ travelling wave. However, the simple approach of symmetrizing the non-linearities about the line $u_1 = u_2$ will not work. In this case the diagonal (though invariant) is not *attracting*, and hence, Theorem 2.1 does not apply. It appears, however, that this can be overcome by generalizing Assumption I, to attracting invariant linear subspaces and then applying more sophisticated tools such as connection matrices and transition matrices.

Predator-prey systems (involving non-linearities which satisfy $\frac{\partial f^i}{\partial u_j} > 0$ and $\frac{\partial f^j}{\partial u_i} < 0$) which have not been discussed in this paper are more difficult. In part, this is due to the fact that the removability of the critical points can no longer be done by a local argument and global constraints become necessary. In particular, integral conditions in some sense similar to that of (B3) appear to be necessary to apply the techniques discussed in this paper (see [3] or [7] for more details). Of course it is not clear when such conditions are actually necessary for the existence of the travelling waves themselves. The predator-prey systems also give rise to simple examples where the choice of isolating neighborhood, i.e. Assumption II, is not clear. Consider Figure 7 and the existence of an A to B travelling wave. If one tries to construct the obvious rectangle for $\mathcal{R}_u$ it is not clear that N^∞ does not intersect the edges of $\mathcal{R}$, i.e. it is not known how to prove the lemmas corresponding to Lemmas 3.2 and 4.2 of this paper. Solutions to problems of this nature will probably require new non-trivial ideas.

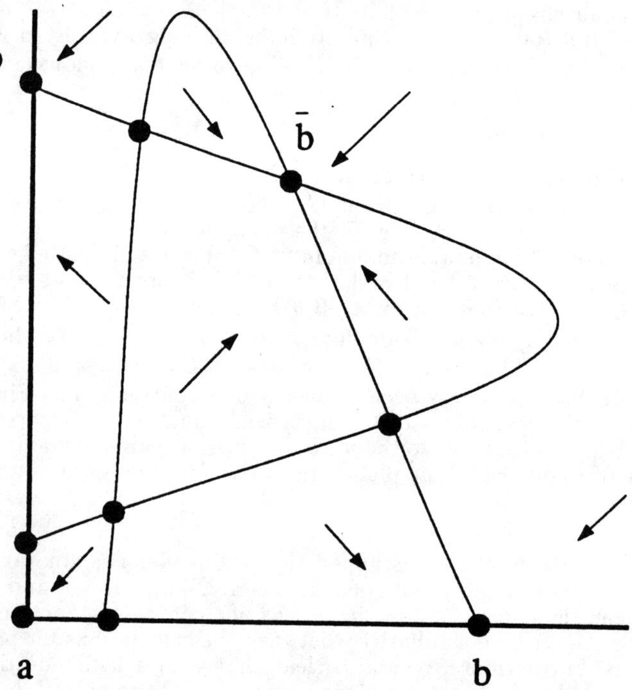

Figure 6

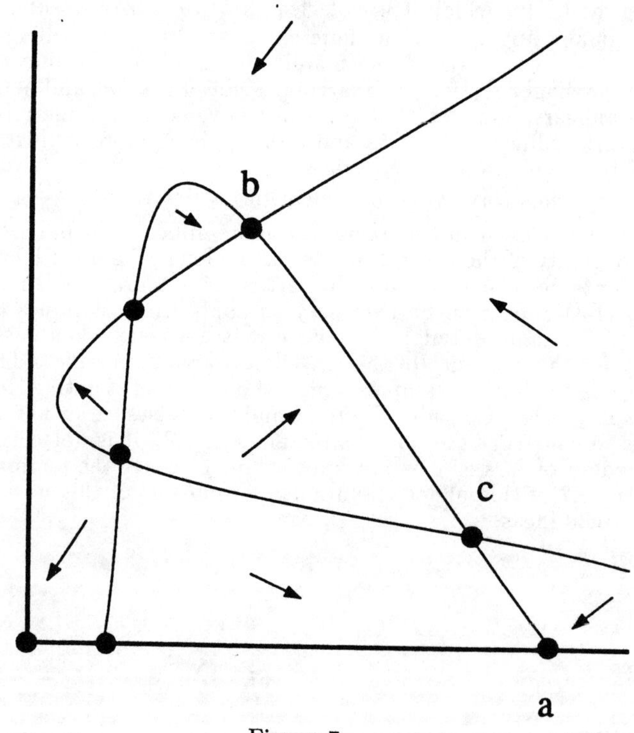

Figure 7

REFERENCES

[1] C. Conley, *Isolated invariant sets and the Morse Index*, CBMS **38**, AMS, Providence, 1978.

[2] C. Conley and R. Gardner, *An application of the generalized Morse index to travelling wave solutions of a competitive reaction-diffusion model*, Indiana Univ. Math. J., 33 (1984), 319-343.

[3] R. Gardner, *Existence of travelling wave solutions of predator-prey systems via the Conley index*, SIAM J. Appl. Math., 44 (1984), 56-79.

[4] V. Hutson and K. Mischaikow, *Travelling Waves for Competing Species* (in progress).

[5] C. McCord and K. Mischaikow, *Connected simple systems, transition matrices, and heteroclinic orbits*, Trans. AMS (to appear).

[6] K. Mischaikow and V. Hutson, *Travelling waves for mutualist species*, CDSNS90-38 preprint series.

[7] K. Mischaikow and J. Reineck, *Travelling waves for predator-prey systems* (in progress).

CHAPTER 13

Transversality for Undercompressive Shocks in Riemann Problems

Stephen Schecter*
Michael Shearer†

Abstract. We describe the natural role of undercompressive shocks in solving Riemann problems for systems of two hyperbolic conservation laws in one space dimension. Undercompressive shocks correspond to saddle-to-saddle trajectories of a vector field. Given one such trajectory, nearby trajectories are described by the zeroes of a separation function. It is natural to assume the separation function has nonzero derivative with respect to shock speed at the given trajectory. Under this transversality assumption and some other assumptions, we construct unique three-wave solutions to Riemann problems using an undercompressive shock and weak slow and fast waves on either side of the shock. The assumptions are verified for an example.

1 Introduction

In this paper we describe the natural role of undercompressive shocks in solving Riemann problems for systems of hyperbolic conservation laws in one space dimension. The Riemann problem for a system of conservation laws

$$U_t + F(U)_x = 0, \tag{1.1}$$

where $U = U(x,t) \in \mathbf{R}^n$ and $F : \mathbf{R}^n \to \mathbf{R}^n$, is the initial value problem for (1.1) with initial data consisting of a single jump between constant values of U:

$$U(x,t) = \begin{cases} U_L & \text{if } x < 0 \\ U_R & \text{if } x > 0. \end{cases} \tag{1.2}$$

*Department of Mathematics, North Carolina State University, Raleigh, NC 27695.
Research supported by National Science Foundation grant DMS 9002803.

†Department of Mathematics, North Carolina State University, Raleigh, NC 27695.
Research supported by National Science Foundation grant DMS 8907736 and by Army Research Office grant DAAL03-88-K-0080.

Equation (1.1) will be assumed to be strictly hyperbolic near U_L and U_R, i.e., the Jacobian matrix $DF(U)$ has real and distinct eigenvalues for each U near U_L and for each U near U_R.

A centered shock wave

$$U(x,t) = \begin{cases} U_- & \text{if } x < st \\ U_+ & \text{if } x > st \end{cases} \tag{1.3}$$

with speed s is **admissible** if there is a trajectory $U(\xi)$ for the vector field

$$BU' = F(U) - F(U_-) - s(U - U_-), \tag{1.4}$$

satisfying

$$U(\pm\infty) = U_\pm, \quad U'(\pm\infty) = 0, \tag{1.5}$$

where $'$ denotes ordinary differentiation with respect to ξ, and B is a constant invertible matrix. (Of course admissibility depends upon the choice of B.) Such a trajectory corresponds to a travelling wave $U = U((x - st)/\mu)$ that approximates the shock wave (1.3) as $\mu \to 0+$ and is a solution of the related system

$$U_t + F(U)_x = \mu B U_{xx}. \tag{1.6}$$

In (1.4), U_- and s are parameters. There is always an equilibrium of (1.4) at $U = U_-$; if (1.3) is an admissible shock, then there is also an equilibrium of (1.4) at $U = U_+$.

An admissible shock is **compressive** if U_- and U_+ are hyperbolic equilibria with stable manifolds of dimension $k-1$ and k respectively for some $k, 1 \le k \le n$. For example, for a system of two equations, $n = 2$. Then for $k = 1$, U_- is an unstable node, U_+ is a saddle point, and (1.3) is a **slow shock**; for $k = 2$, U_- is a saddle point, U_+ is a stable node, and (1.3) is a **fast shock**.

An admissible shock is **undercompressive** if U_- and U_+ are hyperbolic equilibria of the same type. For $n = 2$, U_- and U_+ are necessarily saddle points. Undercompressive shocks occur naturally in nonstrictly hyperbolic conservation laws [7,8], and in systems of quasilinear partial differential equations of mixed hyperbolic and nonhyperbolic type [1]. It has been found that when the equation admits undercompressive shocks, then there are data U_L, U_R in (1.2) for which the solution of the Riemann problem has $n + 1$ waves rather than the n waves that generally suffice for strictly hyperbolic systems.

For the case $n = 2$, we shall show how this structure with an extra wave arises in a way that is generic, in the sense that the solutions of Riemann problems with the three wave structure are stable to perturbation of the initial data or the equations. For $n > 2$, our constructions readily generalize to shocks for which the stable manifold of U_+ has dimension less than or equal to that of the stable manifold of U_-. We plan to deal with this generalization in a subsequent paper.

Let $n = 2$. Suppose there is a saddle-to-saddle heteroclinic orbit $U_o \to U_o^*$ for (1.4) with $U_- = U_o$ and $s = s_o$. Then generically, for $U_- = U_o$ and $s \neq s_o$ near s_o, there will be no saddle-to-saddle heteroclinic orbit . However, since saddle-to-saddle heteroclinic orbits have codimension one, there is a function $s = s(U_-)$, defined on a neighborhood of U_o, such that for each U_- in that neighborhood, (1.4) has a saddle-to-saddle heteroclinic orbit from U_- to some U_-^* near U_o^* at $s = s(U_-)$. The main result, Theorem 2.8, states that, under a condition on B, and assuming certain transversality conditions, for each U_L near U_o and each U_R near U_o^*, the Riemann problem has a solution consisting of three waves. The solution, which is unique in a local sense, consists of a slow compressive shock wave or a slow rarefaction wave from U_L to some U_- near U_o, an undercompressive shock from

U_- to U_-^* with speed $s = s(U_-)$, and a fast compressive shock or fast rarefaction wave from U_-^* to U_R.

In Section 3, we check the conditions for a model equation of mixed type.

2 Main Result

We consider (1.1) with $n = 2$, F of class $C^p, p \geq 3$, and a fixed diffusion matrix B in (1.4). We rewrite (1.4) as

$$U' = G(U, U_-, s) = B^{-1}[F(U) - F(U_-) - s(U - U_-)]. \tag{2.1}$$

We assume:

(H1) For $U_- = U_o$ and $s = s_o$, (2.1) has a heteroclinic orbit $U_o \longrightarrow U_o^*$ joining distinct hyperbolic saddle point equilibria.

Since $D_U G(U_o^*, U_o, s_o)$ is invertible, the implicit function theorem implies the following.

Lemma 2.1 *Suppose (H1) is satisfied. Then there exist neighborhoods A_1 of U_o and A_2 of U_o^*, $\mu > 0$, and a C^p function $\tilde{U} : A_1 \times (s_o - \mu, s_o + \mu) \longrightarrow A_2$, with $\tilde{U}(U_o, s_o) = U_o^*$, such that for $(U_-, s) \in A_1 \times (s_o - \mu, s_o + \mu), U = \tilde{U}(U_-, s)$ is the unique equilibrium of (2.1) in A_2. Moreover, for $(U_-, s) \in A_1 \times (s_o - \mu, s_o + \mu)$, both $U = U_-$ and $U = \tilde{U}(U_-, s)$ are hyperbolic saddles of (2.1).*

The **Hugoniot locus** $H(U_-)$ is the set of points U that are equilibria of (2.1) for some value of s. Lemma 2.1 implies that for each U_- in $A_1, H(U_-)$ includes points in A_2.

The sets A_1 and A_2 and the number μ may need to be shrunk several times in the remainder of this section; we shall not mention this explicitly.

Hypothesis (H1) guarantees the existence of a heteroclinic solution $U = \gamma(t)$ of (2.1) with $(U_-, s) = (U_o, s_o)$. Let Σ be a line segment through $\gamma(0)$ transverse to $\dot{\gamma}(0)$. By Lemma 2.1, for $(U_-, s) \in A_1 \times (s_o - \mu, s_o + \mu)$, (2.1) has hyperbolic saddles U_- and $\tilde{U}(U_-, s)$. A branch of the unstable manifold of U_- meets Σ at a point $x^1(U_-, s)$, and a branch of the stable manifold of $\tilde{U}(U_-, s)$ meets Σ at a point $x^2(U_-, s)$. We define the *separation function* $d(U_-, s)$ to be a convenient multiple of the signed distance between $x^1(U_-, s)$ and $x^2(U_-, s)$:

$$d(U_-, s) = \dot{\gamma}(0) \wedge (x^1(U_-, s) - x^2(U_-, s)),$$

where $v \wedge w = det(v, w)$. (Both the definition and notation are standard [6].) The stable manifold theorem guarantees that the function d is of class C^p. For use in examples, we note that the partial derivatives of d at (U_o, s_o) are computable. Let q denote s or one of the components of U_-. Then

$$\frac{\partial d}{\partial q}(U_o, s_o) =$$

$$\int_{-\infty}^{\infty} exp\left[-\int_0^t divG(\gamma(s), U_o, s_o)ds\right] G(\gamma(t), U_o, s_o) \wedge \frac{\partial G}{\partial q}(\gamma(t), U_o, s_o)dt. \tag{2.2}$$

Here $divG = trace D_U G$ as usual.

We now assume that the saddle-to-saddle heteroclinic orbit $U_o \longrightarrow U_o^*$ is generic in the sense that it is broken by varying s:

(H2) $\frac{\partial d}{\partial s}(U_o, s_o) \neq 0$.

Then the implicit function theorem implies the following.

Lemma 2.2 *Suppose the function d satisfies (H2). Then for each $U_- \in A_1$, the equation*

$$d(U_-, s) = 0$$

has a unique solution $s = s^*(U_-)$ in $(s_o - \mu, s_o + \mu)$. Moreover, s^* is C^p and $s^*(U_o) = s_o$.

Next we make two assumptions that allow us to set up slow and fast wave curves in A_1 and A_2 respectively:

(H3) *The partial differential equation (1.1) is strictly hyperbolic and genuinely nonlinear in $A_1 \cup A_2$.*

(H4) *The diffusion matrix B is strictly stable with respect to $DF(U_o)$ and $DF(U_o^*)$, in the sense of [5].*

Assumption (H4) means that the linearization of (1.6) at U_o and U_o^* is well-posed in L^2, uniformly as $\mu \to 0+$. The following two lemmas are proved in [5]. Let A be a nonsingular 2×2 real matrix with distinct real eigenvalues $\lambda_1 < \lambda_2$. Let ℓ_k, r_k, $k = 1, 2$, be left and right eigenvectors of A corresponding to eigenvalue λ_k, chosen so that

$$\ell_j r_k = \delta_{jk}.$$

Lemma 2.3 *A real 2×2 matrix B is strictly stable with respect to A if and only if*

$$(i) \quad \ell_k B r_k > 0, \qquad k = 1, 2 \quad (no\ summation) \tag{2.3}$$

and

$$(ii) \quad \det B > 0. \tag{2.4}$$

Lemma 2.4 *If (H4) holds, then hyperbolic equilibria of (2.1) near U_o and U_o^* remain hyperbolic equilibria of the same type when B is replaced by the identity matrix.*

A consequence of Lemma 2.4 is that weak Lax shocks [3] are admissible and compressive. In Appendix A we give a simple proof that Lemma 2.3 implies Lemma 2.4. Note also that Lemma 2.3 can be used to check (H4) in examples.

For $k = 1, 2$, let $\lambda_k(U)$, $r_k(U)$, $\ell_k(U)$ denote respectively the kth eigenvalue, right eigenvector, left eigenvector of $DF(U)$, with $\lambda_1(U) < \lambda_2(U)$. By (H3), we can normalize the eigenvectors by

$$D\lambda_k(U) r_k(U) = 1, \qquad \ell_j(U) r_k(U) = \delta_{jk}.$$

Then $\ell_k(U)$, $r_k(U)$ are C^{p-1} functions of U. With the aid of Lemma 2.4, the following result can be proved in the usual way (cf. [3,4]).

Lemma 2.5 *Suppose (H3) and (H4) are satisfied. For each $k = 1, 2$ there are numbers $\delta_k > 0$, $\rho_k > 0$, and C^2 functions $U_k(U_L, \epsilon)$ defined on $\{(U_L, \epsilon) : U_L \in A_k \text{ and } |\epsilon| < \delta_k\}$, and $s_k(U_L, \epsilon)$ defined on $\{(U_L, \epsilon) : U_L \in A_k \text{ and } -\delta_k < \epsilon \leq 0\}$, such that:*

$$(i) \qquad U_k(U_L, 0) = U_L, \quad \frac{\partial}{\partial \epsilon} U_k(U_L, 0) = r_k(U_L). \tag{2.5}$$

$$s_k(U_L,0-) = \lambda_k(U_L), \quad \frac{\partial}{\partial\epsilon} s_k(U_L,0-) = \frac{1}{2}. \tag{2.6}$$

(ii) *For* $-\delta_k < \epsilon \leq 0$, *there is an admissible centered shock from* U_L *to* $U_k(U_L,\epsilon)$ *with speed* $s = s_k(U_L,\epsilon)$.

(iii) *For* $0 \leq \epsilon < \delta_k$, *there is a centered rarefaction wave joining* U_L *to* $U_k(U_L,\epsilon)$ *with fastest wave speed* $\lambda_k(U_k(U_L,\epsilon))$ *and intermediate values* $U(x/t)$ *given by*

$$U(\xi) = U_k(U_L, \xi - \lambda_k(U_L)), \qquad \lambda_k(U_L) < \xi < \lambda_k(U_k(U_L,\epsilon)). \tag{2.7}$$

(iv) *If* $U_L \in A_k, U \in A_k$, *and there is an admissible centered shock from* U_L *to* U *with speed* s *satisfying* $|s - \lambda_k(U_L)| < \rho_k$, *or a centered rarefaction wave from* U_L *to* U *with fastest wave speed* $\lambda_k(U)$, *then* $s = s_k(U_L,\epsilon)$ *and* $U = U_k(U_L,\epsilon)$ *for some* $\epsilon, -\delta_k < \epsilon \leq 0$, *or* $U = U_k(U_L,\epsilon)$ *for some* ϵ, $0 \leq \epsilon < \delta_k$.

We define

$$W_k(U_L) = \{U = U_k(U_L,\epsilon) : |\epsilon| < \delta_k\}.$$

The curve $W_1(U_L)$ is referred to as the **slow wave curve** through U_L, while $W_2(U_L)$ is the **fast wave curve** through U_L.

Now fix U_L in A_1 and consider the equation

$$d(U_1(U_L,\epsilon), s) = 0, \quad |\epsilon| < \delta_k, \tag{2.8}$$

which determines undercompressive shocks with left state on $W_1(U_L)$. By Lemma 2.2, equation (2.8) has the unique solution $s = s^*(U_1(U_L,\epsilon))$. There corresponds a function $U_1^* = U_1^*(U_L,\epsilon)$, taking values near U_o^*, such that (2.1) with $(U_-, s) = (U_1(U_L,\epsilon), s^*(U_1(U_L,\epsilon)))$ has a saddle-to-saddle heteroclinic orbit from $U_1(U_L,\epsilon)$ to $U_1^*(U_L,\epsilon)$. In fact,

$$U_1^*(U_L,\epsilon) = \tilde{U}(U_1(U_L,\epsilon), s^*(U_1(U_L,\epsilon)),$$

where $\tilde{U}$ is given in Lemma 2.1, and s^* in Lemma 2.2. Thus for each U_L in A_1 we may define an *undercompressive shock curve* $\Sigma(U_L)$, which consists of right states of undercompressive shocks with left state on $W_1(U_L)$:

$$\Sigma(U_L) = \{U_1^*(U_L, \epsilon) : \epsilon \text{ small}\}.$$

In order to use our wave curves to construct three-wave solutions of (1.1), we first observe that (H1), (H4), and Lemma A.3 imply that the undercompressive shocks move more slowly than the fast waves, and faster than the slow waves:

$$\lambda_1(U_o) < s_o < \lambda_2(U_o^*). \tag{2.9}$$

Now for each U_L near U_o and $(\epsilon_1, \epsilon_2) \in \mathbf{R}^2$ small, we can construct a solution of (1.1) consisting of three waves: a slow wave from U_L to $U_1(U_L,\epsilon_1) \in W_1(U_L)$ with speed $s_1(U_L, \epsilon_1)$ near $\lambda_1(U_0)$; an undercompressive shock wave from $U_1(U_L, \epsilon_1)$ to $U_1^*(U_L, \epsilon_1) \in \Sigma(U_L)$ with speed $s^*(U_L, \epsilon_1)$ near s_o; and a fast wave from $U_1^*(U_L, \epsilon_1)$ to $U_2(U_1^*(U_L, \epsilon_1), \epsilon_2) \in W_2(U_1^*(U_L, \epsilon_1))$ with speed $s_2(U_1^*(U_L, \epsilon_1), \epsilon_2)$ near $\lambda_2(U_o^*)$. These three waves give a solution of (1.1) by (2.9). Now let U_R be given. Then our three-wave solution of (1.1) solves the Riemann problem with data (U_L, U_R) if and only if

$$\mathcal{F}(U_L, U_R, \epsilon_1, \epsilon_2) = U_R - U_2(U_1^*(U_L, \epsilon_1), \epsilon_2) = 0. \tag{2.10}$$

From Lemmas 2.1, 2.2, and 2.5, $\mathcal{F}$ is C^2.

To show that this construction gives a unique solution to the Riemann problem for each U_L near U_o and U_R near U_o^*, we need a final hypothesis:

(H5) $\frac{\partial}{\partial \epsilon} U_1^*(U_o, 0)$ *and* $r_2(U_o^*)$ *are linearly independent.*

Geometrically, (H5) says that the undercompressive shock curve $\Sigma(U_o)$ and the fast wave curve $W_2(U_o^*)$ meet transversally at U_o^*. Notice that (H5) implies in particular that $\frac{\partial}{\partial \epsilon} U_1^*(U_o, 0) \neq 0$, i.e., that Σ is a nonsingular curve.

For use in examples, we record a formula for $\frac{\partial}{\partial \epsilon} U_1^*(U_o, 0)$

Lemma 2.6 *Suppose (H1),(H2),(H3) are satisfied. Then*

$$\frac{\partial}{\partial \epsilon} U_1^*(U_o, 0) = [DF(U_o^*) - s_o]^{-1} \left\{ (\lambda_1(U_o) - s_o) r_1(U_o) - (U_o^* - U_o) \left[\frac{\partial d}{\partial s}(U_o, s_o)\right]^{-1} D_{U_-} d(U_o, s_o) r_1(U_o) \right\}.$$

Proof This is a straightforward application of the chain rule. We first note that (H1) implies $D_U G(U_o^*, U_o, s_o)$ is invertible, so

$$DF(U_o^*) - s_o I \quad \text{is invertible.} \tag{2.11}$$

Now,

$$\frac{\partial}{\partial \epsilon} U_1^*(U_L, \epsilon) = \frac{\partial}{\partial \epsilon} \tilde{U}(U_1(U_L, \epsilon), s^*(U_1(U_L, \epsilon)) = D_{U_-} \tilde{U} \frac{\partial}{\partial \epsilon} U_1 + \frac{\partial \tilde{U}}{\partial s} D_{U_-} s^* \frac{\partial}{\partial \epsilon} U_1.$$

Setting $U_L = U_o$ and $\epsilon = 0$, and using (2.1), we obtain

$$\frac{\partial}{\partial \epsilon} U_1^*(U_o, 0) = D_{U_-} \tilde{U}(U_o, s_o) r_1(U_o) + \frac{\partial \tilde{U}}{\partial s}(U_o, s_o) D_{U_-} s^*(U_o) r_1(U_o) \tag{2.12}$$

Now $\tilde{U}(U_1, s)$ satisfies $\tilde{U}(U_o, s_o) = U_o^*$ and from

$$F(\tilde{U}) - F(U_-) - s(\tilde{U} - U_-) = 0$$

we obtain

$$DF(\tilde{U}) D_{U_-} \tilde{U} - DF(U_-) - s(D_{U_-} \tilde{U} - I) = 0 \tag{2.13}$$

and

$$DF(\tilde{U}) \frac{\partial \tilde{U}}{\partial s} - (\tilde{U} - U_-) - s \frac{\partial \tilde{U}}{\partial s} = 0. \tag{2.14}$$

Since $\tilde{U}(U_o, s_o) = U_o^*$, evaluating (2.13) and (2.14) at $(U_-, s) = (U_o, s_o)$ and using (2.11) yields

$$D_{U_-} \tilde{U}(U_o, s_o) = (DF(U_o^*) - s_o I)^{-1} (DF(U_o) - s_o I). \tag{2.15}$$

and

$$\frac{\partial \tilde{U}}{\partial s}(U_o, s_o) = (DF(U_o^*) - s_o I)^{-1} (U_o^* - U_o). \tag{2.16}$$

Similarly, since

$$d(U_-, s^*) = 0$$

we obtain

$$D_{U_-}s^*(U_o) = -\left(\frac{\partial d}{\partial s}(U_o, s_o)\right)^{-1} D_{U_-}d(U_o, s_o). \tag{2.17}$$

Substituting (2.15),(2.16) and (2.17) into (2.12) yields the result.□

We now state an existence-uniqueness result for Riemann problems.

Proposition 2.7 *Assume (H1)-(H5). Let $\mathcal{F}$ be given by (2.10). Then there exist neighborhoods M_1 of U_o, M_2 of U_o^*, and M_3 of $(0, 0)$ such that for each $U_L \in M_1$ and $U_R \in M_2$ there is a unique $(\epsilon_1, \epsilon_2) \in M_3$ such that $\mathcal{F}(U_L, U_R, \epsilon_1, \epsilon_2) = 0$.*

Proof. We have $\mathcal{F}(U_o, U_o^*, 0, 0) = 0$, and $D_{(\epsilon_1,\epsilon_2)}\mathcal{F}(U_o, U_o^*, 0, 0)$ is invertible by (2.5) and (H5). The result follows from the implicit function theorem. □

To obtain a more geometric version of Proposition 2.7, let N_1 and N_2 be smaller neighborhoods of U_o and U_o^*. Consider the set $\mathcal{S}$ of functions taking values in $N_1 \cup N_2$ and consisting of three admissible waves: a slow wave joining U_L to a point $U_m \in N_1$, with speed close to $\lambda_1(U_o)$; an undercompressive shock wave joining U_m to a point U_m^* in N_2, with speed close to s_o and corresponding trajectory close to $\gamma(t)$; and a fast wave joining U_m^* to U_R, with speed close to $\lambda_2(U_o^*)$.

Theorem 2.8 *Suppose (H1)-(H5) are satisfied. Then there exist neighborhoods N_1 of U_o, N_2 of U_o^* such that for each $U_L \in N_1$, $U_R \in N_2$, the Riemann problem has a unique solution in $\mathcal{S}$. The solution consists of three waves, as constructed above.*

Proof. First extend the definition of $s_k(U_L, \epsilon)$ of Lemma 2.5 to apply to rarefaction waves:

$$s_k(U_L, \epsilon) = \lambda_k\left(U_k(U_L, \epsilon)\right), \qquad 0 \le \epsilon < \delta_k.$$

Then s_k is continuous on $A_k \times (-\delta_k, \delta_k)$ and C^2 away from $\epsilon = 0$. Since

$$\frac{\partial}{\partial \epsilon} s_k(U_L, 0+) = D\lambda_k(U_L) r_k(U_L) = 1,$$

$s_k(U_L, \epsilon)$ is not C^1 at $\epsilon = 0$ (see (2.6)). We refer to $s_k(U_L, \epsilon)$ as the *speed* of the wave from U_L to $U_k(U_L, \epsilon)$.

Next, we use Proposition 2.7 to construct a solution of the Riemann problem in $\mathcal{S}$, and finally we show that this solution is unique. Let M_1, M_2, M_3 be as in Proposition 2.7. For fixed $U_L \in M_1$, $U_R \in M_2$, let $(\epsilon_1, \epsilon_2) \in M_3$ be the unique solution of

$$\mathcal{F}(U_L, U_R, \epsilon_1, \epsilon_2) = 0.$$

Then a solution of the Riemann problem in $\mathcal{S}$ is constructed as follows. There is a slow wave from U_L to $U_1(U_L, \epsilon_1)$ with speed $s_1(U_L, \epsilon_1)$ close to $\lambda_1(U_o)$; an undercompressive shock from $U_1(U_L, \epsilon_1)$ to $U_1^*(U_L, \epsilon_1)$ with speed $s^*(U_1(U_L, \epsilon_1))$ close to s_o, and trajectory near $\gamma(t)$ (combining Lemmas 2.1 and 2.2); and a fast wave from $U_1^*(U_L, \epsilon_1)$ to $U_R = U_2(U_1^*(U_L, \epsilon_1), \epsilon_2)$ with speed $s_2(U_1^*(U_L, \epsilon_1), \epsilon_2)$ close to $\lambda_2(U_o^*)$.

To prove uniqueness, we choose smaller neighborhoods N_1 of U_o, N_2 of U_o^* so that the uniqueness statements of Lemmas 2.1, 2.2, 2.5 are valid with $A_k = N_k$, $k = 1, 2$. Let $U_L \in N_1$, $U_R \in N_2$, and let $U \in \mathcal{S}$ be a solution of the Riemann problem. By Lemma 2.5, the slow wave from U_L to $U_m \in N_1$ with speed s_{slow} near $\lambda_1(U_o)$ has $s_{\text{slow}} = s_1(U_L, \overline{\epsilon}_1)$ and

$$U_m = U_1(U_L, \overline{\epsilon}_1), \tag{2.18}$$

for some $\bar{\epsilon}_1$, $|\bar{\epsilon}_1| < \delta_1$. Similarly, the fast wave from U_m^* to U_R with speed s_{fast} satisfies $s_{\text{fast}} = s_2(U_m^*, \bar{\epsilon}_2)$ and

$$U_R = U_2(U_m^*, \bar{\epsilon}_2), \tag{2.19}$$

for some $\bar{\epsilon}_2$, $|\bar{\epsilon}_2| < \delta_2$.

The undercompressive shock from U_m to U_m^* has speed s_u close to s_o and trajectory close to $\gamma(t)$. Therefore $d(U_m, s_u) = 0$, and Lemma 2.2 implies $s_u = s^*(U_m)$, and $U_m^* = \tilde{U}(U_m, s^*(U_m))$. From (2.18), we have

$$U_m^* = U_1^*(U_L, \bar{\epsilon}_1). \tag{2.20}$$

Combining (2.19) and (2.20), we find

$$\mathcal{F}(U_L, U_R, \bar{\epsilon}_1, \bar{\epsilon}_2) = 0.$$

Proposition 2.7 then gives $\bar{\epsilon}_1 = \epsilon_1$, $\bar{\epsilon}_2 = \epsilon_2$. □

We can strengthen Theorem 2.6 slightly if we assume

(H6) *There exists $M > 0$ such that if $|s| > M$, U and $U_- \in A_k$, $k = 1$ or 2, and*

$$F(U) - F(U_-) - s(U - U_-) = 0,$$

then $U = U_-$.

Hypothesis (H6) implies that all weak slow waves in A_1 are described by the curves $W_1(U_L)$, $U_L \in A_1$, while all weak fast waves in A_2 are described by $W_2(U_L)$, $U_L \in A_2$. Let $\tilde{\mathcal{S}}$ be a set defined like $\mathcal{S}$ except that the restrictions on the possible speeds of the slow and fast waves are removed. Then we have:

Theorem 2.9 *Suppose (H1)–(H6) are satisfied. Then there exist neighborhoods $\tilde{N}_1$ of U_o, $\tilde{N}_2$ of U_o^* such that for $U_L \in \tilde{N}_1$, $U_R \in \tilde{N}_2$, the Riemann problem has a unique solution in $\tilde{\mathcal{S}}$. The solution consists of three waves as constructed above.*

3 Example

Here, we verify the assumptions of Section 2 for an example. Consider the system

$$\begin{aligned} u_t - v_x &= 0 \\ v_t - \sigma(u)_x &= 0, \end{aligned} \tag{3.1}$$

where $\sigma : I \to \mathbf{R}$ is a C^2 function defined on an interval I, with the following properties. There exist numbers $m < \alpha < u_I < \beta < M$ in I, with

$$\begin{aligned} \sigma'(u) < 0 \qquad & \text{for} \quad \alpha < u < \beta \\ \sigma'(u) > 0 \qquad & \text{for} \quad u < \alpha \quad \text{and} \quad u > \beta \\ (u_I - u)\sigma''(u) < 0 \qquad & \text{for} \quad u \in I \\ \int_m^M (\sigma(u) - \sigma(m))du = 0, \qquad & \sigma(m) = \sigma(M). \end{aligned}$$

Then system (3.1) is of mixed type. It is hyperbolic outside the (u, v) strip $\alpha < u < \beta$, and elliptic inside the strip. System (3.1) is a simple model for dynamic changes of phase in one-dimensional elasticity [2] with a non-monotone stress-strain law $\sigma = \sigma(u)$, and for one-dimensional isentropic gas dynamics with a van der Waals type isotherm, corresponding to the graph of $-\sigma$ [10].

Slemrod [10] proposed adding small viscosity and capillarity to (3.1), producing the system

$$\begin{aligned} u_t - v_x &= 0 \\ v_t - \sigma(u)_x &= \mu v_{xx} - \mu^2 A u_{xxx}. \end{aligned} \tag{3.2}$$

If $0 < A \le \frac{1}{4}$, we define $b_\pm = \frac{1}{2}(1 \pm (1-4A)^{\frac{1}{2}})$, and let $w = v - \mu b_+ u_x$ in (3.2). This converts (3.2) into

$$\begin{aligned} u_t - w_x &= \mu b_+ u_{xx} \\ w_t - \sigma(u)_x &= \mu b_- w_{xx}. \end{aligned} \tag{3.3}$$

Note that (3.3) has the form (1.6), with

$$U = \begin{pmatrix} u \\ w \end{pmatrix}, \quad F(u,w) = \begin{pmatrix} -w \\ -\sigma(u) \end{pmatrix}, \quad B = diag(b_+, b_-).$$

The differential equation (2.1) becomes

$$\begin{aligned} \dot{u} &= b_+^{-1}\left[-(w - w_-) - s(u - u_-)\right] \\ \dot{w} &= b_-^{-1}\left[-(\sigma(u) - \sigma(u_-)) - s(w - w_-)\right]. \end{aligned} \tag{3.4}$$

When $s = 0$, system (3.4) reduces to

$$\begin{aligned} \dot{u} &= -b_+^{-1}(w - w_-) \\ \dot{w} &= -b_-^{-1}(\sigma(u) - \sigma(u_-)), \end{aligned} \tag{3.5}$$

which is Hamiltonian.

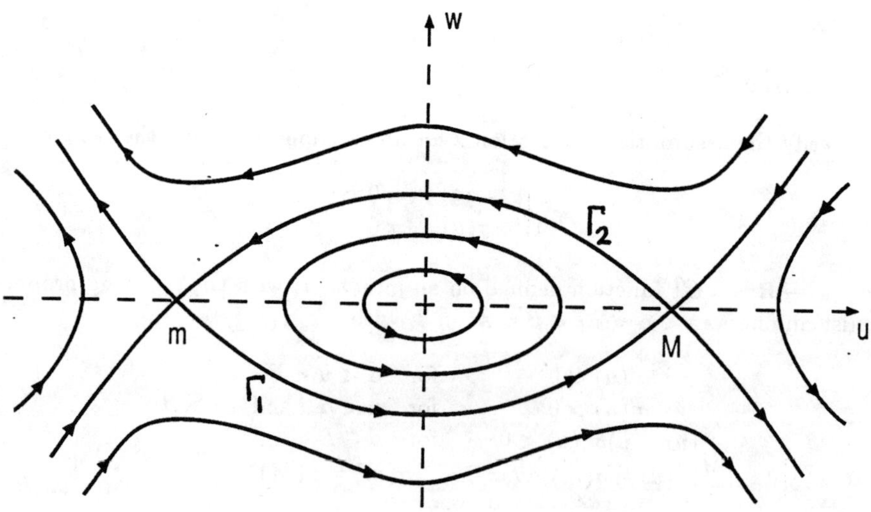

Figure 1. Phase portrait for equation (3.5).

The phase portrait for (3.5) when $(u_-, w_-) = (m, 0)$ or $(u_-, w_-) = (M, 0)$ is shown in Figure 1. There are heteroclinic orbits Γ_1 from $(m,0)$ to $(M,0)$ and Γ_2 from $(M,0)$ to $(m,0)$; moreover, $(m,0)$ and $(M,0)$ are hyperbolic saddles. Thus (H1) holds for $(u_-, w_-, s) = (m, 0, 0)$ and $(u_-, w_-, s) = (M, 0, 0)$. We shall verify the remaining hypotheses only at $(u_-, w_-, s) = (m, 0, 0)$; the verification at $(u_-, w_-, s) = (M, 0, 0)$ is left to the reader.

We define a separation function d for Γ_1 as in Section 2. The function d depends on the parameters in (3.4), but we suppress the dependence on (b_+, b_-), since these parameters are considered fixed. For $(u_-, w_-) = (m, 0)$, let $(u(t), w(t))$ be a heteroclinic solution of (3.5) from $(m,0)$ to $(M,0)$. Let R be the region bounded by Γ_1 and the u-axis, with area $A(R)$. By (2.2) and Green's Theorem

$$\begin{aligned}\frac{\partial d}{\partial s}(m,0,0) &= \int_{-\infty}^{\infty} \begin{bmatrix} \dot{u} \\ \dot{w} \end{bmatrix} \wedge \begin{bmatrix} -b_+^{-1}(u(t) - m) \\ -b_-^{-1} w(t) \end{bmatrix} dt \\ &= \int_{\Gamma_1} b_+^{-1}(u - m)dw - b_-^{-1} w du \\ &= (b_+^{-1} + b_-^{-1})A(R) > 0.\end{aligned} \tag{3.6}$$

Thus (H2) is satisfied at $(u_-, w_-, s) = (m, 0, 0)$. We note for future reference that

$$\frac{\partial d}{\partial u_-}(m,0,0) = \int_{-\infty}^{\infty} \begin{bmatrix} \dot{u} \\ \dot{w} \end{bmatrix} \wedge \begin{bmatrix} 0 \\ b_-^{-1}\sigma'(m) \end{bmatrix} dt = b_-^{-1}\sigma'(m)(M - m); \tag{3.7}$$

$$\frac{\partial d}{\partial w_-}(m,0,0) = \int_{-\infty}^{\infty} \begin{bmatrix} \dot{u} \\ \dot{w} \end{bmatrix} \wedge \begin{bmatrix} b_+^{-1} \\ 0 \end{bmatrix} dt = 0. \tag{3.8}$$

The other basic quantities used in Section 2 depend only on u as follows. The characteristic speeds are

$$\lambda_1(u) = -c(u); \quad \lambda_2(u) = c(u),$$

where $c(u) = (\sigma'(u))^{1/2}$. Normalizing the right eigenvectors as in Section 2, we have

$$r_1(u) = k(u)\begin{pmatrix} 1 \\ c(u) \end{pmatrix}; \quad r_2(u) = -k(u)\begin{pmatrix} 1 \\ -c(u) \end{pmatrix}, \tag{3.9}$$

where $k(u) = -2c(u)/\sigma''(u)$ is positive for $u < u_I$ and negative for $u > u_I$.

We set

$$U_o = (m, 0), \quad U_o^* = (M, 0), \quad s_o = 0. \tag{3.10}$$

Although system (3.1) is of mixed type, it is strictly hyperbolic ($\sigma'(u) \neq 0$) and genuinely nonlinear ($\sigma''(u) \neq 0$) in the neighborhoods $A_1 = \{(u,v) : u < \alpha\}$ of U_o, $A_2 = \{(u,v) : u > \beta\}$ of U_o^*, thus verifying hypothesis (H3). Hypothesis (H4) follows from Lemma 2.3, (3.9), and the fact that B is diagonal with $b_\pm > 0$.

Hypothesis (H5) involves some calculation. From Lemma 2.4,(3.6-3.9), we have

$$\frac{\partial}{\partial\epsilon}U_1^*(U_o,0) = \begin{pmatrix} 0 & -b_+^{-1} \\ -b_-^{-1}\sigma'(M) & 0 \end{pmatrix}^{-1}\Bigg\{-c(m)k(m)\begin{pmatrix} 1 \\ c(m) \end{pmatrix}$$

$$-\begin{pmatrix} M-m \\ 0 \end{pmatrix}(b_+^{-1}+b_-^{-1})^{-1}A(R)^{-1}(b_-^{-1}\sigma'(m)(M-m) \quad 0)k(m)\begin{pmatrix} 1 \\ c(m) \end{pmatrix}\Bigg\}$$

$$= k(m)\begin{pmatrix} 0 & b_-[\sigma'(M)]^{-1} \\ b_+ & 0 \end{pmatrix}\begin{pmatrix} c(m)+(b_+^{-1}+b_-^{-1})^{-1}A(R)^{-1}(b_-^{-1}\sigma'(m)(M-m)^2 \\ \sigma'(m) \end{pmatrix},$$

which has positive slope. But the slope of $r_2 = r_2(M)$ at U_o^* is $-c(M)$, which is negative. Thus, hypothesis (H5) is verified.

Hypothesis (H6) follows from the identity $s^2 = (\sigma(u_+) - \sigma(u_-))/(u_+ - u_-)$ for shocks with $u(x,t) = \begin{cases} u_+ & \text{if } x > st \\ u_- & \text{if } x < st \end{cases}$.

We conclude from Theorem 2.9 that the Riemann problem for (3.1), for U_L near $U_o = (m,0)$ and U_R near $U_o^* = (M,0)$ has a unique solution (within the class $\widetilde{\mathcal{S}}$ of the Theorem) involving three waves, including one undercompressive shock, across which u jumps across the interval $[\alpha,\beta]$. This result was proved in [9] using an ad hoc analysis of the phase plane.

4 Appendix. Strictly Stable 2×2 Matrices.

In this Appendix we give a simple proof that Lemma 2.3 implies Lemma 2.4. Let A, B, λ_k, ℓ_k, r_k be as in Lemma 2.3.

Lemma A.1 *Suppose B is strictly stable with respect to A, and A has j positive eigenvalues and $2-j$ negative eigenvalues, $j = 0, 1$ or 2. Then BA has j eigenvalues with positive real part and $2-j$ eigenvalues with negative real part.*

Proof. The proof follows from two identities:

$$sgn(\det BA) = sgn\,(\det A), \tag{A.1}$$

which follows immediately from (2.4), and

$$\text{trace (BA)} = \sum_{k=1}^{2}\lambda_k\ell_k B r_k. \tag{A.2}$$

To see (A.2), let L be the matrix with rows ℓ_1, ℓ_2, and let R be the inverse of L, having columns r_1, r_2. Then

$$\begin{aligned}\text{trace (BA)} &= \text{trace (LBAR)} \\ &= \text{trace } (LB\Lambda R) \quad (\text{where } \Lambda = \text{diag }(\lambda_1, \lambda_2)) \\ &= \sum_{k=1}^{2}\lambda_k\ell_k B r_k.\end{aligned}$$

Suppose A has eigenvalues of opposite sign. Then $\det A < 0$. From (A.1) we have $\det(BA) < 0$, so that BA also has real eigenvalues of opposite sign.

Alternatively, suppose A has eigenvalues of the same sign. Suppose $\lambda_2 > \lambda_1 > 0$. (The argument is similar if $\lambda_1 < \lambda_2 < 0$.) Then $\det BA > 0$ from (A.1), while trace $BA > 0$ from (2.3) and (A.2). Therefore, both eigenvalues of BA have positive real part. □

Lemma A.2 *B is strictly stable with respect to A if and only if B^{-1} is strictly stable with respect to A.*

Proof. First,

$$\det(B^{-1}) = (\det B)^{-1}. \tag{A.3}$$

Moreover, since $LR = I$,

$$LB^{-1}R = R^{-1}B^{-1}L^{-1} - (LBR)^{-1}$$

Therefore,

$$\ell_k B^{-1} r_k = \frac{1}{\det(LBR)} \ell_{3-k} B r_{3-k} = \frac{1}{\det B} \ell_{3-k} B r_{3-k}, \quad k = 1, 2. \tag{A.4}$$

The result follows immediately from (A.3) and (A.4). □

Define $\lambda_o = -\infty$, $\lambda_3 = \infty$.

Lemma A.3 *Suppose B is strictly stable with respect to A. Then for $\lambda_j < s < \lambda_{j+1}$, $j = 0, 1$ or 2, $B^{-1}(A - sI)$ has j eigenvalues with negative real part and $2 - j$ eigenvalues with positive real part.*

Proof. The eigenvalues of $A - sI$ are $\lambda_k - s$, $k = 1, 2$, with left and right eigenvectors ℓ_k, r_k. Therefore, B is strictly stable with respect to $A - sI$, and the result follows from Lemmas A.1 and A.2. □

Lemma 2.4 follows from Lemma A.3 and the observation that if B is strictly stable with respect to A, then B is strictly stable with respect to any matrix close to A.

References

[1] E. Isaacson, D. Marchesin and B. Plohr, Transitional waves for conservation laws. *SIAM J. Math. Anal.*, **21** (1990), pp. 837-866.

[2] R.D. James, The propagation of phase boundaries in elastic bars. *Arch. Rat. Mech. Anal.*, **73** (1980), pp. 125-158.

[3] P.D. Lax, Hyperbolic systems of conservation laws II. *Comm. Pure Appl. Math.*, **10** (1957), pp. 537-566.

[4] T.-P. Liu, The Riemann problem for general 2x2 conservation laws. *Trans. Amer. Math. Soc.*, **199** (1974), pp. 89-112.

[5] A. Majda and R. Pego, Stable viscosity matrices for systems of conservation laws. *J. Differential Equations*, **56** (1985), pp. 229-262.

[6] F.M.A. Salaam, The Melnikov technique for highly dissipative systems. *SIAM J. Appl. Math.*, **47** (1987), pp. 232-243.

[7] M. Shearer, D.G. Schaeffer, D. Marchesin and P. Paes-Leme, Solution of the Riemann problem for a prototype 2 x 2 system of non-strictly hyperbolic conservation laws. *Arch. Rat. Mech. Anal.*, **97** (1987), pp. 299-320.

[8] M. Shearer, The Riemann problem for 2x2 systems of hyperbolic conservation laws with case I quadratic nonlinearities. *J. Differential Equations*, **80** (1989), pp. 343-363.

[9] M. Shearer, Dynamic phase transitions in a van der Waals gas. *Quart. Applied Math.*, **46** (1988), pp. 631-636.

[10] M. Slemrod, Admissibility criteria for propagating phase boundaries in a van der Waals fluid. *Arch. Rat. Mech. Anal.*, **81** (1983), pp. 301-315.

CHAPTER 14

Recent Results on the Kinetic Theory of Coagulation*

Marshall Slemrod†

ABSTRACT. This report summarizes recent work on dynamic problems arising in the analysis of a kinetic theory of coagulation.

1. Introduction. A classic model going back to Smoluchowski [5] to describe cluster dynamics of discrete particles is the coagulation-fragmentation system. The model has served a variety of purposes: studies of polymerization, cloud formation, the immune system, and even formation of the universe. [6]

In this report I will describe some recent results on coagulation-diffusion equations, especially emphasizing the role of diffusion which has often been neglected in studying coagulation equations. The interested reader can consult my paper [4] for more details.

2. The equations. Let $c_j(\underline{x},t) \geq 0$, $j = 1,2,\cdots$, denote the expected number of clusters of j particles/unit volume at position $\underline{x} \in \mathbb{R}^n (n \in 1,2,3)$ at time $t \geq 0$. The discrete coagulation-fragmentation equations with diffusion are

*Sponsored in part by the Air Force Office of Scientific Research, Air Force Systems Command, USAF, under Contract/Grant No. AFOSR-87-0315. The United States Government is authorized to reproduce and distribute reprints for any governmental purposes not withstanding any copyright herein. Also supported in part by grants from the Graduate School of the University of Wisconsin, the Pearlman Foundation, and a Meyerhoff Visiting Professorship at the Weizman Institute of Science.

†Center for the Mathematical Sciences, University of Wisconsin-Madison, Madison, Wisconsin 53705.

$$
\begin{aligned}
\frac{\partial c_j}{\partial t} - D_j \Delta c_j &= \frac{1}{2}\sum_{k=1}^{j-1}[a_{j-k,k}c_{j-k}c_k - b_{j-k,k}c_j] \\
&- \sum_{k=1}^{\infty}[a_{j,k}c_jc_k - b_{j,k}c_{j+k}], \quad j = 1, 2, \cdots, \\
\underline{c}(\underline{x}, t) &= (c_1(\underline{x}, t), \cdots).
\end{aligned} \tag{2.1}
$$

The diffusivities D_j, coagulation rates $a_{j,k}$ and fragmentation rates $b_{j,k}$, are non-negative constants with $a_{j,k} = a_{k,j}$, $b_{j,k} = b_{k,j}$; Δ is the n dimensional Laplacian. Stokes' law tells us that D_j is proportional to $j^{-1/n}$ though Binder [1] has derived a slightly different relation.

3. Some special solutions. Consider (2.1) with $b_{j,k} \equiv 0$ (pure coagulation) and product coagulation coefficients $a_{j,k} = r_j r_k$; $r_j \geq 0$, $j = 1, 2, \cdots$. In this case (2.1) takes the form

$$
\frac{\partial c_j}{\partial t} - D_j \Delta c_j = \frac{1}{2}\sum_{k=1}^{j-1} r_{j-k}r_k c_{j-k}c_k - r_j c_j \sum_{k=1}^{\infty} r_k c_k, \; 1 \leq j < \infty. \tag{3.1}
$$

Next we recall some results of Leyvraz and Tschudi [2] and Leyvraz [3] for spatially independent non-mass-conserving solutions.

Spatially homogeneous solutions $c_j(\underline{x}, t) = c_j(t)$ will satisfy

$$
\frac{dc_j}{dt} = \frac{1}{2}\sum_{k=1}^{j-1} r_{j-k}r_k c_{j-k}c_k - r_j c_j \sum_{k=1}^{\infty} r_k c_k, \; 1 \leq j < \infty. \tag{3.2}
$$

In [2] it was shown for the special case

$$
r_j = Aj, \quad A > 0,
$$

an exact solution of (3.2) satisfying the monodisperse initial conditions

$$
c_1(0) = 1, \; c_j(0) = 0, \; 2 \leq j < \infty
$$

is given by

$$
\begin{aligned}
c_j(t) &= \frac{j^{j-3}(A^2t)^{j-1}\exp(-A^2jt)}{(j-1)!}, \; 0 \leq t \leq \frac{1}{A^2}, \\
&= \frac{j^{j-3}e^{-j}}{(j-1)!A^2t}, \; \frac{1}{A^2} \leq t < \infty.
\end{aligned}
$$

This solution has the property that the mass density is given by

$$
\begin{aligned}
\rho(t) = \sum_{j-1}^{\infty} jc_j(t) &= 1, \; 0 \leq t \leq \frac{1}{A^2}, \\
&= \frac{1}{A^2t}, \; \frac{1}{A^2} \leq t < \infty.
\end{aligned}
$$

That is mass conservation ceases at *the gelation time* $t_g = \frac{1}{A^2}$. After t_g mass is lost to the infinite particle (the gel) which is not accounted for in the original model (2.1).

The above equations (3.2) with $r_j = j$ correspond to the Flory-Stockmayer theory of gelation. The presence of the infinite cluster make it difficult to say whether this solution is physically relevant after the gelation time t_g. If however $r_j/j \to 0$ as $j \to \infty$ then it has been shown in [2] that the infinite system (3.2) is physically relevant for all time in the following sense: The finite systems

$$\dot{c}_j^{(N)} = \frac{1}{2}\sum_{k=1}^{j-1} r_k r_{j-k}\, c_k^{(N)} c_{j-k}^{(N)} - j c_j^{(N)} \sum_{k=1}^{N-j} k c_k^{(N)}$$

possess solutions $c_j^{(N)}$, $1 \le j < \infty$, where limits as $N \to \infty$ exist and are indeed solutions of (3.2) for all $t > 0$. These limiting solutions will still exhibit gelation when $r_j = j^\alpha$, $\frac{1}{2} < \alpha$, but now the meaning and "admissibility" of solutions to the infinite system is clear: The infinite system is the limit of finite systems which contain interaction between large and small molecules, i.e. between what in the limit of large N will become sol and gel respectively. This limiting procedure is the key for our analysis of boundary value problems for the dynamic diffusion systems (3.1) in Section 5.

In an effort to find other non-mass-conserving post gel solutions for (3.2) Leyvraz [2] suggested the following ansatz: Let

$$c_j(t) = \frac{\alpha_j}{t + C} \tag{3.3}$$

where $\alpha_j > 0$ and C is a constant. Substitution of (3.3) into (3.2) yields the infinite algebraic system

$$-\alpha_j = \frac{1}{2}\sum_{k=1}^{j-1} r_{j-k} r_k \alpha_{j-k} \alpha_k - r_j \alpha_j \sum_{k=1}^{\infty} r_k \alpha_k\,, \quad 1 \le j < \infty\,. \tag{3.4}$$

The assignment $j = 1$ in (3.4) shows

$$r_1^{-1} = \sum_{k=1}^{\infty} r_k \alpha_k \tag{3.5}$$

and hence (3.4) becomes

$$\alpha_j = \frac{r_1}{2(r_j - r_1)} \sum_{k=1}^{j-1} r_{j-k} r_k \alpha_{j-k} \alpha_k,\ 2 \le j < \infty\,. \tag{3.6}$$

Notice that once $\alpha_1 > 0$ is prescribed, (3.6) determines the subsequent α_j, $j \ge 2$. The goal then is to find $\alpha_1 > 0$ so that the constraint (3.5) is satisfied. This problem was successfully solved by Leyvraz [3] who proved the following result.

Proposition. Let r_j satisfy

(i) $r_j > r_1 > 0 \quad (j \geq 2)$,
(ii) $lim_{j\to\infty} r_j^{-1} = 0$.

Then equations (3.5), (3.6) have a positive solution α_j, $1 \leq j < \infty$. Furthermore if

(iii) $r_j > Bj^\alpha$, $\alpha > 0$, $B > 0$ constants then the above solution satisfies
(a) $\sum_{j=1}^{\infty} j\alpha_j < \infty \quad (\alpha > \frac{1}{2})$,
(b) $\sum_{j=1}^{N} j\alpha_j = O(N^{1-2\alpha+\varepsilon})(\alpha \leq \frac{1}{2})$

for any positive ε: Hence there exists a finite mass, spatially homogeneous non-mass-conserving solution $c_j(t) = \frac{\alpha_j}{t+C}$ when $\alpha > \frac{1}{2}$.

Finally consider the possibility of time independent, spherically symmetric solutions of (3.1). First note spherically symmetric solutions $c_j(\underline{x}, t) = c_j(R, t)$, $R = |\underline{x}|$, of (3.1) satisfy

$$\frac{\partial c_j}{\partial t} - D_j\left(\frac{\partial^2 c_j}{\partial R^2} + \frac{(n-1)}{R}\frac{\partial c_j}{\partial R}\right) = \frac{1}{2}\sum_{k=1}^{j-1} r_{j-k}r_k c_{j-k}c_k - r_j c_j \sum_{k=1}^{\infty} r_k c_k \tag{3.7}$$

where $n \in \{1,2,3\}$.

In the same spirit as Leyvraz's solution (3.3) for spatially independent solutions, we make the ansatz

$$c_j(R) = \frac{2(4-n)\alpha_j}{D_j R^2} \tag{3.8}$$

for time independent solutions of (3.7). Substitution of (3.8) into (3.7) yields after some manipulations that the α_j satisfy (3.5), (3.6) with r_j replaced by $r_j D_j^{-1}$. Hence the Proposition may be immediately applied to yield:

Theorem. Let $r_j D_j^{-1}$ satisfy (i), (ii) of the Proposition. Then there exists $\alpha_j > 0$, $1 \leq j < \infty$, so that (3.8) provides a positive solution $c_j(R)$, $1 \leq j < \infty$, of (3.7). Furthermore if (iii) holds with $\alpha > \frac{1}{2}$ then

$$\rho(R) = \sum_{j=1}^{\infty} j c_j(R) = \frac{2(4-n)}{R^2}\sum_{j=1}^{\infty}\frac{\alpha_j}{D_j}$$

for all $R > 0$. Hence there exists a finite density time independent solution of (3.7) when $r_j D_j^{-1} > Bj^\alpha$ and $\alpha > \frac{1}{2}$.

4. Diffuse interface similarity solutions. Take as an ansatz for spherically symmetric solutions to (3.7):

$$c_j(R,t) = \frac{\phi_j(\xi)}{t} \qquad \xi = \frac{R}{\sqrt{t}}. \tag{4.1}$$

Substitution of (4.1) into (3.1) yields the infinite system of ordinary differential equations for $\phi_j(\xi)$:

$$D_j(\phi_j'' + \frac{(n-1)}{\xi}\phi_j') + \frac{\xi}{2}\phi_j' + \phi_j = -\frac{1}{2}\sum_{k=1}^{j-1} r_{j-k}r_k\phi_{j-k}\phi_k + r_j c_j \sum_{k=1}^{\infty} r_k\phi_k \tag{4.2}$$
$$1 \leq j < \infty, \quad n \in \{1,2,3\}.$$

Notice $\phi_j \equiv 0$ represents the no cluster equilibrium. In addition if r_j satisfy (i), (ii), (iii) (with $\alpha > \frac{1}{2}$) of the Proposition, $\phi_j = \alpha_j$, $1 \le j < \infty$, provides a second equilibrium. In the case $n = 1$ a solution $\phi_j(\xi)$, $1 \le j < \infty$, on $-\infty < \xi < \infty$, connecting the equilibria when inserted into (4.1) would provide a structured diffuse interface between the space independent non-mass-conserving solution (3.3) and the far field zero cluster equilibrium. Analytically the goal would be to solve (4.2) on $(-\infty, \infty)$ subject to boundary conditions

$$\begin{aligned} \phi_j &= \alpha_j \quad \text{at} \quad \xi = -\infty\,, \\ &= 0 \quad \text{at} \quad \xi = +\infty,\ 1 \le j < \infty\,. \end{aligned} \tag{4.3}$$

Resolution of the issues of existence, uniqueness, stability of solutions to (4.2), (4.3) are all open problems.

5. Results on boundary value problems for the diffuse interface equations. Since (4.2), (4.3) seem to be rather difficult systems to analyze in [4] I have detailed proofs for some existence theorems for the diffuse interface equations under more restrictive hypotheses. Specifically I avoided the nasty singularity at $\xi = 0$ and worked on finite domains with prescribed boundary data.

The main result is:

Theorem. Assume $D_j = D > 0$ for all j and

$$\frac{r_j}{j} \to 0 \quad \text{as} \quad j \to \infty\,.$$

Then the boundary value problem (4.2) with boundary conditions

$$\phi_j(\xi_0) = \beta_j,\ \phi_j(\xi_1) = \gamma_j,\ 1 \le j < \infty,\ 0 < \xi_0 < \xi_1 < \infty,\ \beta_j > 0,\ \gamma_j > 0\,,$$
$$\sum_{j=1}^{\infty} j\beta_j < \infty,\ \sum_{j=1}^{\infty} j\gamma_j < \infty$$

has a classical C^2 positive solution ϕ_j, $1 \le j < \infty$ for which

$$\sum_{j=1}^{\infty} j\phi_j(\xi) < \infty\,.$$

The proof used heavily the fact that the D_j are all equal and solutions are obtained as the limits of solutions of finite systems of N equations in the same spirit as Leyvraz and Tschudi [2] discussed in Section 3. Hence we may view these solutions as "admissible". It remains an open problems to generalize the theorem to more general D_j non-constant, $D_j \to 0$ as $j \to \infty$, the physically realistic case.

References

[1.] Binder, K., Theory for the dynamics of clusters, II, Critical diffusion in binary systems and the kinetics of phase separation, Phys. Rev. B 15, 4425 - 4447 (1977).

[2.] Leyvraz, F., Tschudi, H. R., Singularities in the kinetics of coagulation processes, J. Physics A. Math. Gen. 14, 3389 -3405 (1981).

[3.] Leyvraz, F., Existence and properties of post – gel solutions for the kinetic equations of coagulation, J. Physics A. Math. Gen. 16, 2861 - 2873 (1983).

[4.] Slemrod, M., Coagulation – Diffusion systems : derivation and existence of solutions for the diffuse interface structure equations, to appear Physica D.

[5.] von Smoluchowski, M., Versuch einer mathematischen Theorie der Kolloiden Losungen, Z. Phys. Chem. 92, 129 - 168 (1917).

[6.] Drake, R., Reviews in aerosol physics and chemistry, in Topics in Current Aerosol Research, vol. 3, eds. Hidy, G. M., Brock, J. R., Oxford: Pergamon Press (1972).

CHAPTER 15

Growth or Decay of Shock Waves in the Generalized Goursat-Riemann Problem*

T. C. T. Ting†
Tankin Wang†

Abstract. For plane waves propagating in an elastic half-space $x \geq 0$ due to an impact at the plane $x = 0$, the problem is that of a generalized Goursat-Riemann problem in which the initial conditions at $t = 0$ are constant but the boundary conditions at $x = 0$ are not. If $\boldsymbol{s}(x,t)$ is the traction vector at any plane x = constant, $\boldsymbol{s}(0,t)$ depends on t. Assuming that $\boldsymbol{s}(0,t)$ is linear in t, i.e., $\boldsymbol{s}(0,t) = \boldsymbol{s}^a + \boldsymbol{s}_t^a\, t$ where $\boldsymbol{s}^a$ and $\boldsymbol{s}_t^a$ are constant, the Goursat-Riemann problem corresponds to the special case in which $\boldsymbol{s}_t^a$ vanishes. We study the generalized Goursat-Riemann problem with non-zero $\boldsymbol{s}_t^a$. We investigate how the solution to the Goursat-Riemann problem is affected by the non-zero $\boldsymbol{s}_t^a$. In particular, we examine whether the shock waves generated in the Goursat-Riemann problem grow or decay due to the non-zero $\boldsymbol{s}_t^a$.

1. Introduction. The equations governing the plane waves in an isotropic elastic solid, with certain restrictions, can be written as a system of 4×4 conservation laws [1,2],

$$\mathbf{U}_x - \mathbf{f}(\mathbf{U})_t = \mathbf{0}, \tag{1}$$

$$\mathbf{U} = (\sigma, \tau, u, v),$$

$$\mathbf{f}(\mathbf{U}) = (\rho u, \rho v, \varepsilon, \gamma).$$

*The work presented here is supported by the U. S. Air Force Office of Scientific Research under Contract AFSOR-89-0013

† Department of Civil Engineering, Mechanics and Metallurgy, University of Illinois at Chicago, Box 4348, Chicago, IL 60680.

In the above, σ, τ are the normal and shear stresses, ε, γ are the normal and shear strains, u, v are the normal and transverse velocities and ρ is the mass density. For hyperelastic solids there exists a potential function, or the complementary strain energy $W(\sigma, \tau)$ such that [3]

$$\varepsilon = W_\sigma, \qquad \gamma = W_\tau, \tag{2}$$

in which the subscript denotes differentiation. Introducing the 2-vectors

$$\boldsymbol{s} = (\sigma, \tau), \quad \mathbf{u} = (u, v), \quad \mathbf{p} = (\varepsilon, \gamma), \tag{3}$$

equation (1) can be written for continuous solutions as

$$\mathbf{U}_x - \mathbf{A}\mathbf{U}_t = \mathbf{0}, \qquad \mathbf{U} = \begin{bmatrix} \boldsymbol{s} \\ \mathbf{u} \end{bmatrix}, \tag{4}$$

$$\mathbf{A} = \begin{bmatrix} \mathbf{0} & \rho\mathbf{I} \\ \mathbf{G} & \mathbf{0} \end{bmatrix},$$

where $\mathbf{I}$ is a unit matrix and

$$\mathbf{G} = \begin{bmatrix} W_{\sigma\sigma} & W_{\sigma\tau} \\ W_{\sigma\tau} & W_{\tau\tau} \end{bmatrix}. \tag{5}$$

The characteristic wave speed c is obtained from [4]

$$(c\mathbf{A} + \mathbf{I})\boldsymbol{\xi} = \mathbf{0}, \quad \boldsymbol{\xi} = \begin{bmatrix} \mathbf{r} \\ -\mathbf{r}/(\rho c) \end{bmatrix}, \tag{6}$$

which can be reduced to

$$\{\mathbf{G} - (\rho c^2)^{-1}\mathbf{I}\}\,\mathbf{r} = \mathbf{0}. \tag{7}$$

Assuming that $\mathbf{G}$ is positive definite, there are two pairs of characteristic wave speeds $\pm c_1, \pm c_3$ and we let

$$c_1 \geq c_3 > 0.$$

Had we considered fully general plane waves, (1) would be a system of 6×6 conservation laws and there would be an additional pair of characteristic wave speed $\pm c_2$ which can be shown to be linearly degenerate [2].

It should be pointed out that (7) is, mathematically, the characteristic equation with characteristic wave speed $(\rho c^2)^{-1}$ for the following 2×2 conservation laws

$$\left.\begin{aligned} &\boldsymbol{s}_t + \mathbf{f}(\boldsymbol{s})_x = \mathbf{0}, \\ &\boldsymbol{s} = (\sigma, \tau), \quad \mathbf{f}(\boldsymbol{s}) = (W_\sigma, W_\tau). \end{aligned}\right\} \tag{8}$$

Equation (8) is a familiar equation in fluid dynamics and hence the results obtained here can be applied to problems in fluid dynamics. Many studies have been made on (8) which is in general not strictly hyperbolic [5-9].

In this paper we consider the following initial and boundary conditions for the generalized Goursat-Riemann problem

$$\left.\begin{aligned} &\boldsymbol{s}(x,0) = \boldsymbol{s}^b, \quad x > 0, \\ &\boldsymbol{s}(0,t) = \boldsymbol{s}^a + \boldsymbol{s}_t^a t, \quad t > 0, \end{aligned}\right\} \tag{9}$$

where $\boldsymbol{s}^b, \boldsymbol{s}^a, \boldsymbol{s}_t^a$ are constant. The superscripts "b" and "a" stand for "before" and "after", respectively. The constants $\boldsymbol{s}^b, \boldsymbol{s}^a, \boldsymbol{s}_t^a$ are 2-vectors defined in $(3)_1$. The problem reduces to the Goursat-Riemann problem when $\boldsymbol{s}_t^a = \mathbf{0}$.

When $\boldsymbol{s}_t^a = \mathbf{0}$, $\mathbf{U}(x,t)$ depends on x/t only and is best expressed in the form of wave curves in the $\mathbf{U}$ space. The wave curve consists of a sequence of simple wave curves and/or shock wave curves [10]. If the boundary conditions are prescribed in terms of stress $\boldsymbol{s}$, it suffices to consider wave curves in the sub-space $\boldsymbol{s}$ [1,2]. This is not possible if the boundary conditions are prescribed in terms of velocity [11,12]. The simple waves are "centered simple waves" because, in the x-t plane, they are represented by a family of straight lines from the origin x=t=0. Along each straight line $\mathbf{U}$ is a constant. The shock wave in the x-t plane is represented by a double line which is a straight line because the shock wave speed is a constant. Between the simple waves and/or the shock waves we have constant state regions. When $\boldsymbol{s}_t^a \neq \mathbf{0}$, we no longer have simple waves and constant state regions. Moreover, the shock wave speed is not a constant and may grow or decay as time increases. We will investigate the perturbation of the solution to the generalized Goursat-Riemann problem from the Goursat-Riemann problem in the region near x = 0 and t = 0. In particular, we examine whether the shock waves grow or decay due to the non-zero

s_t^a. The perturbation is with respect to small x and t, not with s_t^a, so that the value of s_t^a need not be small.

2. Perturbation in the constant state and simple wave regions. For the generalized Goursat-Riemann problem with initial and boundary conditions prescribed in (9) we consider the solution $\mathbf{U}(x,t)$ in the form [13-15]

$$\left.\begin{aligned} &\mathbf{U}(x,t) = \mathbf{U}^0(\lambda) + t\,\mathbf{U}^1(\lambda) + \frac{t^2}{2}\mathbf{U}^2(\lambda) + \cdots, \\ &\lambda = x/t, \end{aligned}\right\} \tag{10}$$

where $\mathbf{U}^0, \mathbf{U}^1, \mathbf{U}^2, \cdots$, are functions of λ only. Clearly, $\mathbf{U}^1, \mathbf{U}^2, \cdots$ vanish identically for the Goursat-Riemann problem. Differentiation of (10) with x and t results in

$$\left.\begin{aligned} \mathbf{U}_x &= \frac{1}{t}\frac{d}{d\lambda}\mathbf{U}^0 + \frac{d}{d\lambda}\mathbf{U}^1 + O(t), \\ \mathbf{U}_t &= -\frac{\lambda}{t}\frac{d}{d\lambda}\mathbf{U}^0 + (\mathbf{U}^1 - \lambda\frac{d}{d\lambda}\mathbf{U}^1) + O(t). \end{aligned}\right\} \tag{11}$$

Since $\mathbf{A}$ in (4) depends on $\mathbf{U}$, we also expand $\mathbf{A}$ as

$$\mathbf{A}(\mathbf{U}) = \mathbf{A}^0(\lambda) + t\mathbf{A}^1(\lambda) + \frac{t^2}{2}\mathbf{A}^2(\lambda) + \cdots \tag{12}$$

$$\left.\begin{aligned} &\mathbf{A}^0(\lambda) = \mathbf{A}(\mathbf{U}^0) \\ &\mathbf{A}^1(\lambda) = \mathbf{U}^1 \cdot \nabla\mathbf{A}^0 \\ &\qquad\vdots \end{aligned}\right\} \tag{13}$$

The operator ∇ is the gradient with respect to the components of $\mathbf{U}$. With (11) and (12), $(4)_1$ leads to

$$(\lambda\mathbf{A}^0 + \mathbf{I})\frac{d}{d\lambda}\mathbf{U}^0 = \mathbf{0}, \tag{14}$$

$$(\lambda\mathbf{A}^0 + \mathbf{I})\frac{d}{d\lambda}\mathbf{U}^1 = \mathbf{A}^0\mathbf{U}^1 - \lambda\mathbf{A}^1\frac{d}{d\lambda}\mathbf{U}^0, \tag{15}$$

$$\vdots$$

We will study the first order perturbation, namely, the solution of $\mathbf{U}^1$.

Let c^0 be a root of the determinant

$$\|c^0\mathbf{A}^0(c^0)+\mathbf{I}\|=0.$$

The solution of $\mathbf{U}^0, \mathbf{U}^1$ depends on whether λ equals to any of the four roots c^0 or not. We call the region "regular" if $\lambda \neq c^0$ and "singular" if $\lambda = c^0$. For the Goursat-Riemann problem the regions $\lambda \neq c^0$ and $\lambda = c^0$ correspond, respectively, to the constant state region and the simple wave region.

2.1 Regular region. When $\lambda \neq c^0$, (14) and (15) yield

$$\left.\begin{aligned}\mathbf{U}^0 &= \text{constant},\\ \mathbf{U}^1 &= \{(\mathbf{A}^0)^{-1}+\lambda\,\mathbf{I}\}\,\mathbf{q},\end{aligned}\right\} \tag{16}$$

where $\mathbf{q}$ is an arbitrary constant. Equation $(11)_2$ then gives

$$\mathbf{U}_t|_{t=0} = (\mathbf{A}^0)^{-1}\,\mathbf{q}, \tag{17}$$

and (16) and (17) provide the relation

$$\mathbf{U}^1 = (\lambda\mathbf{A}^0+\mathbf{I})\,\mathbf{U}_t|_{t=0} \tag{18}$$

which allows us to convert $\mathbf{U}_t|_{t=0}$ to $\mathbf{U}^1$ and vice versa. We see that while $\mathbf{U}^0$, $\mathbf{U}_t|_{t=0}$ are constant, $\mathbf{U}^1$ is linear in λ.

2.2 Singular region. When

$$\lambda = c^0, \tag{19}$$

equations (14) and $(6)_1$ imply that

$$\frac{d}{d\lambda}\mathbf{U}^0 = -\zeta\boldsymbol{\xi}^0$$

where ζ is a proportionality factor. In particular,

$$\frac{d}{d\lambda}s^0 = -\zeta\mathbf{r}^0. \tag{20}$$

The proportionality factor ζ is determined by differentiating (19) with λ and using (20). We have

$$1 = -\zeta \mathbf{r}^0 \cdot \nabla c^0 \tag{21}$$

and (20) is rewritten as

$$\frac{d}{d\lambda}\boldsymbol{s}^0 = \{\mathbf{r}^0 \cdot \nabla c^0\}^{-1}\, \mathbf{r}^0. \tag{22}$$

It is clear that ∇ here is the gradient with respect to the components of $\boldsymbol{s}$ because c depends on $\boldsymbol{s}$, not on $\mathbf{u}$. Integration of (20) or (22) furnishes the simple wave curves in the stress space [1,2].

Inserting $(13)_2$ in (15) we obtain the differential equation for $\mathbf{U}^1$ for the singular region as

$$(\lambda \mathbf{A}^0 + \mathbf{I})\frac{d}{d\lambda}\mathbf{U}^1 = \mathbf{A}^0\mathbf{U}^1 - \lambda\, \mathbf{U}^1 \cdot (\nabla \mathbf{A}^0)\frac{d}{d\lambda}\mathbf{U}^0 \tag{23}$$

which is linear in $\mathbf{U}^1$. Hence if $\mathbf{U}^1$ is a solution, so is $\mathrm{k}\mathbf{U}^1$ where k is an arbitrary constant. Despite the fact that the coefficient of $d\mathbf{U}^1/d\lambda$ in (23) is singular, it is shown in [13] that (23) can be solved for $d\mathbf{U}^1/d\lambda$ and that the solution for $\mathbf{U}^1$ is unique.

3. Discontinuity of $\mathbf{U}^1$ between the singular and regular regions. We consider $\mathbf{U}(x,t)$ on a curve Γ in the x-t plane which is given by

$$x_\Gamma = \lambda_\Gamma t + \frac{1}{2}\kappa_\Gamma t^2 + \cdots$$

where λ_Γ and κ_Γ are constant. The curvature of x_Γ at $t = 0$ is κ_Γ. Hence

$$\lambda = x_\Gamma/t = \lambda_\Gamma + \frac{1}{2}\kappa_\Gamma t + \cdots$$

and, from (10), $\mathbf{U}(x,t)$ on Γ is

$$\mathbf{U}(x_\Gamma, t) = \mathbf{U}^0(\lambda_\Gamma) + t\ \{\mathbf{U}^1 + \frac{1}{2}\kappa_\Gamma \frac{d}{d\lambda}\mathbf{U}^0\}\Big|_{\lambda=\lambda_\Gamma} + O(t^2).$$

If Γ is a boundary between a regular region and a singular region and if Γ is not a shock wave, $\mathbf{U}(x,t)$ and $\mathbf{U}^0(\lambda_\Gamma)$ are continuous across Γ. We then have

$$\{\mathbf{U}^1\}_{\text{regular region}} = \{\mathbf{U}^1 + \frac{1}{2}\kappa_\Gamma \frac{d}{d\lambda}\mathbf{U}^0\}_{\text{singular region}} \tag{24}$$

This shows that $\mathbf{U}^1$ is in general discontinuous across the boundary between a regular region and a singular region even when $\mathbf{U}(x,t)$ is continuous.

4. Perturbation of the shock wave solution. The Rankine-Hugoniont equation for (1) is

$$[\mathbf{U}] + V^*[\mathbf{f}(\mathbf{U})] = \mathbf{0} \tag{25}$$

in which V^* is the shock wave speed and the square brackets denote the jump across the shock wave. Employing the 2-vectors (3), (25) can be written as

$$\left.\begin{aligned} [\boldsymbol{s}] + \rho V^*[\mathbf{u}] &= \mathbf{0} \\ [\mathbf{u}] + V^*[\mathbf{p}] &= \mathbf{0}. \end{aligned}\right\} \tag{26}$$

Let $x_v(t)$ be the locus of the shock wave in the x-t plane and be given by

$$x_v = V\,t + \frac{1}{2}\kappa_v\,t^2 + \cdots$$

where V, κ_v are , respectively, the initial velocity and the initial acceleration of the shock wave at t = 0. The shock wave speed V^* at anytime t is

$$V^* = V + \kappa_v\,t + \cdots \tag{27}$$

Equation (10) written for $\mathbf{u}$ and $\boldsymbol{s}$ are

$$\left.\begin{aligned} \mathbf{u}(x,t) &= \mathbf{u}^0 + t\,\mathbf{u}^1 + \frac{t^2}{2}\mathbf{u}^2 + \cdots \\ \boldsymbol{s}(x,t) &= \boldsymbol{s}^0 + t\,\boldsymbol{s}^1 + \frac{t^2}{2}\boldsymbol{s}^2 + \cdots . \end{aligned}\right\} \tag{28}$$

With the notations $(3)_3$, (2) can be written as

$$p_i = \frac{\partial W}{\partial s_i}$$

and the expression of $\mathbf{p}$ in the form (10) leads to

$$\begin{aligned} \mathbf{p}(\boldsymbol{s}) &= \mathbf{p}^0 + t\,\mathbf{p}^1 + \frac{t^2}{2}\,\mathbf{p}^2 + \cdots, \\ \mathbf{p}^0 &= \mathbf{p}(\boldsymbol{s}^0), \qquad \mathbf{p}^1 = \mathbf{G}^0\boldsymbol{s}^1, \\ p_i^2 &= G_{ij}^0\, s_j^2 + \frac{\partial^3 W}{\partial s_i \partial s_j \partial s_k} s_j^1 s_k^1. \end{aligned} \tag{29}$$

In the above, $\mathbf{G}^0 = \mathbf{G}(\boldsymbol{s}^0)$. Substituting (27) -(29) into (26), we have

$$\left.\begin{aligned} [\boldsymbol{s}^0] + \rho V[\mathbf{u}^0] = \mathbf{0} \\ [\mathbf{u}^0] + V[\mathbf{p}^0] = \mathbf{0} \end{aligned}\right\} \tag{30}$$

for the zero order and

$$\left.\begin{aligned} [\boldsymbol{s}^1] + \rho V[\mathbf{u}^1] + \rho\kappa_v[\mathbf{u}^0] = \mathbf{0} \\ [\mathbf{u}^1] + V[\mathbf{G}^0\boldsymbol{s}^1] + \kappa_v[\mathbf{p}^0] = \mathbf{0} \end{aligned}\right\} \tag{31}$$

for the first order.

From (30) we obtain

$$[\mathbf{u}^0] = -\frac{1}{\rho V}[\boldsymbol{s}^0], \qquad [\mathbf{p}^0] = \frac{1}{\rho V^2}[\boldsymbol{s}^0], \tag{32}$$

and the use of these results in (31) leads to

$$\left.\begin{aligned} \rho V[\mathbf{u}^1] &= -[\boldsymbol{s}^1] + (\kappa_v/V)[\boldsymbol{s}^0], \\ [\mathbf{u}^1] &= -V[\mathbf{G}^0\boldsymbol{s}^1] - (\kappa_v/\rho V^2)[\boldsymbol{s}^0]. \end{aligned}\right\} \tag{33}$$

Elimination of $[\mathbf{u}^1]$ between the two equations in (33) yields

$$\rho V^2[\mathbf{G}^0\boldsymbol{s}^1] - [\boldsymbol{s}^1] = -2(\kappa_v/V)[\boldsymbol{s}^0]. \tag{34}$$

Let $\boldsymbol{s}^+$,$\boldsymbol{s}^-$ be , respectively, the value of $\boldsymbol{s}$ in front of and behind the shock wave. When $\boldsymbol{s}^+$ is given, $(32)_2$ provides a solution for $\boldsymbol{s}^-$ with V as the parameter.

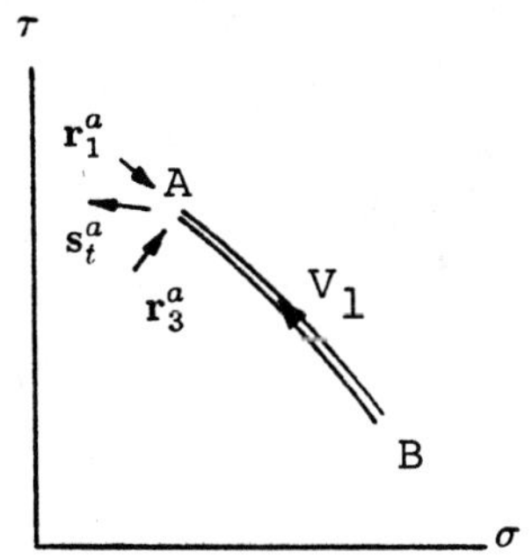

Fig.1 The shock wave curve in the stress space

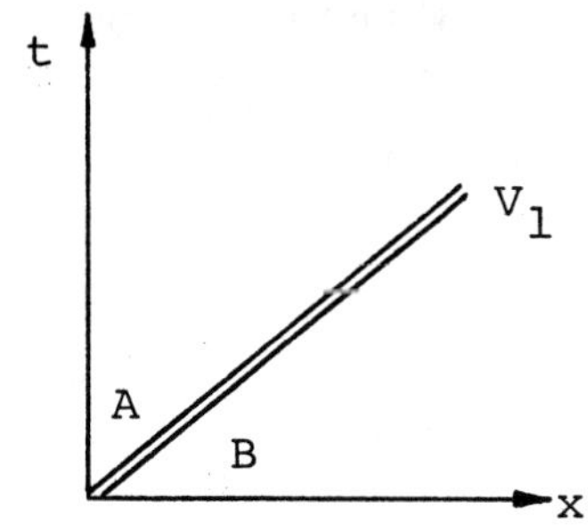

Fig.2 The solution for the Goursat-Riemann problem when $s_t^a = \mathbf{0}$

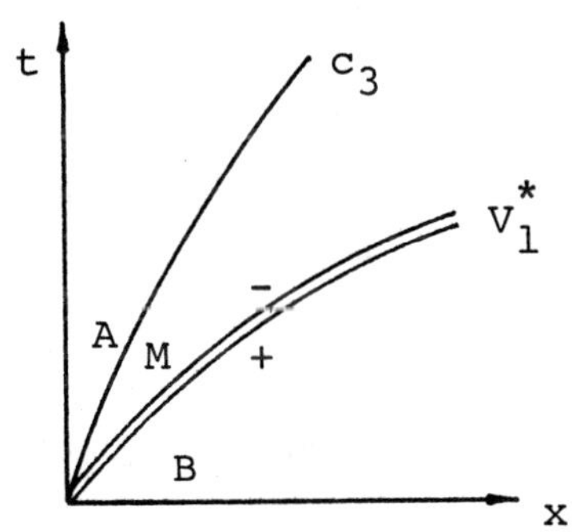

Fig.3 The solution for the generalized Goursat-Riemann problem with $s_t^a \neq \mathbf{0}$

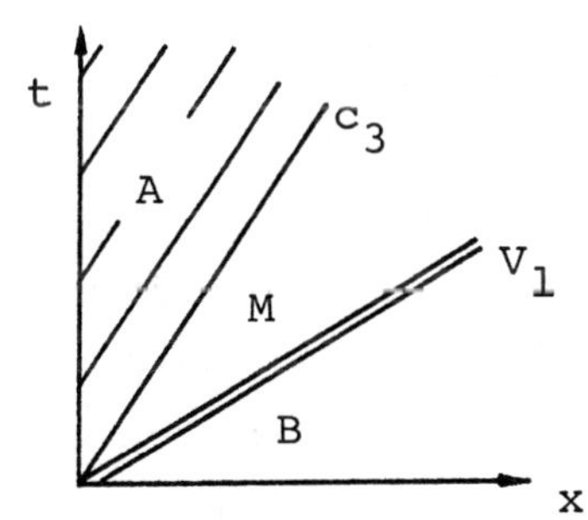

Fig.4 The solution to the special case when $s_t^a = \gamma_3^a r_3^a$

As V varies, $\boldsymbol{s}^-$ traces out a curve in the stress space which is the shock wave curve, Fig. 1. Equation $(32)_1$ then gives the jump in $\mathbf{u}^0$ while (34) and $(33)_1$ or $(33)_2$ give the jump in $\boldsymbol{s}^1$ and $\mathbf{u}^1$ in terms of κ_v.

5. Growth or decay of the shock wave. When the initial acceleration κ_v of the shock wave is non-zero, the shock wave grows or decays as it propagates. If the state ahead of the shock wave is at rest or at a uniform state, the stress amplitude and velocity amplitude in general increase or decrease according to whether the acceleration κ_v is positive or negative. If the state ahead of the shock wave is not a uniform state, the positivity of κ_v may not be accompanied by the growth of the stress amplitude and/or the velociy amplitude. This was pointed out in [16] for one-dimensional shock waves and in [17-19] for three-dimensional shock waves. Thus the definition of growth or decay of a shock wave depends on whether one is looking at the stress amplitude, the velocity amplitude or the acceleration κ_v.

In this paper we define the growth or decay of shock waves according to whether κ_v is positive or negative.

6. The solution to the generalized Goursat-Riemann problem. The perturbation $\mathbf{U}^1$ presented in Sections 2, 4 and the discontinuity condition of $\mathbf{U}^1$ between the singular and regular regions derived in Section 3 are the basis for the constructing the solution $\mathbf{U}^1$ to the generalized Goursat-Riemann problem prescribed in (9). As an illustration, consider the special case in which $\boldsymbol{s}^a, \boldsymbol{s}^b$ in (9) are located on a shock wave curve associated with c_1 which is denoted by V_1 in Fig. 1. If $\boldsymbol{s}_t^a = \mathbf{0}$, the solution in the x-t plane consists of a single shock wave of wave speed V_1, Fig. 2. Since $\boldsymbol{s}_t^a \neq \mathbf{0}$, we would have the wave pattern shown in Fig. 3 in which there is a non-constant shock wave V_1^* followed by an acceleration wave c_3. The acceleration wave c_1 does not appear because, by Lax stability conditions [20], the c_1 wave speed behind the shock wave is larger than the shock wave speed V_1^*. The region between the V_1^* shock wave and the c_3 acceleration wave denoted by M is a regular region which is not a constant state. We will determinate the initial acceleration κ_v of the V_1^* shock and $\boldsymbol{s}_t^m$ which is the value of $\boldsymbol{s}_t$ in the region M as $t \to 0$. In the sequel, the superscripts $a, b, m, \cdots$ denote the value of the quantity concerned in the regions $A, B, M, \cdots$, respectively, as $t \to 0$.

The first step is to find $\mathbf{U}^{1^-}$ from $\mathbf{U}^{1^+}$ across the shock wave V_1^*. Applying (34) to the shock wave V_1^* at $t = 0$ and noticing that the region B is a constant state which means that $\mathbf{U}^{1^+} = \mathbf{0}$, we have

$$\rho V_1^2 \mathbf{G}^a \boldsymbol{s}^{1^-} - \boldsymbol{s}^{1^-} = -2\,(\kappa_v / V_1)\,[\boldsymbol{s}^0]. \tag{35}$$

The regions M and A are both regular regions and hence $\boldsymbol{s}^a = \boldsymbol{s}^m$ which leads to $\mathbf{G}^a = \mathbf{G}^m$. Let

$$\boldsymbol{s}^{1^-} = \gamma_1^- \mathbf{r}_1^a + \gamma_3^- \mathbf{r}_3^a, \tag{36}$$

$$[\boldsymbol{s}^0] = \boldsymbol{s}^a - \boldsymbol{s}^b = -(\gamma_1^0 \mathbf{r}_1^a + \gamma_3^0 \mathbf{r}_3^a), \tag{37}$$

in which $\gamma_1^-, \gamma_3^-, \gamma_1^0, \gamma_3^0$ are constant and $\mathbf{r}_1^a, \mathbf{r}_3^a$ are the right eigenvectors of (7) associated with $c = c_1^a, c_3^a$, i.e.,

$$\left.\begin{aligned} \{\rho(c_1^a)^2\mathbf{G}^a - \mathbf{I}\}\, \mathbf{r}_1^a &= \mathbf{0}, \\ \{\rho(c_3^a)^2\mathbf{G}^a - \mathbf{I}\}\, \mathbf{r}_3^a &= \mathbf{0}. \end{aligned}\right\} \tag{38}$$

In the above it is understood that $c_1^- = c_1^a$ and $c_3^- = c_3^a$. The direction of $\mathbf{r}_1^a$ and $\mathbf{r}_3^a$ are taken to be in the direction of decreasing c_1 and c_3, respectively. With $[\boldsymbol{s}^0]$ given in (9), γ_1^0 and γ_3^0 are determined from (37). For the V_1 shock wave and the $\mathbf{r}_1^a, \mathbf{r}_3^a$ shown in Fig. 1, we have

$$\gamma_1^0 > 0. \tag{39}$$

Using (36) -(38), (35) can be written as

$$\begin{aligned} &\{(V_1/c_1^a)^2 - 1\}\, \gamma_1^- \mathbf{r}_1^a + \{(V_1/c_3^a)^2 - 1\}\, \gamma_3^- \mathbf{r}_3^a \\ &= 2\,(\kappa_v/V_1)\,(\gamma_1^0 \mathbf{r}_1^a + \gamma_3^0 \mathbf{r}_3^a). \end{aligned}$$

Hence γ_1^- and γ_3^- are given by

$$\left.\begin{aligned} \{(V_1/c_1^a)^2 - 1\}\, \gamma_1^- &= 2\,(\kappa_v/V_1)\, \gamma_1^0 \\ \{(V_1/c_3^a)^2 - 1\}\, \gamma_3^- &= 2\,(\kappa_v/V_3)\, \gamma_3^0 \end{aligned}\right\} \tag{40}$$

with κ_v as an unknown parameter. We have thus obtained $\boldsymbol{s}^{1^-}$ of (36). As to $\mathbf{u}^{1^-}$, using (36) and (37), we write $(33)_1$ as

$$\rho V_1 \mathbf{u}^{1^-} = -(\gamma_1^- \mathbf{r}_1^a + \gamma_3^- \mathbf{r}_3^a) - (\kappa_v/V_1)(\gamma_1^0 \mathbf{r}_1^a + \gamma_3^0 \mathbf{r}_3^a).$$

We are interested in $\rho V_1 \mathbf{u}^{1^-} - \boldsymbol{s}^{1^-}$, not in $\rho V_1 \mathbf{u}^{1^-}$. Therefore we employ (36) and (40) to obtain

$$\rho V_1 \mathbf{u}^{1^-} - \boldsymbol{s}^{1^-} = -\frac{(V_1/c_1^a)^2+3}{(V_1/c_1^a)^2-1}\left(\frac{\kappa_v}{V_1}\right)\gamma_1^0\,\mathbf{r}_1^a - \frac{(V_1/c_3^a)^2+3}{(V_1/c_3^a)^2-1}\left(\frac{\kappa_v}{V_1}\right)\gamma_3^0\,\mathbf{r}_3^a. \tag{41}$$

The next step is to convert $\mathbf{U}^{1^-}$ to $\mathbf{U}_t^m$ using (18) which is written in full as

$$\boldsymbol{s}^{1^-} = \boldsymbol{s}_t^m + \rho V_1 \mathbf{u}_t^m,$$

$$\mathbf{u}^{1^-} = V_1\,\mathbf{G}^a \boldsymbol{s}_t^m + \mathbf{u}_t^m.$$

Eliminating $\mathbf{u}_t^m$ leads to

$$\rho V_1 \mathbf{u}^{1^-} - \boldsymbol{s}^{1^-} = (\rho V_1^2 \mathbf{G}^a - \mathbf{I})\,\boldsymbol{s}_t^m. \tag{42}$$

To determine $\boldsymbol{s}_t^m$ explicitly, we let

$$\boldsymbol{s}_t^m = \gamma_1^m\,\mathbf{r}_1^a + \gamma_3^m\,\mathbf{r}_3^a, \qquad \boldsymbol{s}_t^a = \gamma_1^a\,\mathbf{r}_1^a + \gamma_3^a\,\mathbf{r}_3^a, \tag{43}$$

where $\gamma_1^m, \gamma_3^m, \gamma_1^a, \gamma_3^a$ are constant. With $\boldsymbol{s}_t^a$ prescribed in (9), γ_1^a and γ_3^a are determined from $(43)_2$. The boundary between the region A and region M is a c_3 characteristic. This means that the discontinuity in $\boldsymbol{s}_t^a$ and $\boldsymbol{s}_t^m$ must be proportional to $\mathbf{r}_3^a$. Hence

$$\gamma_1^m = \gamma_1^a. \tag{44}$$

Equation (42) with the aids of (38) and $(43)_1$ has the expression

$$\rho V_1 \mathbf{u}^{1^-} - \boldsymbol{s}^{1^-} = \{(V_1/c_1^-)^2 - 1\}\,\gamma_1^m \mathbf{r}_1^a + \{(V_1/c_3^-)^2 - 1\}\,\gamma_3^m \mathbf{r}_3^a$$

and from (41) we obtain

$$\left.\begin{aligned} \gamma_1^a &= -\frac{(V_1/c_1^a)^2+3}{\{(V_1/c_1^a)^2-1\}^2}\left(\frac{\kappa_v}{V_1}\right)\gamma_1^0 = \gamma_1^m, \\ \gamma_3^m &= -\frac{(V_1/c_3^a)^2+3}{\{(V_1/c_3^a)^2-1\}^2}\left(\frac{\kappa_v}{V_1}\right)\gamma_3^0. \end{aligned}\right\} \tag{45}$$

Equation $(45)_1$ provides the initial acceleration κ_v of the shock wave V_1^*. Since γ_1^0 is positive by (39), κ_v and γ_1^a have the opposite signs. In other words, the shock wave V_1^* grows when $\gamma_1^a < 0$ and decays when $\gamma_1^a > 0$. For the $\boldsymbol{s}_t^a$ shown in Fig. 1, $\gamma_1^a < 0$ and the shock wave V_1^* grows as indicated by the positive curvature of V_1^* in Fig. 3. The value of γ_3^a has no effects on the growth or decay of shock wave V_1^*.

If $\gamma_1^a = 0, \kappa_v = 0$ and $\boldsymbol{s}_t^m = \mathbf{0}$. The wave pattern in the x-t plane is then given in Fig. 4 in which the region M is a constant state and the region A has a non-centered simple wave solution.

7. Concluding remarks. The example given in the last section illustrates how one can determine the growth or decay of a shock wave in the generalized Goursat-Riemann problem. The example was one of the simplest cases in which the two stress states $\boldsymbol{s}^a$ and $\boldsymbol{s}^b$ are on a shock wave curve V_1. Other simple cases have the two stress states on a shock wave curve V_3 or on a simple wave curve c_1 or c_3. For the latter, the discontinuity condition (24) for $\mathbf{U}^1$ will be needed. More complicated cases would have the two stress states at arbitrary locations in the stress space in which $\boldsymbol{s}^a$ can be reached from $\boldsymbol{s}^b$ through a series of simple wave curves and/or shock wave curves. More studies on the subject is being undertaken and results will be reported in the future.

Acknowledgements The work presented here is supported by the U.S. Air Force Office of Scientific Research under Contract AFSOR-89-0013.

REFERENCES

[1] Yongchi Li and T. C. T. Ting, *Plane waves in simple elastic solids and discontinuous dependence of solution on boundary conditions*, Int. J. Solids Structures, **19** (1983), 989–1008.

[2] Zhijing Tang and T. C. T. Ting, *Wave curves for the Riemann problem of plane waves in isotropic elastic solids*, Int. J. Eng. Sci., **25** (1987), 1343–1381.

[3] C. Truesdell and W. Noll, "The Nonlinear Field Theories of Mechanics," Handbuch der Physik, Vol.III/3, Springer, Berlin (1965).

[4] R. Courant and D. Hilbert, "Methods of Mathematical Physics," Vol.II, Interscience, New York (1962).

[5] T. -P. Liu, *The Riemann problem for general* 2×2 *conservation laws*, Trans. Amer. Math. Soc., **199** (1974), 89–112.

[6] D. G. Schaeffer and M. Shearer, *The classification of* 2×2 *systems of nonstrictly hyperbolic conservation laws, with application to oil recovery*, Appendix with D. Marchesin and P. J. Paes-Leme, Comm. Pure. Appl. Math., **40** (1987), 141–178.

[7] M. Shearer, D. G. Schaeffer, D. Marchesin and P.J. Paes-Leme, *Solution of the Riemann problem for a prototype* 2×2 *system of nonstrictly hyperbolic conservation laws*, Arch. Rat. Mech. Anal., **97** (1987), 299–320.

[8] D. G. Schaeffer and M. Shearer, *Riemann problems for nonstrictly hyperbolic 2×2 systems of conservation laws*, Trans. Amer. Math. Soc., **304** (1987), 267–306.

[9] Guangshan Zhu and T. C. T. Ting, *Classification of 2×2 nonstrictly hyperbolic systems for plane waves in isotropic elastic solids*, Int. J. Eng. Sci., **27** (1989), 1621–1638.

[10] C. M. Dafermos, "Hyperbolic systems of conservation laws," Brown Univ. Report, LCDS 83-5 (1983).

[11] Xabier Garaizar, *Solution of a Riemann problem for elasticity*, J. Elasticity. (in press).

[12] T. C. T. Ting and Tankin Wang, *The Goursat-Riemann problem for plane waves in isotropic elastic solids with velocity boundary conditions*, in IMA Volumes in Mathematics and Its Applications, **29**, James Glimm and Andrew J. Majda, eds., Springer-Verlag (1991),367-386.

[13] T. C. T. Ting, *The initiation of combined stress waves in a thin walled tube due to impact loadings*, Int. J. Solids Structures, **8** (1972), 269–293.

[14] Ph. Le Floch and P. A. Raviart, *An asymptotic expansion for the solution of the generalized Riemann problem. Part 1: general theory*, Ann. Inst. H. Poincare, Nonlinear Analysis. (in press).

[15] A. Bourgeade, Ph. Le Floch and P. A. Raviart, *An asymptotic expansion for the solution of the generalized Riemann problem. Part 2: application to the equations of gas dynamics*, Ann. Inst. H. Poincare. (in press).

[16] T. C. T. Ting, *Further study on one-dimensional shock waves in nonlinear elastic media*, Q. Appl. Math., **37** (1980), 421–429.

[17] T. C. T. Ting, *Intrinsic description of three-dimensional shock waves in elastic fluids*, Int. J. Eng. Sci., **19** (1981), 629–638.

[18] Yongchi Li and T. C. T. Ting, *Lagrangian description of transport equations for shock waves in three-dimensional elastic solids*, Appl. Math. & Mech., **3** (1982), 491–506.

[19] T. C. T. Ting and Yongchi Li, *Eulerian formulation of transport equations for three-dimensional shock waves in simple elastic solids*, J. Elasticity (1983), 295–310.

[20] P. D. Lax, *Hyperbolic systems of conservation laws II*, Comm. Pure Appl. Math., **10** (1957), 537–566.

CHAPTER 16

A Comparison of Two Numerical Methods for Shocks in One-Dimensional Elastic-Plastic Solids*

John A. Trangenstein†

Abstract. In this paper we compare and contrast a second-order Godunov method with a second-order explicit finite element method for the dynamic response of one-dimensional elastic-plastic solids. We examine the numerical dissipation and dispersion of the two techniques, and compare their performance on problems describing the nonlinear response of an elastic-plastic bar

1. Introduction. The traditional approach to the numerical simulation of shocks in solids is to use piecewise linear finite elements for the displacements (*i.e.*, centered finite differences) in the spatial discretization, coupled with some form of the method of lines in time [3,5,33]. When central differences are used for time integration, the method is also related to the well-established von Neumann-Richtmyer method for hydrodynamics [29]. This approach is based on the success of finite element methods in handling complicated geometry in steady-state calculations, where the finite element method has been rigorously studied [2,6]. However, these methods typically produce significant numerical oscillations in hyperbolic problems near discontinuities, unless stabilized by artificial viscosities.

In a series of papers [24,25,27], we have studied the application of second-order Godunov methods to one-dimensional shocks in elastic-plastic solids. Our confidence in the successful application of second-order Godunov methods to solid mechanics derives from the considerable success this method has already demonstrated in recent years when applied to gas dynamics; see, for example, [32]. Our goal in this paper is to compare and contrast this method with the traditional finite element methods for shocks in one-dimensional solids. Most of the analysis of the methods in this paper is already known, although not collected together. On the other hand, the numerical results at the end of the paper are new, and the analysis at the beginning of the paper helps the reader to understand the results of the

* This work was performed under the auspices of the U.S. Department of Energy by the Lawrence Livermore National Laboratory under contract No. W-7405-Eng-48. Partial support was provided by the Applied Mathematical Sciences Program of the Office of Energy Research, and by the Defense Nuclear Agency.

† Computing and Mathematics Research Division, Lawrence Livermore National Laboratory L-316, P.O. Box 808, Livermore, California 94550. E-mail address: johnt@lll-crg.llnl.gov.

computations.

We would like to remark at the outset that we have not attempted to compare the second-order Godunov methods with all of the variations available to methods based on finite elements in space and method of lines in time, or with other staggered mesh schemes. There is ongoing work in the area of improving the time-integration of finite elements; see [15] for example. We would also like to mention the ongoing work in applying characteristic-based finite element methods (see, for example, [19]), and streamline diffusion (see, for example, [16]). However, most of these modifications of the finite element method are implicit, and can become computationally inferior to explicit methods for problems involving shock propagation [33].

Throughout this paper, we will apply the methods to the equations of motion for longitudinal deformation of a one-dimensional solid. In order to describe these equations, we let v be the velocity, σ be the stress, ϵ be the displacement gradient, ρ be the density at rest, τ denote time and x denote distance in the original configuration. Then conservation of momentum and equality of mixed partial derivatives of the displacement yield the following first-order system of conservation laws:

$$\frac{\partial}{\partial \tau}\begin{bmatrix} \rho \mathrm{v} \\ \epsilon \end{bmatrix} - \frac{\partial}{\partial x}\begin{bmatrix} \sigma \\ \mathrm{v} \end{bmatrix} = 0 . \tag{1.1}$$

We will assume that the stress is a function of the displacement gradient

$$\epsilon = \frac{\partial \mathrm{d}}{\partial x} ,$$

and possibly a history parameter π:

$$\sigma = \sigma(\epsilon, \pi) . \tag{1.2}$$

We also assume that for fixed history, stress is a nondecreasing function of strain:

$$\frac{\partial \sigma}{\partial \epsilon} \geq 0 .$$

When it becomes necessary, we will present more details of the material models. Note that this formulation of the problem is equivalent to the second-order formulation

$$\rho \frac{\partial^2 \mathrm{d}}{\partial \tau^2} - \frac{\partial}{\partial x}\sigma(\frac{\partial \mathrm{d}}{\partial x}, \pi) = 0 , \tag{1.3}$$

provided that the initial value of the displacement gradient ϵ in (1.1) is actually the gradient of the initial displacement d.

It is easy to see that (1.1) is hyperbolic. We can write this equation in the form

$$\frac{\partial}{\partial \tau}\begin{bmatrix} \rho \mathrm{v} \\ \epsilon \end{bmatrix} - \begin{bmatrix} 0 & \frac{\partial \sigma}{\partial \epsilon} \\ \frac{1}{\rho} & 0 \end{bmatrix} \frac{\partial}{\partial x}\begin{bmatrix} \rho \mathrm{v} \\ \epsilon \end{bmatrix} = 0 ,$$

where the eigenvectors and eigenvalues of the matrix of flux derivatives are given by

$$\begin{bmatrix} 0 & \frac{\partial \sigma}{\partial \epsilon} \\ \frac{1}{\rho} & 0 \end{bmatrix}\begin{bmatrix} \lambda\rho & \lambda\rho \\ -1 & 1 \end{bmatrix} = \begin{bmatrix} \lambda\rho & \lambda\rho \\ -1 & 1 \end{bmatrix}\begin{bmatrix} -\lambda & 0 \\ 0 & \lambda \end{bmatrix} .$$

Since stress is an increasing function of displacement gradient, the characteristic speed

$$\lambda = \sqrt{\frac{1}{\rho}\frac{\partial \sigma}{\partial \epsilon}}$$

is real.

There are interesting theoretical results regarding this problem. For example, the sum of the kinetic and strain energy per volume,

$$v = \frac{1}{2}\rho \mathrm{v}^2 + \int_{\epsilon_0}^{\epsilon} \sigma d\epsilon \ ,$$

serves as an "entropy function" [9] for (1.1). By this we mean that

$$\frac{\partial v}{\partial \begin{bmatrix} \rho \mathrm{v} \\ \epsilon \end{bmatrix}} \frac{\partial \begin{bmatrix} -\sigma \\ -\mathrm{v} \end{bmatrix}}{\partial \begin{bmatrix} \rho \mathrm{v} \\ \epsilon \end{bmatrix}} = \frac{\partial \phi}{\partial \begin{bmatrix} \rho \mathrm{v} \\ \epsilon \end{bmatrix}} ,$$

where the "entropy flux" ϕ is the rate of work per volume,

$$\phi = -\mathrm{v}\sigma \ .$$

Since stress is a nondecreasing function of strain, the entropy function is convex.

In physical situations, the motion is subject to some (typically small) amount of physical viscosity:

$$\frac{\partial}{\partial \tau}\begin{bmatrix} \rho \mathrm{v} \\ \epsilon \end{bmatrix} - \frac{\partial}{\partial x}\begin{bmatrix} \sigma \\ \mathrm{v} \end{bmatrix} = \delta \frac{\partial^2}{\partial x^2}\begin{bmatrix} \mathrm{v} \\ 0 \end{bmatrix} . \tag{1.4}$$

Since the entropy function is convex, it can be shown [18] that the solution to (1.4) obtained in the limit of vanishing viscosity δ satisfies the inequality

$$\frac{\partial v}{\partial \tau} + \frac{\partial \phi}{\partial x} \leq 0 \tag{1.5}$$

at points where the solution is smooth. This implies that the total energy, $\int v dx$, can only decrease in time.

At a discontinuity, the solution to (1.1) satisfies the jump conditions [8]

$$-\begin{bmatrix} 0 \\ \sigma_R - \sigma_L \\ \mathrm{v}_R - \mathrm{v}_L \end{bmatrix} = \begin{bmatrix} \rho_R - \rho_L \\ \rho_R \mathrm{v}_R - \rho_L \mathrm{v}_L \\ \epsilon_R - \epsilon_L \end{bmatrix} s \ , \tag{1.6}$$

where the subscripts label the states on the two sides of the discontinuity, and s is the speed of the discontinuity. Thus, at a zero-speed discontinuity (*i.e.*, a contact discontinuity), the velocity and stress are continuous while the density, strain and displacement may have jumps. At a moving discontinuity the density is continuous while the velocity, stress, strain and displacement may jump. In this case, we can write the shock speed in the form

$$s = \pm\sqrt{\frac{1}{\rho}\frac{\sigma_R - \sigma_L}{\epsilon_R - \epsilon_L}} \ .$$

At a discontinuity achieved in the limit of vanishing viscosity, the entropy function satisfies

$$(\phi_R - \phi_L) \leq (v_R - v_L)s\ .$$

Using the entropy condition (1.5), the jump conditions (1.6) and the mean-value theorem, it is possible to show that the chord condition

$$0 \leq s\left\{\frac{\sigma_* - \sigma_L}{\epsilon_* - \epsilon_L} - \frac{\sigma_R - \sigma_L}{\epsilon_R - \epsilon_L}\right\} \tag{1.7}$$

must be satisfied for some ϵ_* between ϵ_L and ϵ_R. In fact, this condition holds for all such ϵ_* if and only if the entropy condition is satisfied [9].

Using the conservation law and the chord condition (1.7), it is possible to develop an analytic solution for the time evolution of an initial discontinuity [26]. This is called a Riemann problem, and its solution forms an important part of the Godunov method.

The outline of the paper is as follows. In §2, we will describe the second-order finite element method and the second-order Godunov method. In the two sections after that, we will study the modified equation analysis and Fourier mode analysis for these two methods. Finally, in §5 we will present computational results on a model for an elastic-plastic bar.

2. Algorithms. The traditional development of finite element methods for the solution of (1.3) involves the selection of piecewise-polynomial basis functions, quadrature rules for the integration of the weak form of the partial differential equation in space, and a technique for time integration. Rather than go through this development, we will summarize our selections in words and proceed directly to the resulting difference equations. We have used piecewise-linear elements in space, with the midpoint rule for the integrals involving the stresses, and the trapezoidal rule for the mass matrix (*i.e.*, lumping). Also, we have used central differences for the time discretization. Other common, but implicit, choices for time discretization are due to Houbolt, to Wilson and to Newmark [3,5]. The method we will discuss is equivalent to Wilkin's finite difference method [31], which was in turn based on the von Neumann-Richtmyer method [29] for gas dynamics. Note that a large variety of alternative basis elements and formulation of the equations are possible; we have selected what appears to be the most common form of the finite element method for shock propagation problems.

The finite difference equations representing the finite element method for solving (1.3) are given by

$$\rho_i \frac{\mathrm{d}_i^{k+1} - 2\mathrm{d}_i^k + \mathrm{d}_i^{k-1}}{\Delta\tau^2} - \frac{\sigma_{i+\frac{1}{2}}^k - \sigma_{i-\frac{1}{2}}^k}{\Delta x} = 0\ . \tag{2.1}$$

Here, the stress is defined by

$$\begin{aligned} \epsilon_{i+\frac{1}{2}}^k &= \frac{\mathrm{d}_{i+1}^k - \mathrm{d}_i^k}{\Delta x}\ , \\ \sigma_{i+\frac{1}{2}}^k &= \sigma(\epsilon_{i+\frac{1}{2}}^k\ ,\ \pi_{i+\frac{1}{2}}^k)\ . \end{aligned} \tag{2.2}$$

We have used the subscript i to denote the index of the cell edges, and the superscript k to denote the timestep number. This method (2.1) is usually written in the alternate form

$$\mathrm{v}_i^{k+\frac{1}{2}} = \mathrm{v}_i^{k-\frac{1}{2}} + \Delta\tau\frac{\sigma_{i+\frac{1}{2}}^k - \sigma_{i-\frac{1}{2}}^k}{\rho_i \Delta x}\ ,$$

(2.3)
$$d_i^{k+1} = d_i^k + \Delta\tau v_i^{k+\frac{1}{2}} .$$

One difference between (2.1) and (2.3) is that the former stores two time levels of displacements, while the latter stores an array of displacements and an array of velocities. Another difference is that it is easier to incorporate artificial viscosity using (2.3).

One of the fundamental distinctions between the finite element method and the Godunov method lies in the centering of the computed quantities. For example, the velocities and displacements in the finite element method are at the cell edges and half-time levels, while the strains and stresses are located at the cell centers and full-time levels. Thus, if we attempt to write the finite element method in a form analogous to (1.1), we obtain

(2.4)
$$\begin{aligned} \rho_i \frac{v_i^{k+\frac{1}{2}} - v_i^{k-\frac{1}{2}}}{\Delta\tau} - \frac{\sigma_{i+\frac{1}{2}}^k - \sigma_{i-\frac{1}{2}}^k}{\Delta x} &= 0 , \\ \frac{\epsilon_{i+\frac{1}{2}}^{k+1} - \epsilon_{i+\frac{1}{2}}^k}{\Delta\tau} - \frac{v_{i+1}^{k+\frac{1}{2}} - v_i^{k+\frac{1}{2}}}{\Delta x} &= 0 . \end{aligned}$$

Note that the former equation in (2.4) is centered at a full-time level and cell edge, while the latter is centered at a half-time level and cell center. The former of the equations in (2.4) shows that the finite element method conserves discrete momentum.

The continuous-in-time form of the finite element method also conserves the sum of kinetic and strain energy, possibly in violation of the entropy inequality at shocks [22]. However, it is more difficult to show that for smooth motion the fully discrete method conserves some discrete measure of energy at each timestep, or that at discontinuities the discrete method observes some discrete form of the entropy inequality.

Note that the finite element method requires a startup procedure to initialize the displacements (or velocities). One simple alternative is to take

$$v_i^{\frac{1}{2}} = v_i^0 + \frac{1}{2}\Delta\tau \frac{\sigma_{i+\frac{1}{2}}^0 - \sigma_{i-\frac{1}{2}}^0}{\rho_i \Delta x} ,$$

$$d_i^1 = d_i^0 + \Delta\tau v_i^{\frac{1}{2}} .$$

It is also useful to employ a similar formula to compute velocities at full-time levels,

$$v_i^k = v_i^{k-\frac{1}{2}} + \frac{1}{2}\Delta\tau \frac{\sigma_{i+\frac{1}{2}}^k - \sigma_{i-\frac{1}{2}}^k}{\rho_i \Delta x} = \frac{v_i^{k+\frac{1}{2}} + v_i^{k-\frac{1}{2}}}{2} ,$$

in order to graph the velocities during the simulation.

Since it is difficult to verify that the numerical method satisfies the entropy inequality (1.5) or the chord condition (1.7), the most common technique for encouraging convergence to the correct discontinuities is to incorporate an artificial viscosity. In its time-centered form, the viscous terms would modify the stress as follows [3]:

(2.5)
$$\tilde{\sigma}_{i+\frac{1}{2}}^k = \sigma_{i+\frac{1}{2}}^k + \frac{1}{2}\alpha\lambda_{i+\frac{1}{2}}^k \rho_{i+\frac{1}{2}} \Delta x \left[\frac{v_{i+1}^{k+\frac{1}{2}} - v_i^{k+\frac{1}{2}}}{\Delta x} + \frac{v_{i+1}^{k-\frac{1}{2}} - v_i^{k-\frac{1}{2}}}{\Delta x} \right] .$$

This modified stress $\tilde{\sigma}^k_{i+\frac{1}{2}}$ then replaces $\sigma^k_{i+\frac{1}{2}}$ in (2.3). Here, α is a viscosity parameter that may vary with the motion, and λ is some estimate of the local wavespeed. Note that the viscous term in (2.5) (which is analogous to the introduction of a frictional force) does not introduce viscosity at a contact discontinuity. In order to avoid making the equations implicit, Wilkins [31] suggested using solely old information in computing the viscosity:

$$\tilde{\sigma}^k_{i+\frac{1}{2}} = \sigma^k_{i+\frac{1}{2}} + \alpha\lambda^k_{i+\frac{1}{2}}\rho_{i+\frac{1}{2}}\Delta x \frac{\mathrm{v}^{k-\frac{1}{2}}_{i+1} - \mathrm{v}^{k-\frac{1}{2}}_i}{\Delta x} . \tag{2.6}$$

In either of these forms (2.5) or (2.6), the modification of the stress has no effect on the conservation of momentum. On the other hand, it does destroy conservation of energy, by tending to enforce the entropy inequality (1.5).

In practice, the artificial viscosity is usually modified to be a nonlinear function of the velocities. A common example is the following [20]:

$$\tilde{\sigma}^k_{i+\frac{1}{2}} = \begin{cases} \sigma^k_{i+\frac{1}{2}} + \rho_{i+\frac{1}{2}}(\mathrm{v}^{k-\frac{1}{2}}_{i+1} - \mathrm{v}^{k-\frac{1}{2}}_i)\left[\alpha_1\lambda^k_{i+\frac{1}{2}} + \alpha_2 \mid \mathrm{v}^{k-\frac{1}{2}}_{i+1} - \mathrm{v}^{k-\frac{1}{2}}_i \mid\right] & , \quad \mathrm{v}^{k-\frac{1}{2}}_{i+1} < \mathrm{v}^{k-\frac{1}{2}}_i \\ \sigma^k_{i+\frac{1}{2}} & , \quad \mathrm{v}^{k-\frac{1}{2}}_{i+1} > \mathrm{v}^{k-\frac{1}{2}}_i \end{cases} .$$

For example, a previous version of the DYNA finite element code [12] took $\alpha_1 = 0.06$ and $\alpha_2 = 1.5$ as its defaults. There are a large number of variations of these ideas, far more than we can reasonably present in this work.

Without artificial viscosity, the timestep is chosen so that

$$\lambda\Delta\tau \leq \Delta x , \tag{2.7}$$

where λ is the largest elastic wavespeed in the problem. With artificial viscosity, it may be desirable to decrease the size of this timestep somewhat, as we will see in §3 and 4. For expansions, this wavespeed can be found by sampling the characteristic speeds at the cell centers; however, for compressions, the largest speed develops in the middle of the wave and must be estimated in order to start the calculation.

The second-order Godunov method for solving (1.1) can be written in the form

$$\begin{aligned} \rho_j \frac{\mathrm{v}^{k+1}_j - \mathrm{v}^k_j}{\Delta\tau} - \frac{\sigma^{k+\frac{1}{2}}_{j+\frac{1}{2}} - \sigma^{k+\frac{1}{2}}_{j-\frac{1}{2}}}{\Delta x} &= 0 , \\ \frac{\epsilon^{k+1}_j - \epsilon^k_j}{\Delta\tau} - \frac{\mathrm{v}^{k+\frac{1}{2}}_{j+\frac{1}{2}} - \mathrm{v}^{k+\frac{1}{2}}_{j-\frac{1}{2}}}{\Delta x} &= 0 . \end{aligned} \tag{2.8}$$

Here, the variables are subscripted with the cell index $j = i - \frac{1}{2}$, and superscripted with the timestep number k. The stress and velocity at the cell edge and half-time level are computed by a combination of characteristic tracing and the solution of a Riemann problem, both to be described below. Note that the second-order Godunov method conserves momentum, but not energy. In problems involving temperature effects, an additional equation for energy can be included in the system of conservation laws, and included in the system of difference equations for the Godunov method [25].

Note that (2.8) shows that discrete traveling discontinuities computed by the Godunov method satisfy analogues of the jump conditions. On the other hand, if we attempt to write the second-order Godunov method in the form (1.3), we obtain

$$\rho_j \frac{\mathrm{d}_j^{k+\frac{3}{2}} - 2\mathrm{d}_j^{k+\frac{1}{2}} + \mathrm{d}_j^{k-\frac{1}{2}}}{\Delta\tau^2} - \frac{\sigma_{j+\frac{1}{2}}^{k+\frac{1}{2}} - \sigma_{j-\frac{1}{2}}^{k+\frac{1}{2}}}{\Delta x} = 0 .$$

The displacements in this formula are not usually computed by the method, and the stresses in this formula are not given by a formula analogous to (2.2), but by the characteristic tracing and Riemann problem solution process to be discussed next. In this regard, note that the finite element method constructs differences of stresses that are evaluated only by the equation of state, while the stresses $\sigma_{j+\frac{1}{2}}^{k+\frac{1}{2}}$ used by the Godunov method may be computed by other means.

The second-order Godunov method begins a timestep with cell-centered values of the velocities, displacement gradients and stresses. If desired, the displacements at the cell edges could be carried, and computed so that their discrete gradients give the cell-centered values for ϵ:

$$\frac{\mathrm{d}_{j+\frac{1}{2}}^{k+1} - \mathrm{d}_{j+\frac{1}{2}}^{k}}{\Delta\tau} - \mathrm{v}_{j+\frac{1}{2}}^{k+\frac{1}{2}} = 0 ,$$

$$\epsilon_j^{k+1} = \frac{\mathrm{d}_{j+\frac{1}{2}}^{k+1} - \mathrm{d}_{j-\frac{1}{2}}^{k+1}}{\Delta x} .$$

Unlike the finite element method, no startup procedure is needed, and the maximum stable timestep is given by the CFL condition (2.7), as we will see in §4.

In order to determine the stresses and velocities required at the cell edges by the spatial gradients in (2.8), the second-order Godunov method employs a combination of characteristic tracing and the solution of a Riemann problem. The first part of the characteristic tracing step is to compute monotonized slopes in the characteristic expansion coefficients. First, we rewrite the equations of motion as a quasilinear form in the "flux variables" (*c.f.* [14]):

$$\text{(2.9)} \qquad \frac{\partial}{\partial\tau}\begin{bmatrix} \mathrm{v} \\ \sigma \end{bmatrix} - \begin{bmatrix} 0 & \frac{1}{\rho} \\ \frac{\partial\sigma}{\partial\epsilon} & 0 \end{bmatrix} \frac{\partial}{\partial x}\begin{bmatrix} \mathrm{v} \\ \sigma \end{bmatrix} = 0 .$$

Here the "linearized coefficient matrix" is

$$A = -\begin{bmatrix} 0 & \frac{1}{\rho} \\ \frac{\partial\sigma}{\partial\epsilon} & 0 \end{bmatrix} = \begin{bmatrix} 1 & -1 \\ \lambda\rho & \lambda\rho \end{bmatrix} \begin{bmatrix} -\lambda & 0 \\ 0 & \lambda \end{bmatrix} \begin{bmatrix} 1 & -1 \\ \lambda\rho & \lambda\rho \end{bmatrix}^{-1} .$$

Note that the second equation in (2.9) is derived from the equation of state (1.2) by differentiating it in time and using the second equation in (1.1). Next, the jumps in cell averages of the primitive variables are expanded in terms of eigenvectors of A:

$$\begin{bmatrix} \Delta c^- \\ \Delta c^+ \end{bmatrix}_j^R = \begin{bmatrix} 1 & -1 \\ \lambda\rho & \lambda\rho \end{bmatrix}^{-1} \begin{bmatrix} \mathrm{v}_{j+1}^k - \mathrm{v}_j^k \\ \sigma_{j+1}^k - \sigma_j^k \end{bmatrix} ,$$

$$\begin{bmatrix} \Delta c^- \\ \Delta c^+ \end{bmatrix}_j^L = \begin{bmatrix} 1 & -1 \\ \lambda\rho & \lambda\rho \end{bmatrix}^{-1} \begin{bmatrix} \mathrm{v}_j^k - \mathrm{v}_{j-1}^k \\ \sigma_j^k - \sigma_{j-1}^k \end{bmatrix} ,$$

$$\begin{bmatrix} \Delta c^- \\ \Delta c^+ \end{bmatrix}_j^C = \frac{1}{2}\left(\begin{bmatrix} \Delta c^- \\ \Delta c^+ \end{bmatrix}_j^R + \begin{bmatrix} \Delta c^- \\ \Delta c^+ \end{bmatrix}_j^L \right) .$$

Afterward, the expansion coefficients $\Delta c^{\pm}$ are adjusted to prevent the introduction of new extrema [7]:

$$\Delta c^{\pm} = \min\{|\,(\Delta c^{\pm})^L\,|, |\,(\Delta c^{\pm})^R\,|, \frac{1}{2}\,|\,(\Delta c^{\pm})^C\,|\} \times \{\text{sign}\{(\Delta c^{\pm})^L\} + \text{sign}\{(\Delta c^{\pm})^R\}\} .$$

This step is analogous to a local decoupling of the equations of motion into a set of scalar evolution equations. Furthermore, the slope limiting procedure has the effect of introducing a selective numerical viscosity in individual wave families, as we shall see in the next section.

Note that the contribution of the numerical viscosity to the momentum equation is analogous to a viscous force, and the contribution of viscosity to the stress-rate equation is analogous to the introduction of a visco-elastic term in the constitutive law.

In plasticity models, the characteristic speeds change discontinuously from elastic to plastic behavior. This discontinuity destroys the ability of the slope construction to obtain second-order accuracy, and may require the numerical scheme to separate an initial discontinuity into two discontinuities traveling at different speeds in the same direction. As a result, at all cell edges where the neighboring cells are undergoing different loading conditions, we set the slopes $\mathrm{v}_{j+1}^k - \mathrm{v}_j^k$ and $\sigma_{j+1}^k - \sigma_j^k$ to zero; that is, we introduce extra numerical viscosity by falling back to the first-order Godunov method.

We also note that near strong shocks the slope limiting process above is not sufficient to prevent oscillations. Strong shocks can be so self-sharpening that they overwhelm the numerical diffusion of the second-order method, leading to traveling discontinuities with one or no cells in the middle, and numerical oscillations nearby. Thus, when shocks become too sharp to travel as discrete discontinuities, it is useful to "flatten" slopes near the discontinuity [7,27]. Sharp discontinuities can be detected by jumps in stress or velocity 2 cells apart that are essentially the same as jumps 4 cells apart. (In problems involving shear, we look at jumps in pressure and normal velocity.) Accordingly, we compute

$$\omega_j = \max\left\{0\,,\ \min\left\{1\,,\ \frac{0.85 - \frac{\sigma_{j+1}^k - \sigma_{j-1}^k}{\sigma_{j+2}^k - \sigma_{j-2}^k}}{0.1}, \frac{0.85 - \frac{\mathrm{v}_{j+1}^k - \mathrm{v}_{j-1}^k}{\mathrm{v}_{j+2}^k - \mathrm{v}_{j-2}^k}}{0.1}\right\}\right\}, \tag{2.10}$$

where the parameters 0.85 and 0.1 have been determined experimentally, and are never adjusted in practice. Afterward, we multiply the limited slopes $\Delta c_j^{\pm}$ by $\min\{\omega_{j-1}, \omega_j, \omega_{j+1}\}$ in order to introduce additional numerical diffusion in cells near a sharp discontinuity.

Our next step is to use the limited slopes in the characteristic families to construct values for the primitive variables at the left- and right-hand sides of the cell edges at the half-time level. This is accomplished by performing a Taylor expansion about the cell centers in order to approximate the values at the cell edges and half-time levels, using the quasilinear form to replace the time derivative in the Taylor expansion and employing the eigenvectors of A to find the characteristic quantities, and then throwing away characteristic information that goes the wrong way:

$$\begin{bmatrix} \mathrm{v} \\ \sigma \end{bmatrix}_{j+\frac{1}{2}}^L = \begin{bmatrix} \mathrm{v} \\ \sigma \end{bmatrix}_j^k + \begin{bmatrix} -1 \\ \lambda\rho \end{bmatrix}_j^k (1 - \lambda_j^k \frac{\Delta\tau}{\Delta x}) \Delta c_j^+ \frac{1}{2} ,$$

$$\begin{bmatrix} \mathrm{v} \\ \sigma \end{bmatrix}^R_{j+\frac{1}{2}} = \begin{bmatrix} \mathrm{v} \\ \sigma \end{bmatrix}^k_{j+1} - \begin{bmatrix} 1 \\ \lambda\rho \end{bmatrix}^k_{j+1} (1 - \lambda^k_{j+1}\frac{\Delta\tau}{\Delta x})\Delta c^-_{j+1}\frac{1}{2} \, .$$

Afterward, we resolve the interaction of the waves between the left and right states at the cell edges through the approximate solution of a Riemann problem. Note that there are good reasons for developing an approximate Riemann solver. Although an analytic Riemann solver is available for the plasticity model used in our numerical examples in §5, it is far too expensive to use in practice. Another serious problem is that for general equations of state, the analytic solution of the Riemann problem is unknown. A variety of approximate Riemann solvers are possible [1,4,10,13]. Our particular choice is described in [27].

The idea in the approximate Riemann solver is to use the previous characteristic information to construct an approximate path in flux-variable space for the wave interaction. First, we decompose the jump in terms of averaged characteristic directions:

$$(\overline{\lambda\rho})_{j+\frac{1}{2}} = \frac{1}{2}(\lambda^k_j\rho_j + \lambda^k_{j+1}\rho_{j+1}) \, ,$$

$$\begin{bmatrix} \bar{a}^- \\ \bar{a}^+ \end{bmatrix} = \begin{bmatrix} 1 & -1 \\ (\overline{\lambda\rho})_{j+\frac{1}{2}} & (\overline{\lambda\rho})_{j+\frac{1}{2}} \end{bmatrix}^{-1} \begin{bmatrix} \mathrm{v}^R_{j+\frac{1}{2}} - \mathrm{v}^L_{j+\frac{1}{2}} \\ \sigma^R_{j+\frac{1}{2}} - \sigma^L_{j+\frac{1}{2}} \end{bmatrix} .$$

Once we have solved for the wave strengths $\bar{a}^\pm$, we construct a "strength-weighted average" of the acoustic impedances:

$$(\lambda\rho)_{j+\frac{1}{2}} = \frac{(\lambda\rho)^k_j|\bar{a}^-| + (\lambda\rho)^k_{j+1}|\bar{a}^+|}{|\bar{a}^-| + |\bar{a}^+|} \, .$$

Then, we recompute the expansion coefficients:

$$\begin{bmatrix} a^- \\ a^+ \end{bmatrix} = \begin{bmatrix} 1 & -1 \\ (\lambda\rho)_{j+\frac{1}{2}} & (\lambda\rho)_{j+\frac{1}{2}} \end{bmatrix}^{-1} \begin{bmatrix} \mathrm{v}^R_{j+\frac{1}{2}} - \mathrm{v}^L_{j+\frac{1}{2}} \\ \sigma^R_{j+\frac{1}{2}} - \sigma^L_{j+\frac{1}{2}} \end{bmatrix} .$$

Finally, we approximate the state that remains stationary in the solution to the Riemann problem, by either of two equivalent expressions.

$$\begin{bmatrix} \mathrm{v} \\ \sigma \end{bmatrix}^{k+\frac{1}{2}}_{j+\frac{1}{2}} = \begin{bmatrix} \mathrm{v} \\ \sigma \end{bmatrix}^L_{j+\frac{1}{2}} + \begin{bmatrix} 1 \\ (\lambda\rho)_{j+\frac{1}{2}} \end{bmatrix}_j a^- = \begin{bmatrix} \mathrm{v} \\ \sigma \end{bmatrix}^R_{j+\frac{1}{2}} - \begin{bmatrix} -1 \\ (\lambda\rho)_{j+\frac{1}{2}} \end{bmatrix}_{j+1} a^+ \, .$$

This approximate Riemann solver is very simple to implement, and has been shown to work well in practice, even on strong shocks [27]. Note that the Riemann solver provides the stresses and velocities at the cell edges and half-time levels that were needed in (2.8).

3. Modified Equation Analysis. In the first two sections we presented the equations of motion and two numerical methods for their solution. In this section, we will examine the modified equation analysis of the methods presented in the previous section. The goal of this analysis is to determine the modifications to the equations of motion that are implied by the truncation errors of the difference equations [30]. Often, this analysis can give quick qualitative insight into the accuracy of the numerical method. The modified equation analysis can also give quantitative information about the width of numerical shocks [11]. All that is required for this analysis is straightforward application of Taylor expansions.

Our principal goal in performing this analysis is to show that for linear constant-coefficient differential equations, the difference equations for the Godunov method decouple into individual wave families. This will simplify the discussion of the Fourier analysis below. Also, we will see from the modified equation that the slope-limiting process in the Godunov method contributes a diffusive term to the difference equations. The modified equation analysis of the finite element method is not very instructive; it principally tells us the order of the method with artificial viscosity.

In order to simplify the analysis, we will assume a linear stress-strain relationship,

$$\sigma = \lambda^2 \rho \epsilon , \tag{3.1}$$

where λ and ρ are constants. We begin with the Godunov method. We will show that the equations decouple into two wave families. To do so, we define the expansion coefficients of the two wave families by

$$\begin{bmatrix} u^- \\ u^+ \end{bmatrix}_j^k = \begin{bmatrix} 1 & -1 \\ \lambda\rho & \lambda\rho \end{bmatrix}^{-1} \begin{bmatrix} \mathrm{v} \\ \sigma \end{bmatrix}_j^k = \begin{bmatrix} \lambda\rho & -\lambda\rho \\ 1 & 1 \end{bmatrix}^{-1} \begin{bmatrix} \rho \mathrm{v} \\ \epsilon \end{bmatrix}_j^k \lambda .$$

Then the steps in the second-order Godunov method give us

$$\begin{bmatrix} \Delta c^- \\ \Delta c^+ \end{bmatrix}_j^k = \begin{bmatrix} \beta^- & 0 \\ 0 & \beta^+ \end{bmatrix} \begin{bmatrix} (u^-)_{j+1}^k - (u^-)_{j-1}^k \\ (u^+)_{j+1}^k - (u^+)_{j-1}^k \end{bmatrix} \frac{1}{2} ,$$

$$\begin{bmatrix} u^- \\ u^+ \end{bmatrix}_{j+\frac{1}{2}}^L = \begin{bmatrix} (u^-)_j^k \\ (u^+)_j^k + \frac{1}{2}(1 - \lambda\frac{\Delta\tau}{\Delta x})(\Delta c^+)_j^k \end{bmatrix} ,$$

$$\begin{bmatrix} u^- \\ u^+ \end{bmatrix}_{j+\frac{1}{2}}^R = \begin{bmatrix} (u^-)_{j+1}^k + \frac{1}{2}(1 - \lambda\frac{\Delta\tau}{\Delta x})(\Delta c^-)_{j+1}^k \\ (u^+)_{j+1}^k \end{bmatrix} ,$$

$$\begin{bmatrix} a^- \\ a^+ \end{bmatrix}_{j+\frac{1}{2}}^{k+\frac{1}{2}} = \begin{bmatrix} (u^-)_{j+\frac{1}{2}}^R - (u^-)_{j+\frac{1}{2}}^L \\ (u^+)_{j+\frac{1}{2}}^R - (u^+)_{j+\frac{1}{2}}^L \end{bmatrix} ,$$

$$\begin{bmatrix} u^- \\ u^+ \end{bmatrix}_{j+\frac{1}{2}}^{k+\frac{1}{2}} = \begin{bmatrix} (u^-)_{j+\frac{1}{2}}^L + (a^-)_{j+\frac{1}{2}}^{k+\frac{1}{2}} \\ (u^+)_{j+\frac{1}{2}}^L \end{bmatrix} = \begin{bmatrix} (u^-)_{j+\frac{1}{2}}^R \\ (u^+)_{j+\frac{1}{2}}^R - (a^+)_{j+\frac{1}{2}}^{k+\frac{1}{2}} \end{bmatrix} ,$$

$$\begin{bmatrix} u^- \\ u^+ \end{bmatrix}_j^{k+1} = \begin{bmatrix} u^- \\ u^+ \end{bmatrix}_j^k + \begin{bmatrix} (u^-)_{j+\frac{1}{2}}^{k+\frac{1}{2}} - (u^-)_{j-\frac{1}{2}}^{k+\frac{1}{2}} \\ (u^+)_{j+\frac{1}{2}}^{k+\frac{1}{2}} - (u^+)_{j-\frac{1}{2}}^{k+\frac{1}{2}} \end{bmatrix} \frac{1}{2} \lambda \frac{\Delta\tau}{\Delta x} .$$

In each wave family, $0 \le \beta^\pm \le 1$ represents the factor by which slope limiting reduces the centered slopes: $\beta = 1$ for unlimited slopes, and $\beta = 0$ for zero slopes (*i.e.*, first-order Godunov).

Since the wave families decouple, it is sufficient to consider the modified equation analysis for

$$\frac{\partial u}{\partial \tau} + \lambda \frac{\partial u}{\partial x} = 0 .$$

Again, Taylor expansions give us

$$\triangle c_j = \beta \frac{\partial u}{\partial x} \triangle x + O(\triangle x^3) ,$$

$$\triangle c_{j-1} = \beta \frac{\partial u}{\partial x} \triangle x - \beta \frac{\partial^2 u}{\partial x^2} \triangle x^2 + O(\triangle x^3) ,$$

$$u_{j+\frac{1}{2}}^{k+\frac{1}{2}} - u_{j-\frac{1}{2}}^{k+\frac{1}{2}} = \frac{\partial u}{\partial x} \triangle x - \frac{\partial^2 u}{\partial x^2} \frac{1}{2} \triangle x^2 + (1 - \lambda \frac{\triangle \tau}{\triangle x}) \beta \frac{\partial^2 u}{\partial x^2} \frac{1}{2} \triangle x^2 + O(\triangle x^3) .$$

From these, we derive that the modified equation is

$$\frac{\partial u}{\partial \tau} + \lambda \frac{\partial u}{\partial x} \doteq \frac{1}{2} \triangle x (1 - \beta) \lambda (1 - \lambda \frac{\triangle \tau}{\triangle x}) \frac{\partial^2 u}{\partial x^2} .$$

In a similar fashion, we find that the modified equation analysis for the other wave family,

$$\frac{\partial u}{\partial \tau} - \lambda \frac{\partial u}{\partial x} = 0 ,$$

leads to

$$\frac{\partial u}{\partial \tau} - \lambda \frac{\partial u}{\partial x} \doteq \frac{1}{2} \triangle x (1 - \beta) \lambda (1 - \lambda \frac{\triangle \tau}{\triangle x}) \frac{\partial^2 u}{\partial x^2} .$$

Now we can substitute the results for the two wave families back into the original system:

$$\begin{aligned} 0 \doteq\ & \frac{1}{\lambda} \begin{bmatrix} \lambda\rho & -\lambda\rho \\ 1 & 1 \end{bmatrix} \left\{ \frac{\partial}{\partial \tau} \begin{bmatrix} u^- \\ u^+ \end{bmatrix} + \begin{bmatrix} -\lambda & 0 \\ 0 & \lambda \end{bmatrix} \frac{\partial}{\partial x} \begin{bmatrix} u^- \\ u^+ \end{bmatrix} \right. \\ & \left. - \frac{1}{2} \triangle x \lambda (1 - \lambda \frac{\triangle \tau}{\triangle x}) \begin{bmatrix} 1-\beta^- & 0 \\ 0 & 1-\beta^+ \end{bmatrix} \frac{\partial^2}{\partial x^2} \begin{bmatrix} u^- \\ u^+ \end{bmatrix} \right\} \\ =\ & \frac{\partial}{\partial \tau} \begin{bmatrix} \rho \mathrm{v} \\ \epsilon \end{bmatrix} - \frac{\partial}{\partial x} \begin{bmatrix} \sigma \\ \mathrm{v} \end{bmatrix} \\ & - \frac{1}{2} \triangle x \lambda (1 - \lambda \frac{\triangle \tau}{\triangle x}) \begin{bmatrix} \lambda\rho & -\lambda\rho \\ 1 & 1 \end{bmatrix} \begin{bmatrix} 1-\beta^- & 0 \\ 0 & 1-\beta^+ \end{bmatrix} \begin{bmatrix} \lambda\rho & -\lambda\rho \\ 1 & 1 \end{bmatrix}^{-1} \frac{\partial^2}{\partial x^2} \begin{bmatrix} \rho \mathrm{v} \\ \epsilon \end{bmatrix} . \end{aligned}$$

Thus, the slope limiting in each of the characteristic families contributes a diffusive term to the system, provided that the timestep obeys the CFL condition (2.7). Note that if the slopes are unlimited (*i.e.*, if $\beta = 1$), or if the timestep is at its stability limit (*i.e.*, $\lambda \triangle \tau = \triangle x$), then the diffusive term is zero. In practice we keep CFL less than one, so the scheme is second-order accurate if $\beta^{\pm} = 1$; otherwise, its truncation error is first-order.

Now we consider the finite element method using explicit artificial viscosity (2.6):

$$\begin{aligned} 0 \ = \ & \frac{d_i^{k+1} - 2d_i^k + d_i^{k-1}}{\Delta\tau^2} - \lambda^2 \frac{d_{i+1}^k - 2d_i^k + d_{i-1}^k}{\Delta x^2} \\ & - \frac{\alpha\lambda\Delta x}{\Delta\tau}\left[\frac{d_{i+1}^k - 2d_i^k + d_{i-1}^k}{\Delta x^2} - \frac{d_{i+1}^{k-1} - 2d_i^{k-1} + d_{i-1}^{k-1}}{\Delta x^2}\right] . \end{aligned}$$

From Taylor expansions, we derive

$$\frac{d_i^{k+1} - 2d_i^k + d_i^{k-1}}{\Delta\tau^2} = \frac{\partial^2 d}{\partial\tau^2} + \frac{\partial^4 d}{\partial\tau^4}\frac{1}{12}\Delta\tau^2 + O(\Delta\tau^4) ,$$

$$\frac{d_{i+1}^k - 2d_i^k + d_{i-1}^k}{\Delta x^2} = \frac{\partial^2 d}{\partial x^2} + \frac{\partial^4 d}{\partial x^4}\frac{1}{12}\Delta x^2 + O(\Delta x^4) ,$$

$$\begin{aligned} & \frac{d_{i+1}^k - 2d_i^k + d_{i-1}^k}{\Delta x^2} - \frac{d_{i+1}^{k-1} - 2d_i^{k-1} + d_{i-1}^{k-1}}{\Delta x^2} \\ & = \frac{\partial^3 d}{\partial x^2 \partial\tau}\Delta\tau - \frac{1}{2}\frac{\partial^4 d}{\partial x^2 \partial\tau^2}\Delta\tau^2 + \text{4th-order terms} . \end{aligned}$$

From the differential equation (1.3) we derive

$$\frac{\partial^4 d}{\partial\tau^4} = \lambda^4 \frac{\partial^4 d}{\partial x^4} .$$

These can be substituted into the difference equation to derive the modified equation

$$\left[\frac{\partial^2 d}{\partial\tau^2} - \lambda^2\frac{\partial^2 d}{\partial x^2}\right] \doteq \alpha\lambda\Delta x \frac{\partial^3 d}{\partial x^2\partial\tau} + \frac{\lambda^2\Delta x^2}{12}\left[1 - (\frac{\lambda\Delta\tau}{\Delta x})^2 - 6\alpha\frac{\lambda\Delta\tau}{\Delta x}\right]\frac{\partial^4 d}{\partial x^4} .$$

This result shows that the finite element method has a second-order truncation error in the absence of artificial viscosity; the presence of linear artificial viscosity makes the method first-order accurate. This explains why some forms of artificial viscosity are not used in expansions, since these occur in smooth motion and allow high-order resolution. We also note that the first-order term on the right-hand side is dissipative for $\alpha > 0$. However, we will see in the next section that the overall finite element scheme can be *antidissipative* for timesteps near the CFL limit.

4. Fourier Analysis. While the modified equation analysis can give some quick insight into the behavior of numerical methods for small frequency waves, a Fourier mode analysis can give us detailed quantitative information about the numerical diffusion and dispersion in constant-coefficient equations for all wave numbers. Thus, one goal in studying the Fourier analyses of the two methods is to determine rigorously the stability of each scheme. We shall also be interested in the balance of dissipation and dispersion. In practical calculations, it is important that whenever dispersion is present there must also be sufficient dissipation to damp the dispersed waves; otherwise, annoying numerical oscillations can develop and persist. We will see that in this regard there are significant differences between the balance of dissipation and dispersion in the two methods, and the results of the Fourier analysis will help us to understand the numerical examples in the next section.

The fundamental observation underlying the Fourier mode analysis is that the equation

$$\frac{\partial^2 \mathrm{d}}{\partial \tau^2} - \lambda^2 \frac{\partial^2 \mathrm{d}}{\partial x^2} = 0$$

is satisfied by the Fourier modes $\exp i\xi(x \pm \lambda\tau)$. Thus, it is common to substitute

$$\mathrm{d}_m^n = \exp i\xi([m-j]\Delta x + \mu[n-k]\Delta\tau)$$

for the values of the displacement in the finite element difference equations (and to make a similar substitution for the characteristic expansion coefficients in the Godunov difference equations). The magnitude of the imaginary part of μ determines the modulus of the Fourier mode, and thus the amount of dissipation in the method; the dependence of the real part of μ on ξ tells us the dispersion [21]. For stability, we want the Fourier mode to keep its modulus less than or equal to one, so that the wave decays in time. To eliminate dispersion, the scheme must produce $\mu = \lambda$ for all values of xi.

We begin with the analysis of the finite element method using explicit artificial viscosity. Let

$$z = \exp(i\xi\mu\Delta\tau)\ ,\ \theta = \xi\Delta x\ ,\ \gamma = \frac{\lambda\Delta\tau}{\Delta x}\ .$$

Then the difference scheme (2.1), (2.6) and (3.1) leads to the following equation for the Fourier mode:

$$0 = \frac{z - 2 + \frac{1}{z}}{\Delta\tau^2} + 2\lambda^2\frac{1-\cos\theta}{\Delta x^2} + 2\frac{\alpha\lambda\Delta x}{\Delta\tau}\left[1 - \frac{1}{z}\right]\frac{1-\cos\theta}{\Delta x^2}\ .$$

This can be rewritten

$$0 = z^2 - \{2 - 2(1-\cos\theta)\gamma(\gamma+\alpha)\}\, z + 1 - 2\alpha\gamma(1-\cos\theta)\ .$$

Solving for z, we obtain

$$z = 1 - (1-\cos\theta)\gamma(\gamma+\alpha) \pm\sqrt{[1-(1-\cos\theta)\gamma(\gamma+\alpha)]^2 - 1 + 2\alpha\gamma(1-\cos\theta)}\ . \tag{4.1}$$

The remainder of our discussion will be directed toward determining the circumstances under which $|z| \le 1$, so that the finite element scheme is stable.

First, let us consider the case when the discriminant in (4.1) is negative, so that z is complex. Upon inspection, we see that this occurs if

$$(\gamma+\alpha)^2(1-\cos\theta) < 2\ . \tag{4.2}$$

(Note that $\gamma+\alpha < 1$ implies that the discriminant is negative for all Fourier modes; however, this restriction on the timestep is more than is necessary for stability, and is impossible to satisfy for viscosity parameters $\alpha \ge 1$.) Since the constant term in the quadratic form is the product of the roots, we have

$$|\, z\,|^2 = 1 - 2\alpha\gamma(1-\cos\theta) \le 1\ ,$$

and the scheme is dissipative in this case.

On the other hand, if inequality (4.2) is violated, then z is real and we have two cases. If we write

$$z = b \pm \sqrt{b^2 - c}\,,$$

the first case occurs when $b \geq 0$. In this case, we require for stability that

$$z \leq 1 \Longleftrightarrow 0 \leq 1 - 2b + c\,.$$

Otherwise, $-b < 0$ and we want

$$z \geq -1 \Longleftrightarrow 0 \leq 1 + 2b + c\,.$$

From the quadratic form at hand, we see that the former case occurs when

$$1 - (1 - \cos\theta)\gamma(\gamma + \alpha) \geq 0\,,$$

in which case we want

$$0 \leq 1 - 2 + 2(1 - \cos\theta)\gamma(\gamma + \alpha) + 1 - 2\alpha\gamma(1 - \cos\theta) = 2\gamma^2(1 - \cos\theta)\,,$$

which is obviously true. Otherwise, we want

$$0 \leq 1 + 2 - 2\gamma(\gamma + \alpha)(1 - \cos\theta) + 1 - 2\alpha\gamma(1 - \cos\theta) = 4 - 2\gamma(\gamma + 2\alpha)(1 - \cos\theta)\,.$$

In particular, $\cos\theta = -1$ makes this the hardest to satisfy, requiring

$$\gamma(\gamma + 2\alpha) \leq 1\,.$$

This implies a reduced stable timestep for the finite element method with explicit artificial viscosity:

$$\gamma = \frac{\lambda\Delta\tau}{\Delta x} \leq \frac{1}{\alpha + \sqrt{1 + \alpha^2}}\,.$$

A similar Fourier analysis of the finite element method with implicit artificial viscosity (2.3) and (2.5) leads to the stability restriction

$$\frac{\lambda\Delta\tau}{\Delta x} \leq \frac{1}{\frac{1}{2}\alpha + \sqrt{1 + (\frac{1}{2}\alpha)^2}}\,.$$

This implies that the explicit viscosity requires a smaller stable timestep than the implicit method with the same viscosity parameter. We also note that a Fourier analysis can be used to show that if we had used exact quadrature for the mass matrix (*i.e.*, no lumping), then in the absence of artificial viscosity we would have obtained an implicit method with a maximum stable timestep of $\sqrt{1/3}$ times the CFL timestep [22,23].

Next, we consider the second-order Godunov method for

$$\frac{\partial u}{\partial \tau} + \lambda\frac{\partial u}{\partial x} = 0\,.$$

Upon substitution of the Fourier mode

$$u_m^n = \exp i\xi([m - j]\Delta x + \mu[n - k]\Delta\tau)$$

into the difference equation, we obtain

$$0 = \frac{z-1}{\Delta\tau} + \lambda\frac{1-\exp^{-i\theta}}{\Delta x}\left[1 + i\frac{1}{2}(1-\gamma)\beta\sin\theta\right] .$$

(Here, we are assuming a constant level β of slope limiting.) This implies that

$$z = 1 - \gamma\left[(1-\cos\theta) - \frac{1}{2}(1-\gamma)\beta\sin^2\theta\right] - i\gamma\sin\theta\left[1 + \frac{1}{2}\beta(1-\cos\theta)(1-\gamma)\right] ,$$

which in turn implies that the modulus of z is

$$\mid z\mid^2 = 1 - 4\gamma(1-\gamma)\sin^2\frac{\theta}{2}\left[1 - \beta\cos^2\frac{\theta}{2} - \beta^2\gamma(1-\gamma)\sin^2\frac{\theta}{2}\cos^2\frac{\theta}{2}\right] .$$

The quantity in the square brackets is a quadratic form, similar to

$$q(y) = 1 - \beta y - \beta^2\eta y(1-y) ,$$

where $0 \le \beta \le 1$ and $0 < \eta = \gamma(1-\gamma) \le 1/4$. We need only show that this quadratic form is nonnegative over the argument range $0 \le y = \cos^2\theta/2 \le 1$. The minimum value of q occurs at

$$y^* = \frac{1+\beta\eta}{2\beta\eta} .$$

If $0 \le y^* \le 1$, then we have

$$q(y^*) = 1 - (y^*)^2\beta^2\eta > 1 - \frac{1}{4} > 0 .$$

Otherwise, the minimum value of q in the interval $0 \le y \le 1$ could occur at $y = 1$ or $y = 0$. Since $q(0) = 1$ and $q(1) = 1 - \beta \ge 0$, the quadratic form is nonnegative over the desired range. This shows that the Godunov method is dissipative for all Fourier modes and all amounts of slope limiting, provided that the timestep satisfies the CFL condition (2.7).

While the discussion above has been useful in determining the restrictions on the timesteps under which the methods are dissipative, the best way to understand the amount of dissipation and dispersion introduced by the methods is to look at pictures. In Figures 1 to 7 we have drawn the dissipation and dispersion curves for the finite element method with explicit viscosity, and for the Godunov method. The horizontal axis is $\theta = \xi\Delta x$ over the range $0 \le \theta \le 2\pi$; this is sufficient because z is an even 2π-periodic function of *theta* for both methods. Note that $\arg z$ is the frequency of the numerical wave times the numerical timestep, so $\arg(z^{1/\gamma})$ is the frequency of the numerical wave times the time required for the wave to cross a cell. This would be the same as the wave number times the cell width, $\theta = \xi\Delta x$, if there were no dispersion. Thus, to examine the numerical dispersion we plot the relative error in the numerical frequency, $(\arg z^{1/\gamma} - \theta)/\theta$, as a function of θ, which is proportional to the wave number. Also note that $|z|$ tells the amplitude of the numerical wave after one timestep, so $|z^{1/\gamma}|$ tells the amplitude of the numerical wave after the time interval required by the wave to cross the cell. If there were no dissipation in the method, then we would have $\mid z^{1/\gamma}\mid = 1$ for all θ; if there were no dispersion, we would have $(\arg z^{1/\gamma} - \theta)/\theta = 0$ for all θ.

Let us begin by considering the schemes with "no viscosity," *i.e.*, $\alpha = 0$ for the finite element method, and $\beta = 1$ for Godunov. Under these circumstances, the finite element scheme has no dissipation, so we show only its dispersion in figure 1. Since the Godunov method may still have some dissipation, we show its dissipation and dispersion in figures 2 and 3. (Similar dissipation plots are displayed using polar coordinates in [28].) Note that the Godunov method is most dissipative for high wave numbers at CFL = $\frac{1}{2}$. Also note that neither of the two schemes is dispersive at CFL = 1. At high wave numbers with CFL between $\frac{1}{2}$ and 1, the second-order Godunov method is less dispersive than the finite element method, and causes waves to travel faster than they should. On the other hand, the finite element method will cause all wave numbers to travel slower than they should.

Next, we consider the dissipation and dispersion in the methods with viscosity. For the Godunov method we chose $\beta = 0$, corresponding to the first-order version. We have chosen a linear viscosity with $\alpha = 1/2$ for the finite element method. The dispersion plot for the finite element method, figure 5, shows that at high values of CFL the high wavenumbers do not travel. On the other hand, the first-order Godunov method allows all waves to travel, and has no dispersion for CFL = 1 or $\frac{1}{2}$. For values of CFL between $\frac{1}{2}$ and 1, the first-order Godunov method is much less dispersive than the finite element method with this choice of linear artificial viscosity. At CFL = $\frac{1}{2}$, both the first-order Godunov method and the finite element method have zero dispersion; this explains why we chose $\alpha = \frac{1}{2}$ in order to compare the two viscous forms of the methods. The dissipation plots are also interesting. The Godunov method is dissipative for all wave numbers and all values of CFL between 0 and 1, while the finite element method is *anti-dissipative* for higher values of CFL. This indicates that the finite element method with significant amounts of linear viscosity must reduce its timestep in order to be stable.

Let us examine some of the implications of our Fourier analysis. Clearly, we would prefer to run the Godunov method at CFL close to one in order to minimize both dissipation and dispersion, and to reduce the computational cost by taking as few timesteps as possible. As we remarked above, this may not always be possible, especially when the motion leads to significantly higher characteristic speeds than are present in the initial data. On the other hand, in order to reduce the computational cost by running at high CFL with the finite element method, it is useful to use small values of the artificial viscosity α. Furthermore, the only way to reduce the dissipation with the finite element method is to reduce the artificial viscosity. However, for high values of CFL and low values of the artificial viscosity, the finite element method introduces dispersion without dissipation; as a result, high wave numbers will travel slower than they should without sufficient damping to make the dispersed waves unnoticeable.

Our Fourier analysis can also give us some insight into the performance of the methods near shocks. Because shocks are self-sharpening, they introduce high wave number information into the computation. Near shocks the Godunov method will limit the slopes by selecting $\beta = 0$; then for values of CFL between one-half and one, high wave numbers will travel ahead of the shock and will be dissipated. With the finite element method, at high values of CFL the high wave numbers will travel behind the shock, but will not be dissipated strongly enough to be unnoticeable (or may even be antidissipated). For both methods, low values of CFL will strongly disperse high wave numbers to travel behind the shock, where they will be dissipated.

It is not clear whether it is more useful to study the dissipation / timestep, or dissipation / time to cross the cell in using Fourier analysis to understand behavior near shocks. Shocks will inject high wave number information with each timestep, and the method must produce sufficient dissipation of dispersed waves to damp the resulting oscillations. In this regard, we note that the dissipation / timestep at low CFL is nearly zero with the Godunov method.

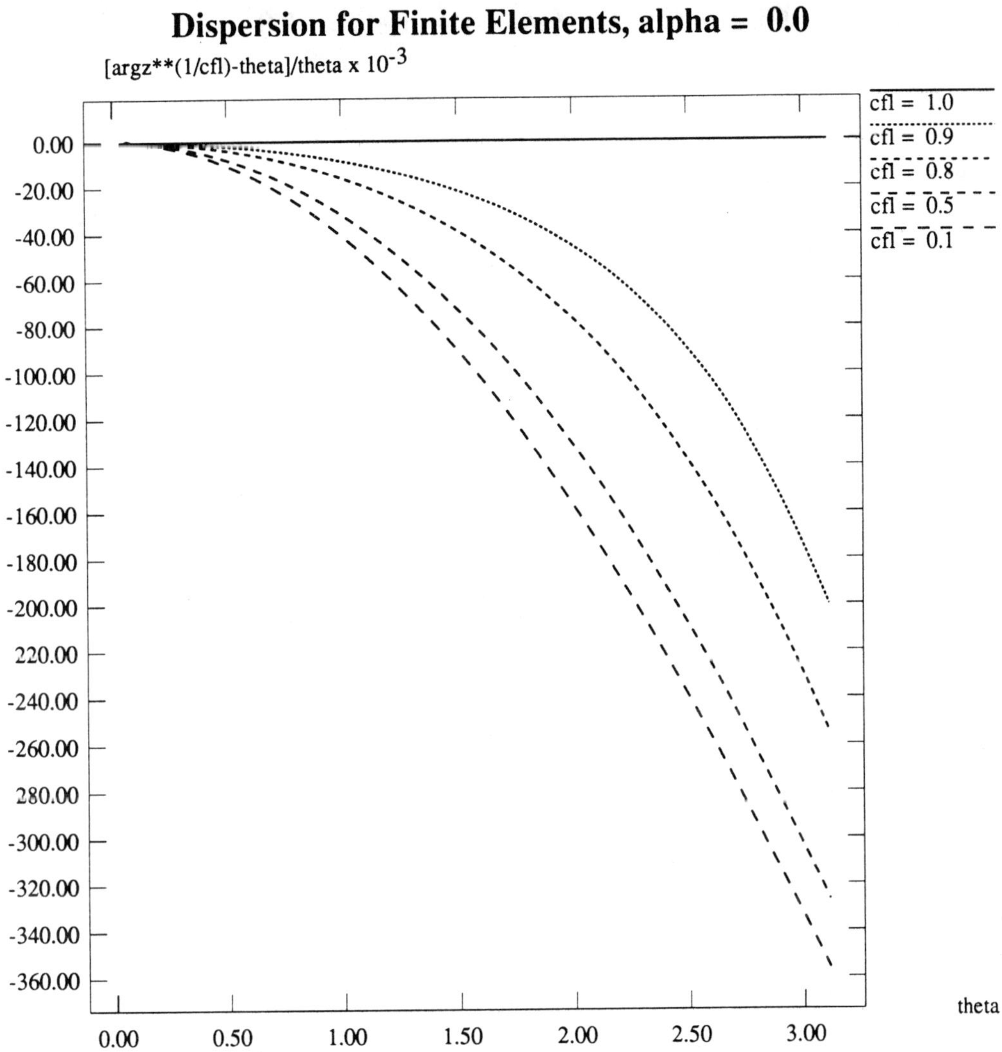
Dispersion for Finite Elements, alpha = 0.0
[argz**(1/cfl)-theta]/theta x 10-3
cfl = 1.0
cfl = 0.9
cfl = 0.8
cfl = 0.5
cfl = 0.1
0.00
-20.00
-40.00
-60.00
-80.00
-100.00
-120.00
-140.00
-160.00
-180.00
-200.00
220.00
-240.00
-260.00
280.00
-300.00
-320.00
-340.00
-360.00
0.00
0.50
1.00
1.50
2.00
2.50
3.00
theta

FIG. 1.

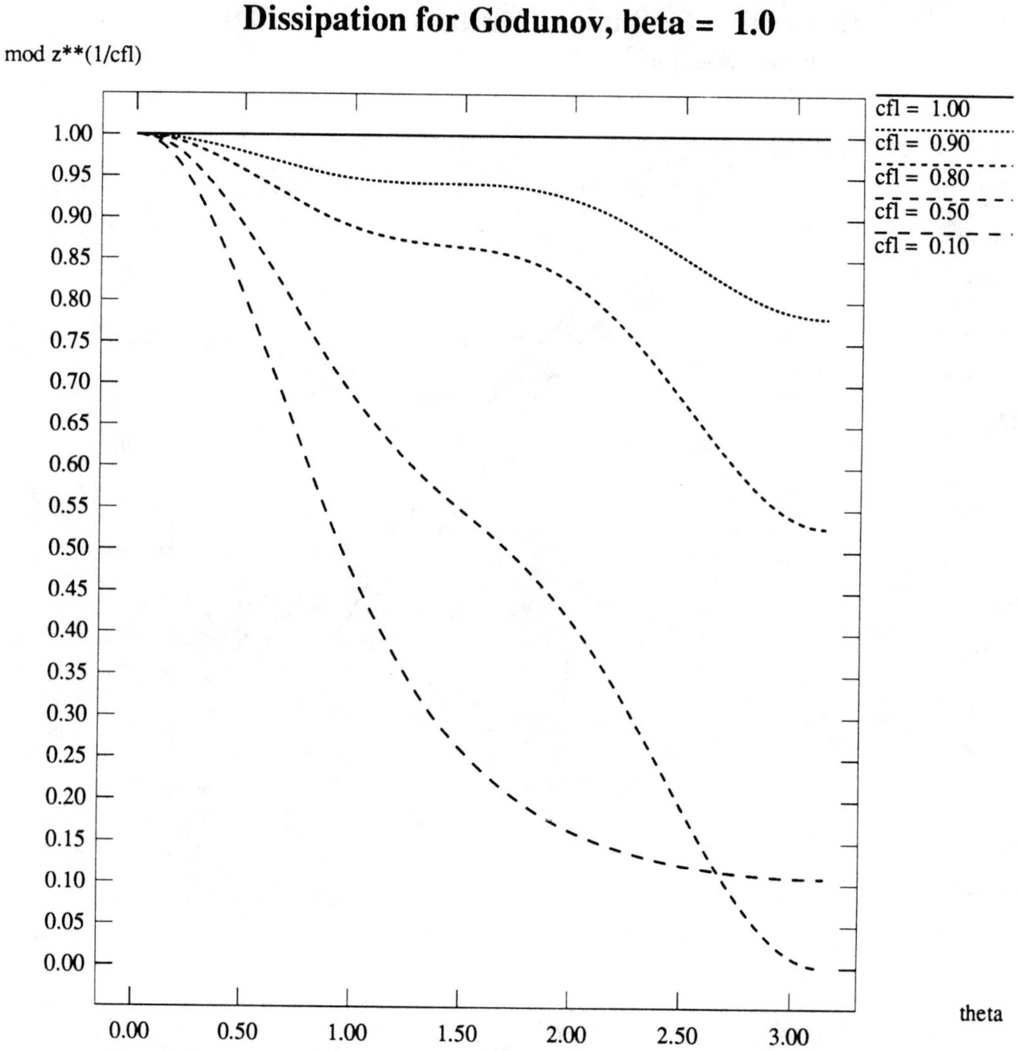
Dissipation for Godunov, beta = 1.0
mod z**(1/cfl)
1.00
0.95
0.90
0.85
0.80
0.75
0.70
0.65
0.60
0.55
0.50
0.45
0.40
0.35
0.30
0.25
0.20
0.15
0.10
0.05
0.00
cfl = 1.00
cfl = 0.90
cfl = 0.80
cfl = 0.50
cfl = 0.10
0.00
0.50
1.00
1.50
2.00
2.50
3.00
theta

FIG. 2.

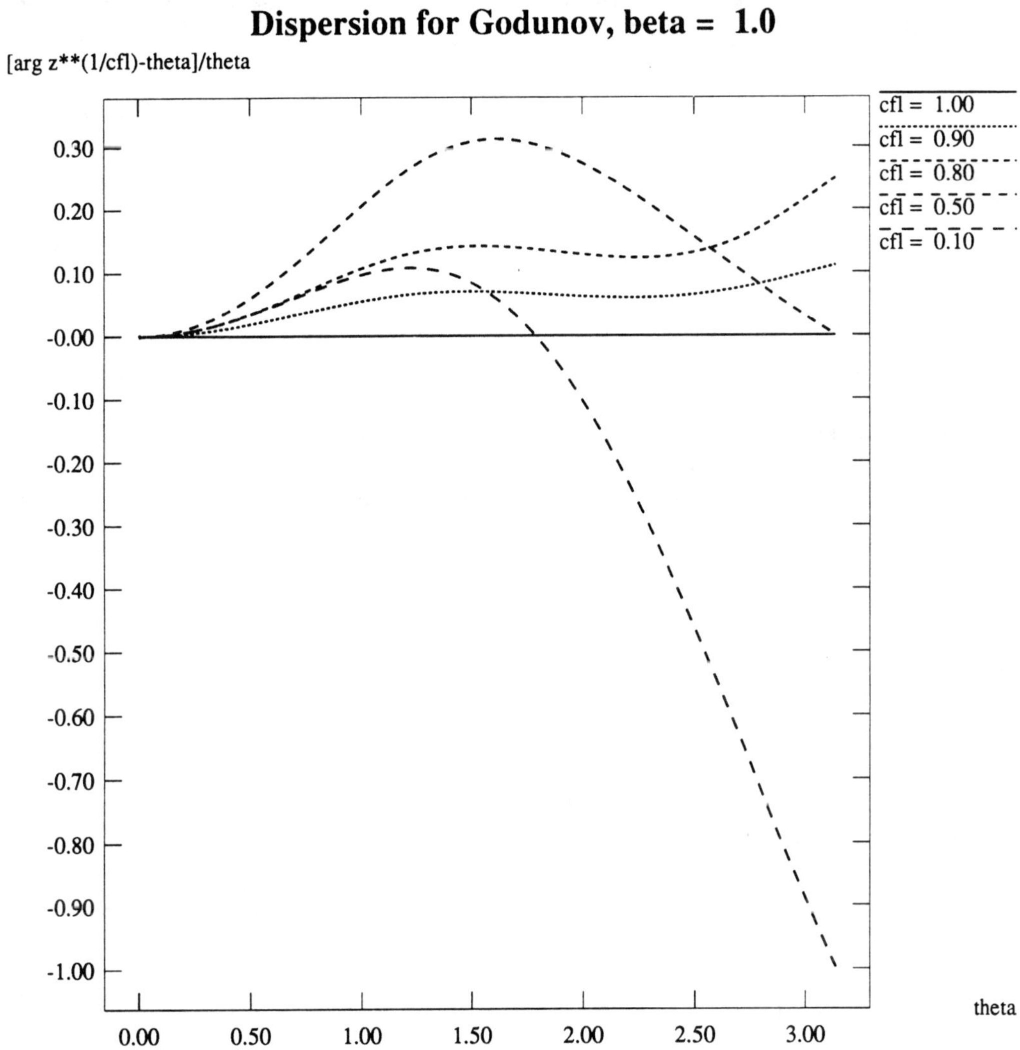
Dispersion for Godunov, beta = 1.0
[arg z**(1/cfl)-theta]/theta
cfl = 1.00
cfl = 0.90
cfl = 0.80
cfl = 0.50
cfl = 0.10
0.30
0.20
0.10
-0.00
-0.10
-0.20
-0.30
-0.40
-0.50
-0.60
-0.70
-0.80
-0.90
-1.00
0.00
0.50
1.00
1.50
2.00
2.50
3.00
theta

FIG. 3.

This may explain why strong slowly moving discontinuities can produce oscillations that are not sufficiently damped by the the Godunov method; in such a case, it may be necessary to include an artificial viscosity in the Godunov method. However, none of the examples in this paper have nearly stationary discontinuities, so no artificial viscosity has been added to the Godunov method.

5. Numerical Examples. The artificial viscosities in the finite element and Godunov methods are designed to work on nonlinear problems involving shocks. While the Fourier analysis of the previous section gives us some idea of how these schemes might perform on linear problems, the more practical issue is how they perform on nonlinear waves. For this reason, we have chosen to compare the methods on a model of an elastic-plastic bar, suggested by Antman and Szymczak [1]. The advantage of this model for numerical purposes is that its Riemann problem can be solved analytically, and involves some interesting wave behavior [26].

In the Antman-Szymczak model the stress σ is a function of the displacement gradient ϵ and a history parameter π representing plasticity. For any physically realistic displacement gradient ϵ there are upper and lower bounds on the stress σ, given by the functions $c(\epsilon)$ and $t(\epsilon)$, representing the plastic compression and tension curves, respectively. Between these bounds, the stress is given by the elastic curves $\sigma = e(\epsilon, \pi)$. Plastic compression occurs if and only if the material is at yield, and the time rate of change of the displacement gradient is negative:

$$c(\epsilon) = e(\epsilon, \pi) \text{ and } \left.\frac{\partial \epsilon}{\partial \tau}\right|_x < 0 \text{ during plastic compression.}$$

Similarly,

$$t(\epsilon) = e(\epsilon, \pi) \text{ and } \left.\frac{\partial \epsilon}{\partial \tau}\right|_x > 0 \text{ during plastic tension.}$$

There are other constraints placed on these functions and variables in order to make a consistent and realistic model; please see [1] or [26] for details. For the examples in this paper, we have chosen

$$\begin{aligned} e(\epsilon, \pi) &= -\frac{0.49}{0.7 + \epsilon - \pi} + 0.7 + \epsilon - \pi \ , \\ c(\epsilon) &= -0.1 - \frac{1}{(1+\epsilon)^2} \ , \\ t(\epsilon) &= 1.1 - \frac{1}{\sqrt{2+\epsilon}} \ . \end{aligned}$$

We would like to note that it is very inexpensive to compute the stress in the Antman-Szymczak model. The process begins with the old value of the history parameter π and the new value of the displacement gradient ϵ. Then we perform the following computations:

$$\begin{gathered} \sigma := \max\{c(\epsilon), \min\{e(\epsilon, \pi), t(\epsilon)\}\} \ ; \\ \text{if } \sigma = c(\epsilon) \text{ solve } \sigma = e(\epsilon, \pi) \text{ for } \pi \ , \\ \text{elseif } \sigma = t(\epsilon) \text{ solve } \sigma = e(\epsilon, \pi) \text{ for } \pi \ . \end{gathered}$$

This produces new values for the stress σ and the history parameter π. Because of the simple form for $e(\epsilon, \pi)$, the computation of π during plastic loading can be performed analytically.

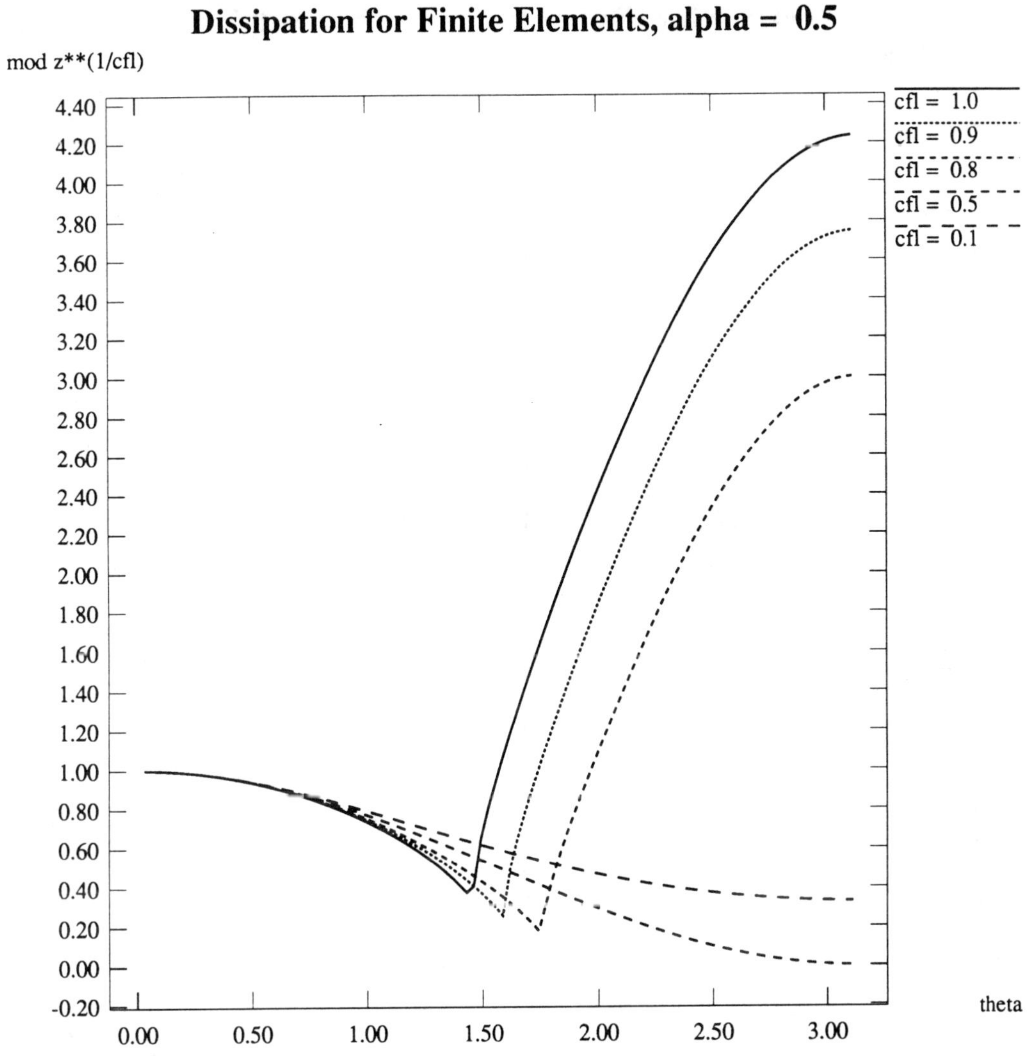
Dissipation for Finite Elements, alpha = 0.5
mod z**(1/cfl)
4.40
4.20
4.00
3.80
3.60
3.40
3.20
3.00
2.80
2.60
2.40
2.20
2.00
1.80
1.60
1.40
1.20
1.00
0.80
0.60
0.40
0.20
0.00
-0.20
0.00
0.50
1.00
1.50
2.00
2.50
3.00
theta
cfl = 1.0
cfl = 0.9
cfl = 0.8
cfl = 0.5
cfl = 0.1

FIG. 4.

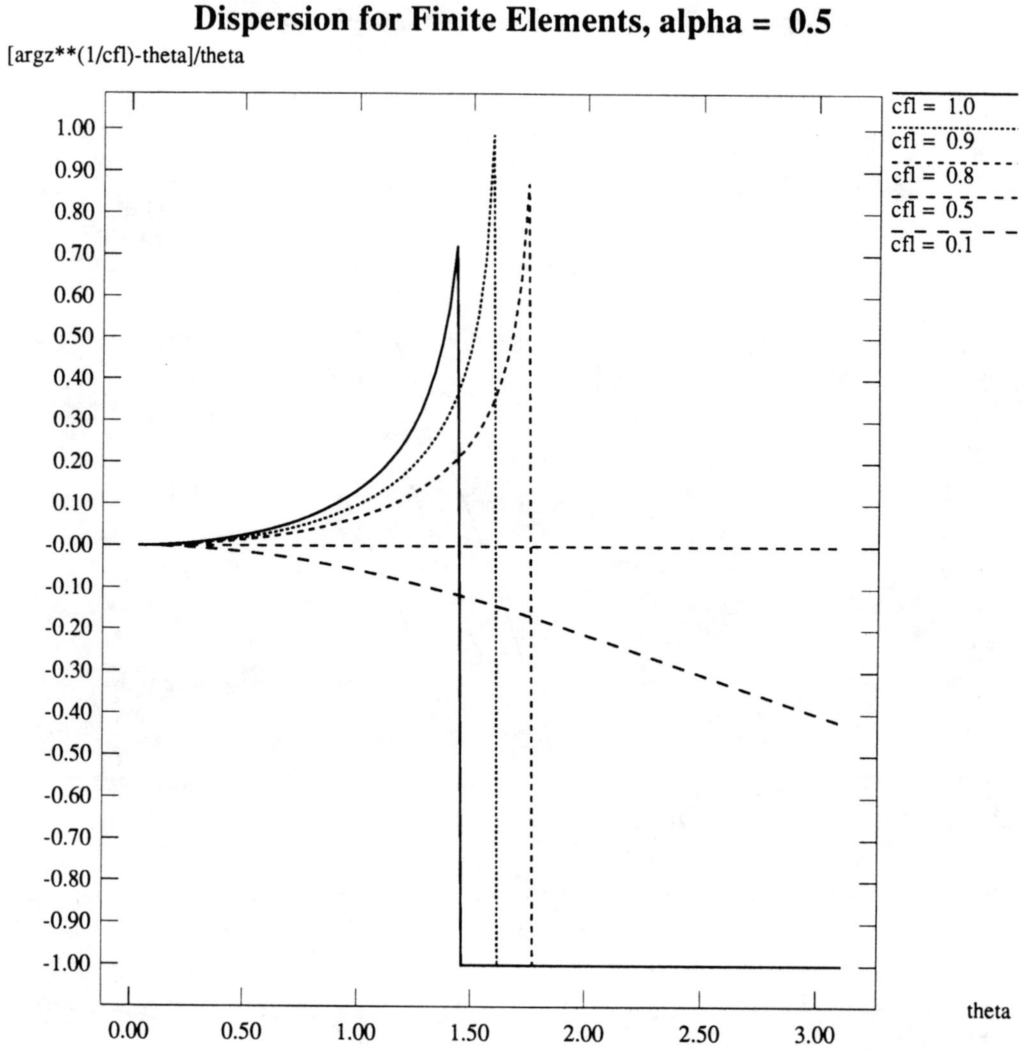
Dispersion for Finite Elements, alpha = 0.5
[argz**(1/cfl)-theta]/theta
cfl = 1.0
cfl = 0.9
cfl = 0.8
cfl = 0.5
cfl = 0.1
1.00
0.90
0.80
0.70
0.60
0.50
0.40
0.30
0.20
0.10
-0.00
-0.10
-0.20
-0.30
-0.40
-0.50
-0.60
-0.70
-0.80
-0.90
-1.00
0.00
0.50
1.00
1.50
2.00
2.50
3.00
theta

FIG. 5.

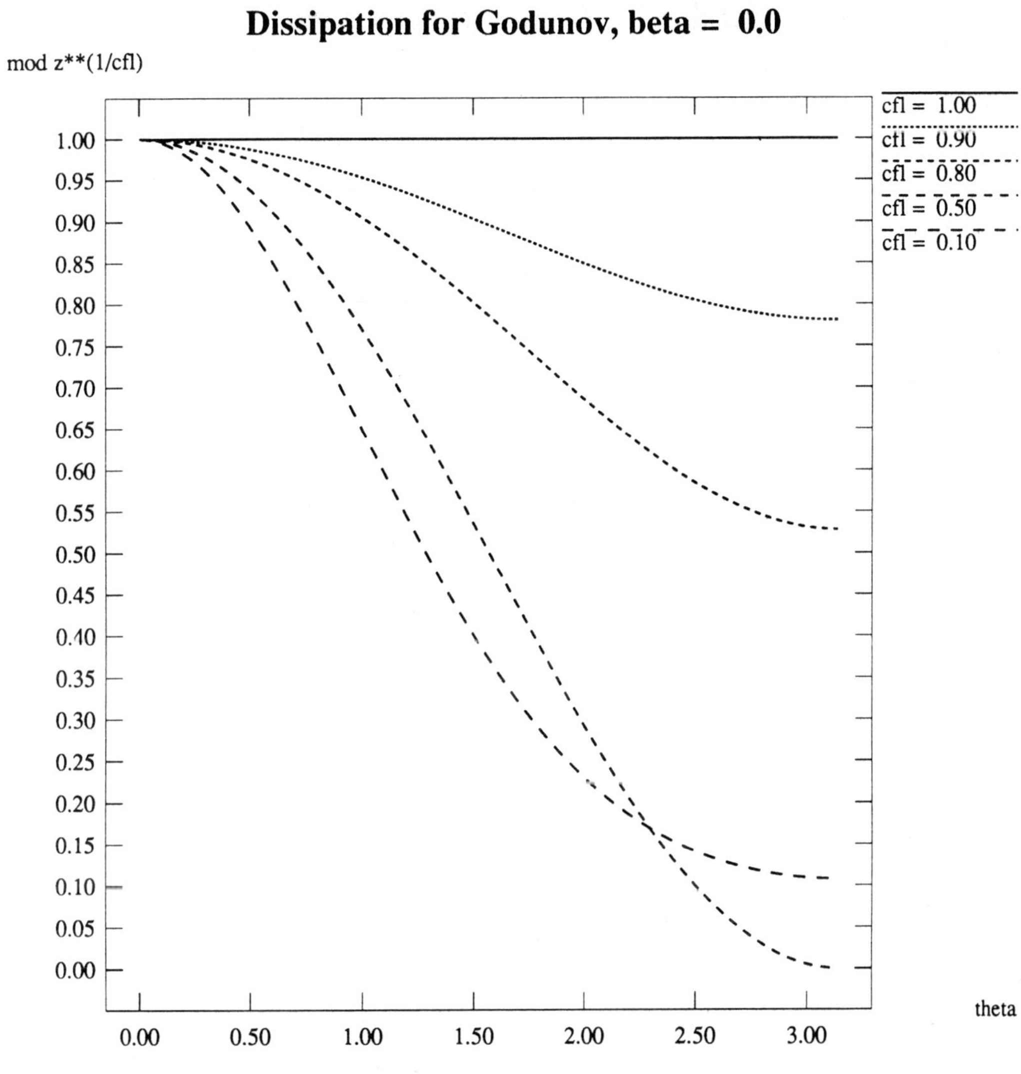
Dissipation for Godunov, beta = 0.0
mod z**(1/cfl)
cfl = 1.00
cfl = 0.90
cfl = 0.80
cfl = 0.50
cfl = 0.10
1.00
0.95
0.90
0.85
0.80
0.75
0.70
0.65
0.60
0.55
0.50
0.45
0.40
0.35
0.30
0.25
0.20
0.15
0.10
0.05
0.00
0.00
0.50
1.00
1.50
2.00
2.50
3.00
theta

FIG. 6.

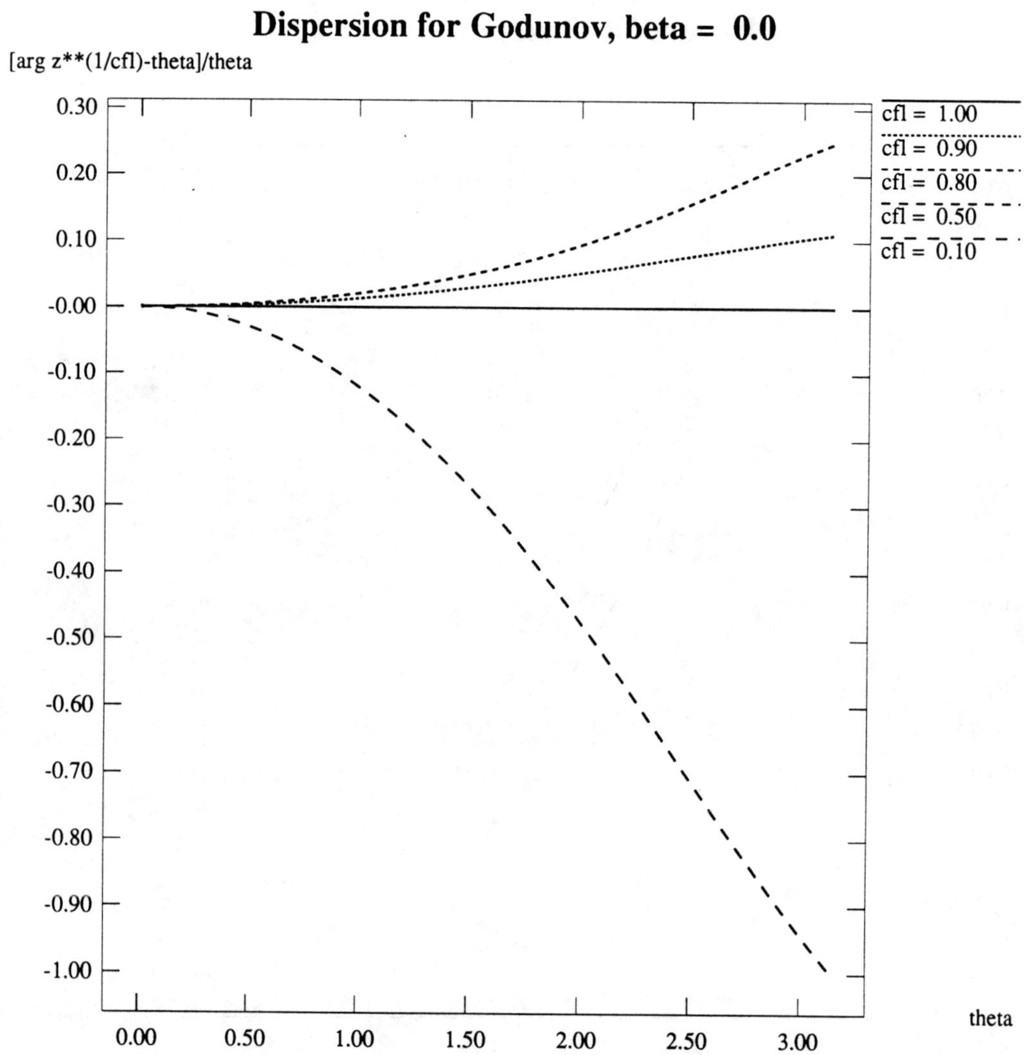
Dispersion for Godunov, beta = 0.0
[arg z**(1/cfl)-theta]/theta
0.30
0.20
0.10
-0.00
-0.10
-0.20
-0.30
-0.40
-0.50
-0.60
-0.70
-0.80
-0.90
-1.00
0.00
0.50
1.00
1.50
2.00
2.50
3.00
theta
cfl = 1.00
cfl = 0.90
cfl = 0.80
cfl = 0.50
cfl = 0.10

FIG. 7.

We consider three examples with this model. In each case, we present the results with the finite element method and the Godunov method. The results with 1000 cells are shown with a solid line, while the results with 100 cells are illustrated with plus signs. In the upper left-hand plot of each figure, the path chosen by the solution of the Riemann problem is shown in velocity-stress space, while the stress-strain relationship is shown in the top right-hand plot. Since the solution of a Riemann problem is self-similar, in the lower four figures we plot velocity, stress, displacement gradient and characteristic speed versus the similarity variable (*i.e.*, distance from the initial discontinuity divided by elapsed time). Thus the horizontal axis of these figures can be used to read the shock speeds, and the plot of the characteristic speeds can be used to verify the Lax entropy condition [17].

In order to plot the finite element velocities, we updated the velocities to the same (full-time level) as the stresses. In order to plot σ versus v, we further averaged the cell edge values for v to obtain cell centered values. Also note that we consider initial discontinuities to occur at cell edges. Thus, we average the velocities on either side of the discontinuity to provide initial data for the velocity at the cell edge in the finite element method. All of the examples involve specifying initial nonzero values for the displacement gradient; for the finite element method, this requires that we integrate the displacement gradient in space to obtain the initial displacements.

The first example involves elastic precursors and plastic shocks moving left and right from an impact. In this case, the initial discontinuity is at the center of the grid, with $\pi = 0.93$, $\epsilon = 1.16$ and $\mathrm{v} = -0.1$ in the left-hand material, and $\pi = 0.56$, $\sigma = 0.196$ and $\mathrm{v} = -3.063$ in the right-hand material. The results with the finite element method using the DYNA 1987 version of the artificial viscosity are shown in figure 8, and the second-order Godunov results are shown in figure 9. In both cases, the timestep has been chosen with CFL = 0.8, in order to allow for the higher characteristic speeds that eventually develop in the middle of the waves. (When we selected CFL = 0.9, small but noticeable noise was detected in both methods near the point of initial impact.) In this example, both methods resolved the waves well.

The second example involves an elastic shock and a strong plastic shock moving left and an elastic-plastic shock moving right. In this case, the initial discontinuity is displaced to the left 5% of the length of the grid from the center. The initial data takes $\pi = 3.25$, $\epsilon = 3.5$ and $\mathrm{v} = 0$ in the left-hand material, and $\pi = 0.217$, $\sigma = 0.207$ and $\mathrm{v} = -5.081$ in the right-hand material. The results with the finite element method are shown in figure 10, and the second-order Godunov results are shown in figure 11. Both computations were run at CFL = 0.5 in order to allow for the large wavespeeds that develop in the center of the grid. Note that the finite element method has significant medium frequency oscillations in the stress. Also note that the Godunov method does a better job of resolving the elastic precursor. On the other hand, the Godunov method has one extra point in the elastic-plastic shock moving to the right; this is due to setting the slopes to zero when changes in the loading conditions are detected (in this case, because the response changes from elastic to plastic compression in the middle of the shock). For this example we could have produced an elastic-plastic shock with only three intermediate cells by removing the reduction to first-order Godunov (*i.e.*, zero slopes) at changes in loading conditions; however, such a modified algorithm would not produce separate elastic and plastic shocks moving to the right in the previous example.

The final example involves an elastic shock moving right toward an elastic rarefaction moving left. Here, there were two initial discontinuities, each at 10% of the length of the grid from either end. The history parameter π was zero throughout the grid. On the left,

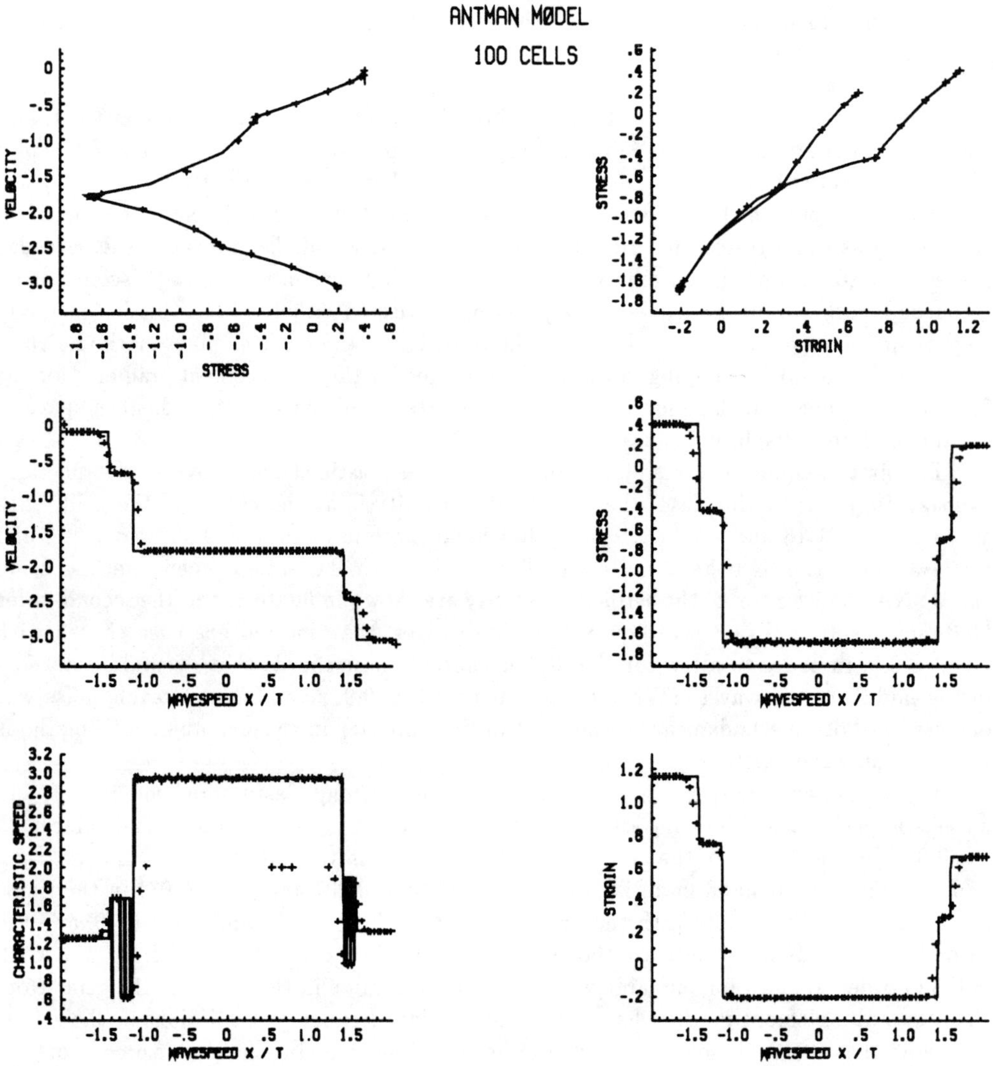

FIG. 8. Finite element method for impact problem

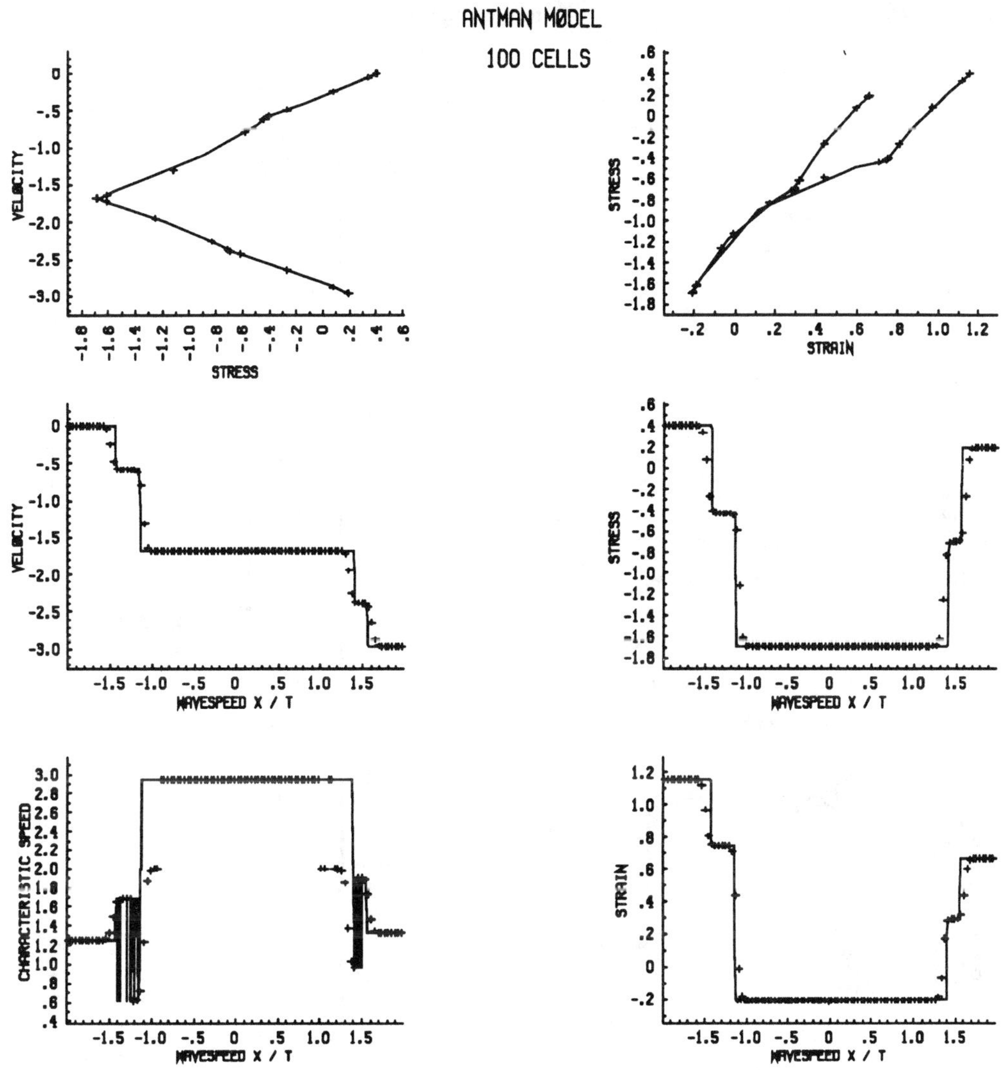

FIG. 9. Second-order Godunov method for impact problem

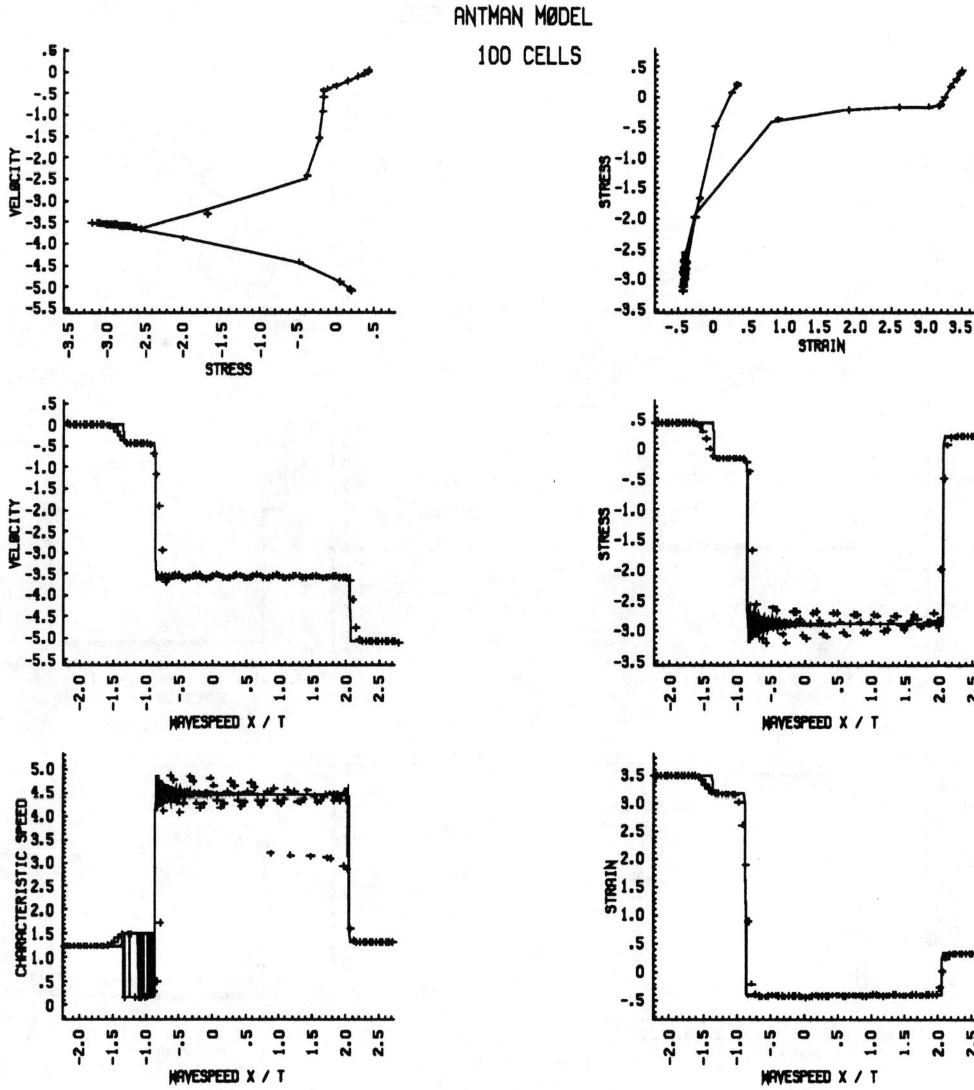

FIG. 10 Finite element method for strong shock

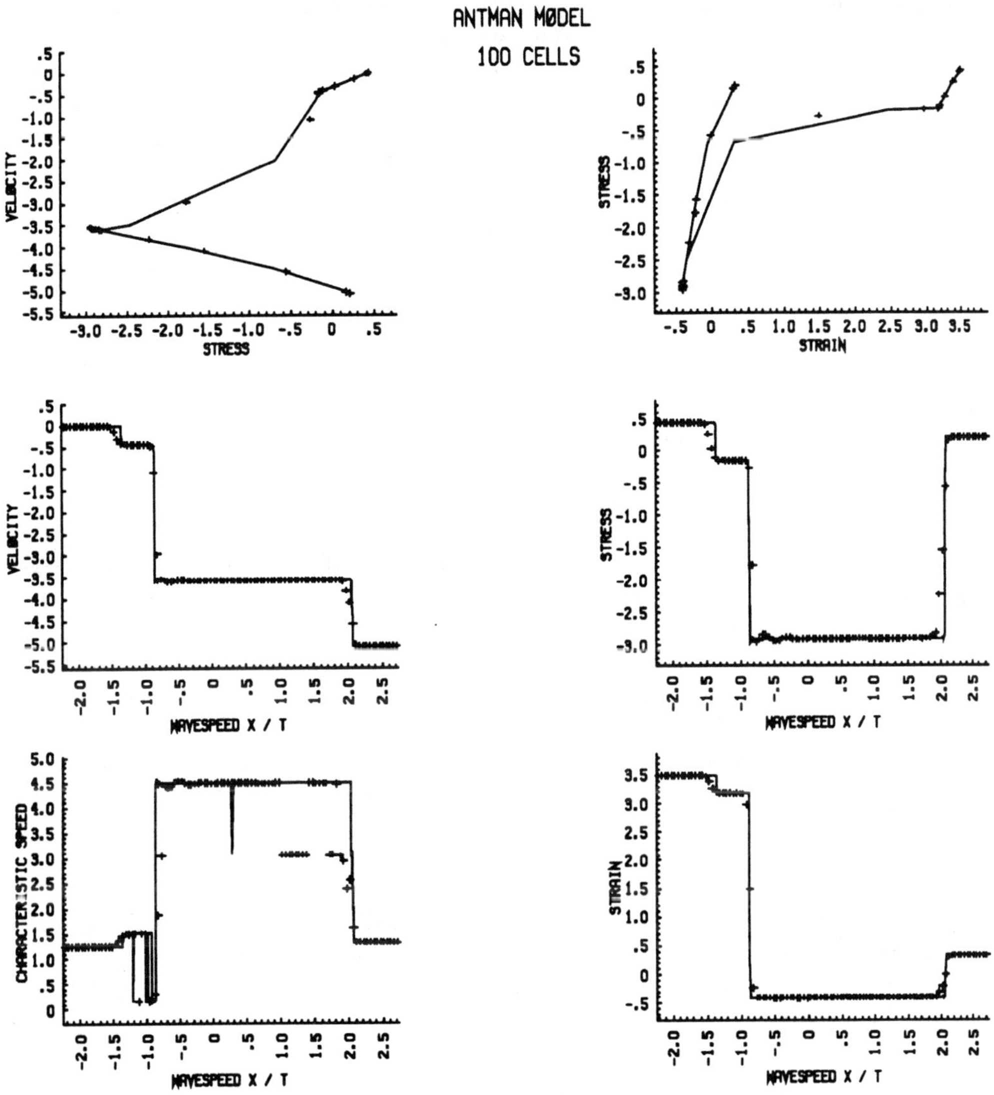

FIG. 11 Second-order Godunov method for strong shock

we took $\sigma = -6$ and $\mathrm{v} = 0.473424$. In the center, our initial data was $\sigma = -3$ and $\mathrm{v} = 0$. On the right, we took $\sigma = 0$ and $\mathrm{v} = 1.20366$. This initial data was determined by an analytic solution to the Riemann problem. Since the largest characteristic speeds in the problem are present in the initial data, we solved the problem with CFL $= 0.9$. The results with the finite element method are shown in figure 12 and the second-order Godunov results are shown in figure 13, at a time before the shock and rarefaction interact. Note that the finite element method has significant oscillations behind this shock, and some noticeable oscillations behind the rarefaction. (Reducing CFL for the finite element method reduced the oscillations behind the shock slightly, but not significantly.) On the other hand, the Godunov method resolved both waves very well. We chose this problem because it appeared that the finite element method should have trouble with shocks connecting into rarefactions: once the waves collide, the expansion of the rarefaction should turn off the artificial viscosity, leaving no dissipation in the method on that side of the shock. However, we found that the difficulties with the finite element method occured much earlier than expected.

6. Summary. It is easy to see from the descriptions that the finite element method with explicit artificial viscosity is less complicated than the Godunov method. In particular, the finite element method does not use characteristics or Riemann problem solutions, and (except for some versions of the artificial viscosity) forms only linear differences of the flow variables. On the other hand, through extra work the Godunov method produces a very nice balance of dispersion and dissipation: whenever it disperses a wave, it includes dissipation to damp the error. In contrast, the finite element method without artificial viscosity makes dispersive errors without providing sufficient dissipation; as a result, the dispersive errors fall behind as oscillations that do not go away.

Although the Godunov method can be significantly more expensive than the finite element method for very simple material models, both models require only one evaluation of the stress per cell and timestep. Thus, for complicated material models both models will be essentially equal in cost per cell and timestep. To make this statement more quantitative, we compared the timing of both methods for 1000 timesteps with 1000 cells on the strong shock of §5 (example number 2). The computations were run on a CRAY XMP, and were fully vectorized. The finite element method was approximately 7.5 times faster than the Godunov method, if we included only the time for the integration method and excluded the time for the stress update and determination of the characteristic speeds. If the latter costs are included, then the finite element method was only twice as fast as the second-order Godunov method for the Antman-Szymczak model, which is a very simple model to evaluate. For more complicated material models, the two methods are essentially equal in cost per timestep.

The true measure of the relative performance of the two methods should be based on some determination of the relative cost required to achieve a desired level of accuracy. The numerical results in this paper show that such a comparison could be highly problem-dependent. This is because the finite element method used in this paper performs better on some problems than on others.

Another way to state the results of this study is that the Godunov method does a better overall job of controlling numerical oscillations and resolving discontinuities without requiring adjustments in the parameters. While the coefficients in slope flattening (2.10) can viewed as adjustable parameters in the Godunov method, in practice these parameters do not need to be adjusted, and have only been introduced to improve the performance on sharp discontinuities, not discontinuities in general. The bulk of the artificial viscosity for the Godunov method enters naturally through the slope-limiting process, without any user

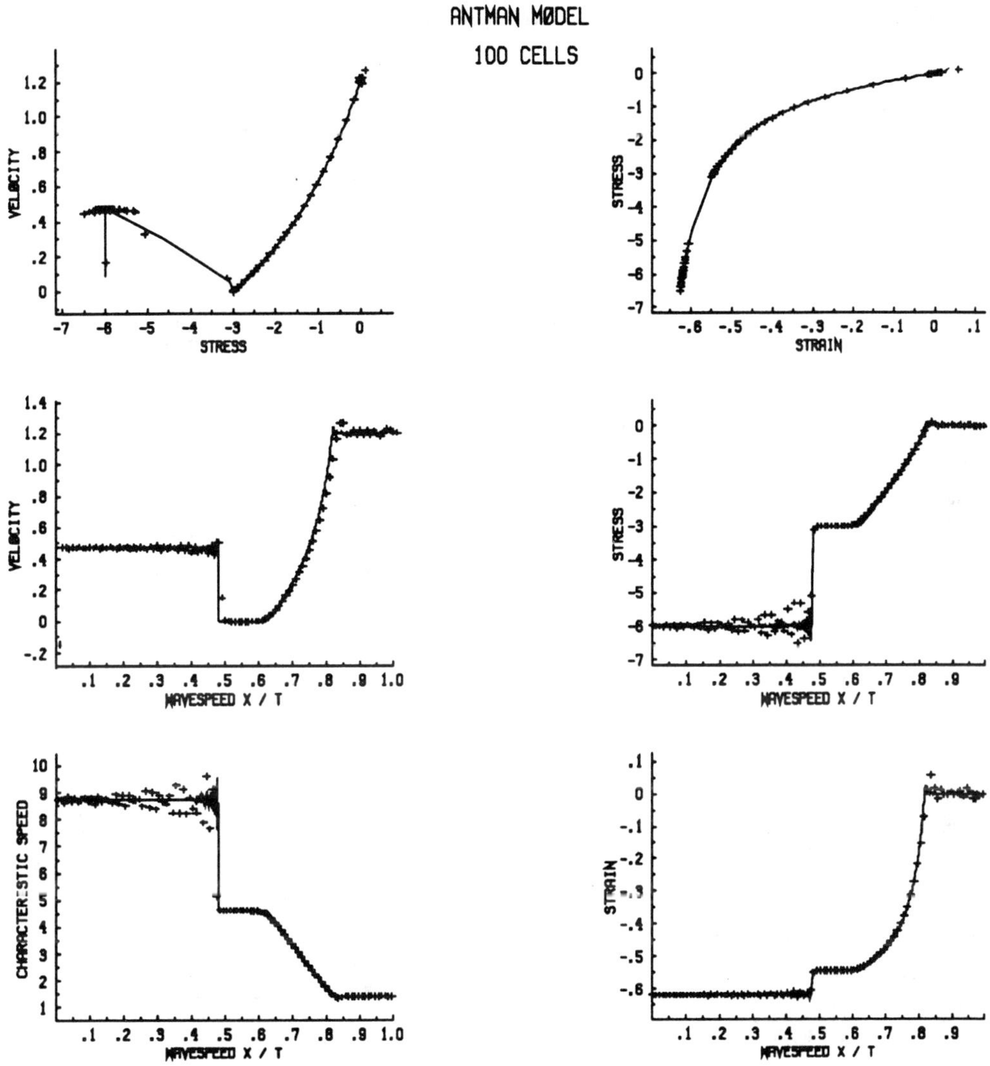

FIG. 12 Finite element method for elastic shock moving toward elastic rarefaction

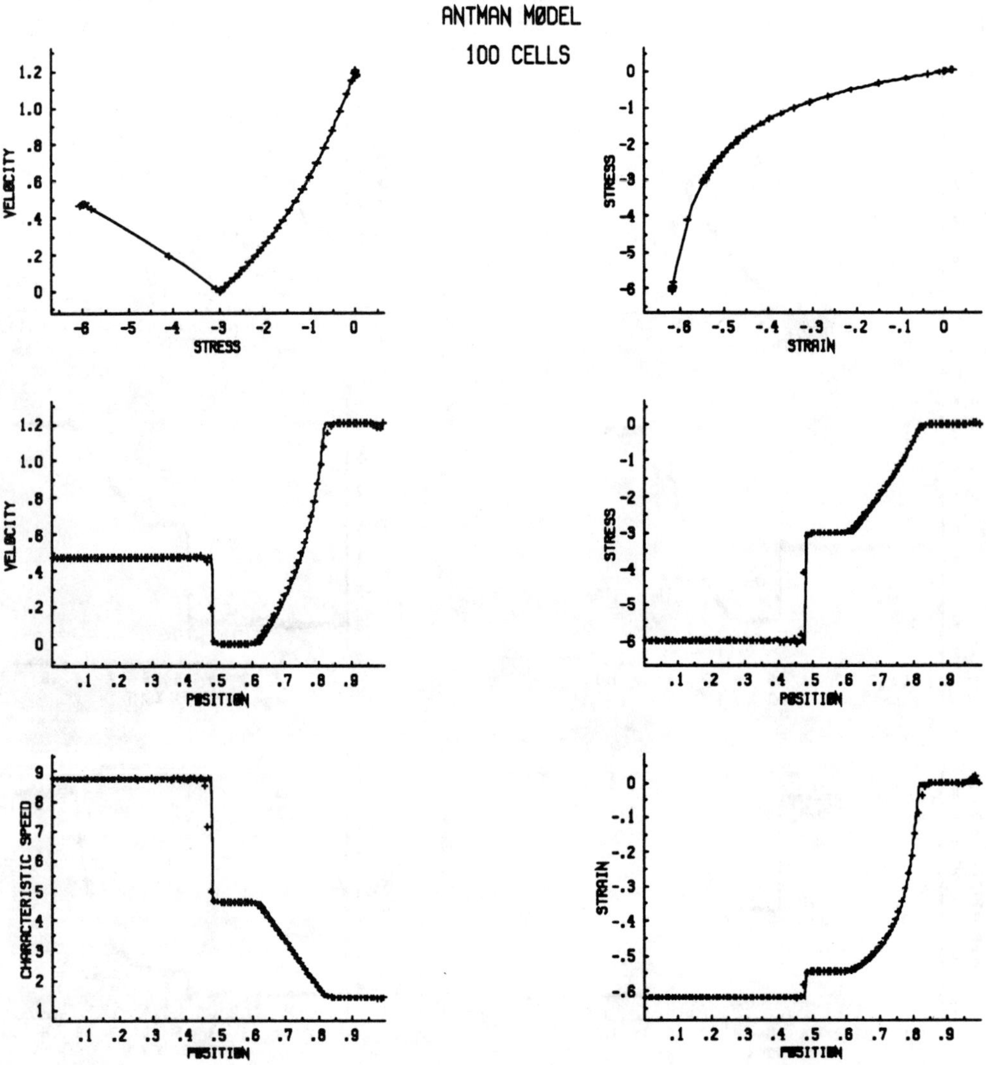

FIG. 13 Second-order Godunov method for elastic shock moving toward elastic rarefaction

interaction.

In summary, the numerical results indicate that the finite element method either needs a different form of the artificial viscosity than the one used in this paper, or it must vary its artificial viscosity to suit the problem at hand. The latter solution to the problem of controlling numerical oscillations is not very satisfactory, however. In problems with several different waves, it might not be possible to adjust the artificial viscosity to handle all of the waves well. For these reasons, the issue of designing appropriate forms of artificial viscosity for the finite element method is still a topic of research at LLNL and other institutions.

7. Acknowledgments. We would like to thank Louis Howell of LLNL for reading a preliminary draft of this document and making several useful suggestions.

REFERENCES

[1] S. S. Antman and W. G. Szymczak. Nonlinear elastoplastic waves. *Contemporary Mathematics*, submitted 1988.

[2] A.K. Aziz, editor. *The Mathematical Foundations of the Finite Element Method with Applications to Partial Differential Equations*. Academic, 1972.

[3] K.-J. Bathe and E. L. Wilson. *Numerical Methods in Finite Element Analysis*. Prentice-Hall, 1976.

[4] J.B. Bell, P. Colella, and J.A. Trangenstein. Higher-order Godunov methods for general systems of hyperbolic conservation laws. *J. Comp. Phys.*, 82:362–397, 1989.

[5] T. Belytschko and T.J.R. Hughes, editors. *Computational Methods for Transient Analysis*, volume 1 of *Computational Methods in Mechanics*. North-Holland, 1983.

[6] P.G. Ciarlet. *The Finite Element Method for Elliptic Problems*. North-Holland, 1978.

[7] P. Colella. A direct eulerian MUSCL scheme for gas dynamics. *SIAM J. Sci. Stat. Comput.*, 6:104–117, 1985.

[8] N. Cristescu. *Dynamic Plasticity*. North-Holland, 1967.

[9] C. M. Dafermos. The entropy rate admissibility criterion for solutions of hyperbolic conservation laws. *J. Diff. Eqns.*, 14:202–212, 1973.

[10] B. Engquist and S. Osher. Stable and entropy satisfying approximations for transonic flow calculations. *Math. Comp.*, 34:45–75, 1980.

[11] J. Goodman and A. Majda. The validity of the modified equation for nonlinear shock waves. *J. Comp. Phys.*, 58:336–348, 1985.

[12] J.O. Hallquist. DYNA3D course notes. Technical Report UCRL-19899, Rev. 2, University of California, Lawrence Livermore National Laboratory, 1987.

[13] A. Harten, P.D. Lax, and B. van Leer. On upstream differencing and Godunov-type schemes for hyperbolic conservation laws. *SIAM Review*, 25:35–61, 1983.

[14] T. J. R. Hughes and J.E. Marsden. Classical elastodynamics as a linear symmetric hyperbolic system. *J. Elasticity*, 8:97–110, 1978.

[15] G.M. Hulbert. *Space-Time Finite Element Methods for Second-Order Hyperbolic Equations*. PhD thesis, Stanford University, Department of Mechanical Engineering, 1989.

[16] C. Johnson, A. Szepessy, and P. Hansbo. On the convergence of shock-capturing streamline diffusion finite element methods for hyperbolic conservation laws. *Math. Comp.*, 54:107–129, 1990.

[17] P.D. Lax. Hyperbolic systems of conservation laws II. *Comm. Pure Appl. Math.*, 10:537–566, 1957.

[18] P.D. Lax. *Hyperbolic Systems of Conservation Laws and the Mathematical Theory of Shock Waves*, volume 11 of *Regional Conference Series in Applied Mathematics*. SIAM, 1973.

[19] K.W. Morton and P.K. Sweby. A comparison of flux limited difference methods and characteristic galerkin methods for shock modeling. *J. Comp. Phys.*, 73:203–230, 1987.

[20] W.F. Noh. Numerical methods in hydrodynamic calculations. Technical Report UCRL-52112, University of California, Lawrence Livermore National Laboratory, 1976.

[21] G.A. Sod. *Numerical Methods in Fluid Dynamics*. Cambridge University, 1985.

[22] G. Strang and G.J. Fix. *An Analysis of the Finite Element Method*. Prentice-Hall, 1973.

[23] P. Tong. On the numerical problems of the finite element methods. In G.L.M. Gladwell, editor, *SMD Symposium on Computer-Aided Engineering*. University of Waterloo, 1971.

[24] J.A. Trangenstein. A second-order algorithm for the dynamic response of soils. *Impact of Computing in Science and Engineering*, 2:1–39, 1990.

[25] J.A. Trangenstein and P. Colella. A higher-order Godunov method for modeling finite deformation in elastic-plastic solids. *Comm. Pure Appl. Math.*, to appear.

[26] J.A. Trangenstein and R.B. Pember. The Riemann problem for longitudinal motion in an elastic-plastic bar. *SIAM J. Sci. Stat. Comput.*, to appear.

[27] J.A. Trangenstein, R.B. Pember, and P. Colella. Numerical algorithms for strong discontinuities in elastic-plastic solids. Submitted to J. Comp. Phys., 1990.

[28] B. van Leer. Towards the ultimate conservative difference scheme. IV. a new approach to numerical convection. *J. Comp. Phys.*, 23:276–299, 1977.

[29] J. von Neumann and R.D. Richtmyer. A method for the numerical calculation of hydrodynamic shocks. *J. Appl. Phys.*, 21:232–237, 1950.

[30] R.F. Warming and B.J. Hyett. The modified equation approach to the stability and accuracy analysis of finite-difference methods. *J. Comp. Phys.*, 14:159–179, 1974.

[31] M.L. Wilkins. Calculation of elastic-plastic flow. *Methods of Computational Physics*, 3:211–263, 1964.

[32] P. Woodward and P. Colella. The numerical simulation of two-dimensional fluid flow with strong shocks. *J. Comp. Phys.*, 54:115–173, 1984.

[33] J.A. Zukas, T. Nicholas, H.F. Swift, L.B. Greszczuk, and D.R. Curran. *Impact Dynamics*. Wiley-Interscience, 1982.

CHAPTER 17

New Theory of MHD Shock Waves*

C. C. Wu†

Abstract. Recent results on MHD shock waves are presented. The studies are based on one-dimensional MHD Navier-Stokes equations. The main result is that whenever a rotation of the magnetic field occurs, the conventional theory, which is a discontinuity theory, is not a good approximation to the dissipative MHD. In particular, we show that intermediate shocks can exist, but rotational discontinuities can not. We show that the evolution of intermediate shocks depends on dissipative mechanism and the Riemann problems are not well-posed in the usual sense. We also show the existence of time-dependent shocks, which do not satisfy Rankine-Hugoniot jump conditions. In addition, we discuss the shock waves for the Modified KdV-Burgers equation, and show that, as in MHD, global solutions of this equation depend on the shock structure.

1. Introduction

There are two fundamental approaches to the shock waves in fluid models: the discontinuity theory and the viscosity method. In the discontinuity theory, the approach is based on conservation laws. Shocks are treated as discontinuities and they must satisfy the Rankine-Hugoniot jump relations. On the other hand, in the viscosity method, Navier-Stokes equations for gas dynamics and dissipative MHD equations for MHD are used; the solutions are smooth everywhere and shock waves are represented by thin layers. In comparing these two approaches, one notices that the viscosity method takes precedence over the discontinuity theory. This is because the physical flows such as in gas dynamics can be properly described by the Navier Stokes equations. On the other hand, the discontinuity theory is only an approximation to the viscous method. The advantage of the discontinuity theory is that it is not required to resolve the shock layers. One can easily see that in order for the discontinuity theory to be a good approximation, shock waves have to have shock structures independent of the dissipation mechanism, and a Riemann problem,

*This work was supported by the NSF grant ATM-89-22133 and by the NASA grant NAGW-1649. Computing was provided by the San Diego Supercomputer Center, which is sponsored by the National Science Foundation.
†Department of Physics, University of California at Los Angeles, Los Angeles, CA 90024.

which is an initial value problem whose initial state consists of two constant states separated by a diaphragm, has to have a unique solution. For gas dynamics, the discontinuity theory is a good approximation to the viscosity method; hence, one can solve the gas dynamics problems without considering complex Navier-Stokes equations.

The conventional MHD shock wave theory is basically a discontinuity theory. The conventional theory ruled out a whole class of intermediate shocks because they do not satisfy certain requirements of the discontinuity theory. The intermediate shocks are shocks with shock frame fluid velocities greater than the Alfvén speed ahead and less than the Alfvén speed behind, or, equivalently, across intermediate shocks the sign of the transverse component of the magnetic field changes.

However, in a series of numerical and analytical studies of the one-dimensional dissipative MHD equations [1-4], we have shown (1) that stable intermediate shocks can evolve from smooth initial conditions; (2) that the evolution of intermediate shocks depends on dissipative mechanism; (3) that a rotational discontinuity, which has no structure in dissipative MHD, is unstable; (4) that there exists a class of time-dependent intermediate-shock-like structures which do not obey the Rankine-Hugoniot conditions and which, in particular, violate coplanarity; (5) that Riemann problems are not well-posed in the traditional sense and can have a family of solutions whenever there is a rotation in the magnetic field; (6) that there are free parameters in the structures of intermediate shocks, and these parameters are related to the shock stability.

In this paper, these results will be presented. In addition, we will point out the differences between the conventional theory and ours. In Section 2, we will review the conventional MHD shock wave theory and present the reasons why intermediate shocks were considered extraneous and not-physical. We will present the new theory in Section 3.

Our approach in studying the MHD shock waves is largely based on numerical examples. Except in a few cases, general proofs of the results are not yet available. However, for a simpler system of CKB (Cohen-Kulsrud-Burgers) equations, which involves only intermediate and fast waves or intermediate and slow waves, one can prove rigorously some features of the intermediate shocks and solve the Riemann problems more or less analytically. The details can be found in the paper by Kennel et al. [5]. In our new theory of MHD shock waves, it is essential to take into account the dissipation mechanism in considering shock waves. This may be a general property for any system which possesses more than two stationary points and whose shock structure depends on a dissipation mechanism. In Section 4, we will give another example to show this property. We will discuss the shock waves for the Modified KdV-Burgers (MKdVB) equation and show how global solutions of this equation depend on the shock structure which in turn depends on the ratio of the dispersive scale length to the dissipative scale length. We will also show the existence of under-compressive shocks in this equation.

2. Conventional Theory of Shock Waves in MHD

The dissipative MHD equations with resistivity, viscosity, and thermal conductivity, are [6]

$$\frac{\partial \rho}{\partial t} = -\nabla \cdot (\rho \mathbf{v}) \tag{1}$$

$$\frac{\partial (\rho \mathbf{v})}{\partial t} = -\nabla \cdot (\rho \mathbf{v}\mathbf{v} + \mathbf{I}(p + \frac{B^2}{2}) - \mathbf{B}\mathbf{B}) + \nu \nabla^2 \mathbf{v} + (\mu + \frac{\nu}{3}) \nabla (\nabla \cdot \mathbf{v}) \tag{2}$$

$$\frac{\partial \mathbf{B}}{\partial t} = -\nabla \times \mathbf{E} \tag{3}$$

$$\frac{\partial \epsilon}{\partial t} = -\nabla \cdot ((\frac{\rho v^2}{2} + \frac{p}{\gamma - 1} + p)\mathbf{v} + \mathbf{E} \times \mathbf{B}) + \nabla \cdot \sigma \cdot \mathbf{v} + \kappa \nabla^2 (\frac{p}{\rho}) \tag{4}$$

Here, σ is the viscous tensor, $\sigma_{ik} = \nu(\partial v_i/\partial x_k + \partial v_k/\partial x_i - 2/3\delta_{ik}\,\partial v_l/\partial x_l) + \mu\delta_{ik}\partial v_l/\partial x_l$. The mass density, pressure, velocity, and magnetic field are denoted by $\rho, p, \mathbf{v}$, and $\mathbf{B}$, respectively; γ is the ratio of the specific heats, and the energy density (ϵ) is given by $\epsilon = \rho v^2/2 + B^2/2 + p/(\gamma - 1)$. The electric field $\mathbf{E}$ is given by Ohm's law, $\eta\mathbf{J} = \mathbf{E} + \mathbf{v} \times \mathbf{B}$, with the current density $\mathbf{J} = \nabla \times \mathbf{B}$ and η constant resistivity. ν and μ are the two coefficients of viscosity, and κ is the thermal conductivity. In addition to these equations, we require $\nabla \cdot \mathbf{B} = 0$. In these equations, we have used units so that factors such as 4π and c do not appear. In this paper, we only consider one-dimensional flow along the x direction, then $\partial/\partial y = \partial/\partial z = 0$, and B_x is a constant. When $\eta = \nu = \mu = 0$, these equations are ideal MHD equations.

The conventional view of shock waves in MHD is summarized here; for details see Kantrowitz and Petschek [7] or Jeffrey and Taniuti [8]. It is basically a discontinuity theory, in which the solution is composed of smooth parts which satisfy (1-4) without the dissipation terms, connected by discontinuities which satisfy shock jump conditions. Therefore it is required that the MHD shock waves satisfy the MHD Rankine-Hugoniot jump relations. As in gas dynamics, the entropy condition in MHD requires that entropy increases across a shock, and so shock waves are compressive, which means both density and pressure increase across a shock.

Besides the entropy wave, where only density changes, there are three characteristic wave modes in MHD. According to the magnitude of the wave speeds, they are called fast, intermediate, and slow waves. Depending on the direction of the magnetic field, these wave speeds may coincide. Thus the system of the MHD equations are not strictly hyperbolic. In general, fast and slow waves involve changes in density, pressure, fluid velocity and the magnitude of the magnetic field. However, when the transverse magnetic field is zero, either the slow or fast wave becomes degenerate. Thus, both fast and slow waves are nonconvex [9]. For both fast and slow waves, the fluid velocity and the magnetic field vary only in a plane and do not rotate across the wave front. On the other hand, intermediate waves are linearly degenerate (non-compressive). Across an intermediate wave, density, pressure, the normal fluid velocity, and the magnitude of the magnetic field remain constant. Only the fluid velocity and the magnetic field rotate across the wave. This means that an intermediate wave is three-dimensional. Thus intermediate waves are not included in the case when we set $B_z = v_z = 0$, as we will do in some computations later.

Let us define state 1 as the state whose shock frame fluid velocity is superfast, state 2 whose fluid velocity is subfast and super-Alfvénic, state 3 whose fluid velocity is sub-Alfvénic and superslow, and state 4 whose fluid velocity is subslow. (In our definition, Alfvén speed is the same as intermediate wave speed.) The entropy-satisfying Rankine-Hugoniot jump relations then include $1 \to 2$, $1 \to 3$, $1 \to 4$, $2 \to 3$, $2 \to 4$, and $3 \to 4$ transitions. The $1 \to 2$ and $3 \to 4$ shocks are fast and slow shocks, respectively. The other four transitions are called intermediate shocks. In Figure 1 the shock classification is summarized according to the converging characteristics. For a fast shock, there is a pair of converging fast waves into the shock. Similarly, there is a pair of slow waves converging into a slow shock. For a $2 \to 3$ intermediate shock, one has a pair of intermediate waves. For a $1 \to 3$ intermediate shock, one can have both fast and intermediate waves. A $2 \to 4$ intermediate shock can have both slow and intermediate waves. Finally, it can have all fast, intermediate and slow waves converging into a $1 \to 4$ intermediate shocks. There are limiting cases when one of the converging characteristics has the same speed as the shock. Switch-on shocks, which will be denoted by $1 \to 2 = 3$ and whose downstream intermediate wave speed is the same as the shock speed, are limiting cases of $1 \to 2$ fast shocks and $1 \to 3$ intermediate shocks. Similarly, $2 = 3 \to 4$ switch-off shocks, whose upstream intermediate wave speed is the same as the shock speed, are limiting cases of $3 \to 4$ slow shocks and $2 \to 4$ intermediate shocks. In addition, when the downstream slow wave speed is the same as the shock speed, one has limiting cases $1 \to 3 = 4$ and $2 \to 3 = 4$. When the upstream fast wave speed is the same as the shock speed, one has limiting cases $1 = 2 \to 3$ and $1 = 2 \to 4$. Note that

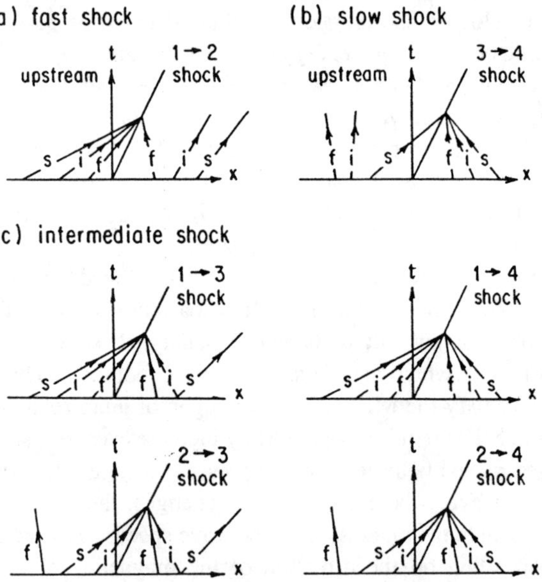

Figure 1: MHD shock classification according to converging characteristics. Shown are for shocks due to $v_x - C_F$, $v_x - C_I$ and $v_x - C_S$ characteristics with upstream states at $x = -\infty$. Symbols f, i, and s denote fast, intermediate, and slow characteristics, respectively.

$1 \to 3$, $1 \to 4$, and $2 \to 4$ intermediate shocks are over-compressive, and that when $B_z = v_z = 0$ identically (namely, no intermediate waves), the $2 \to 3$ shocks are under-compressive.

All these shock waves according to the Rankine-Hugoniot jump relations are coplanar. It means that if both B_z and v_z vanish at the upstream side then they also vanish at the downstream side, and vice versa. The magnitude of the transverse magnetic field increases across a fast shock, and decreases across a slow shock. However, the direction of the transverse magnetic field remains the same across either fast or slow shocks. On the other hand, across an intermediate shock, the direction of the transverse magnetic field changes and the magnitude of the transverse magnetic field can either increase or decrease.

Besides the shock waves, there are other discontinuities in the MHD Rankine-Hugoniot relations. In the case that the normal component of the magnetic field is nonzero, i.e. $B_x \neq 0$, one such discontinuity is called the contact discontinuity, which moves with the fluid velocity and the only discontinuity across which is the density. Another discontinuity is called the rotational discontinuity. It involves rotation of a transverse magnetic field and a transverse velocity without changing the magnitude of the magnetic field, p, ρ, v_x, and entropy. The speed of the discontinuity is given by the intermediate characteristic speed $v_x \pm B_x/\sqrt{\rho}$, which is constant across the discontinuity. A rotational discontinuity is three-dimensional in general. In the limit that the transverse magnetic field reverses its sign, it is called a 180° Alfvén wave, which can also be considered as a limiting case of intermediate shock. In the case that the normal component of the magnetic field is zero, one can have in the Rankine-Hugoniot relations a tangential discontinuity, across which plasma parameters can change under the constraint that the total pressure, $p + B^2/2$, is constant.

In the conventional view, intermediate shocks were considered extraneous, and not thought to exist in physical reality. There had been numerous discussions on the inadmissibility of the intermediate shock. Overall there are three arguments. The first argument was provided by Germain [10], who proved that when μ is much larger than ν or η there is no structure for the intermediate shocks. Thus, an intermediate shock may or may not have a shock structure depending on the dissipation mechanism. This contradicts one requirement of the discontinuity theory that shock structure must exits, independent of dissipation mechanism.

The second argument for not considering intermediate shocks physical was provided by Akhiezer et al. [11], and by Taniuti [8]. It was based on linear stability considerations of ideal MHD equations. By considering a small perturbation of a shock wave, it was reasoned that a stable shock should have $(n-1)$ outgoing waves, where n is the number of equations ($n = 7$ in MHD). For intermediate shocks, the number of outgoing waves is less than 6, and thus there is no solution to the linearized problem. (For the $2 \rightarrow 3$ intermediate shocks, there are 6 outgoing waves, but they were considered unphysical due to an additional requirement explained later.) Based on this "evolutionary condition," intermediate shocks were disregarded.

The third argument was proposed by Kantrowitz and Petschek [7]. Their argument is like this. Suppose an intermediate shock exists. Since the shock frame fluid velocity is super-Alfvénic ahead and sub-Alfvénic behind the intermediate shock, the intermediate waves are convergent on the intermediate shock. Thus the intermediate wave cannot propagate ahead of the shock, neither can it stay behind the shock. Furthermore, since the MHD shock waves according to the MHD Rankine-Hugoniot jump conditions are planar, the plane of the magnetic field cannot rotate across the shock wave, and therefore the intermediate wave can also not exist in the middle of the shock wave. Thus a flow containing an intermediate shock does not have any neighboring solutions corresponding to a small change in the boundary conditions which requires an arbitrarily small angle of rotation of the plane of the magnetic field. Since the magnetic field is expected not to be purely planar, Kantrowitz and Petschek concluded that the intermediate shocks can not occur in nature.

The above arguments indicate that certain properties of intermediate shocks are inconsistent with the requirements of the discontinuity theory and that the intermediate shock should thus be disregarded. But, as was mentioned in Section 1, a discontinuity theory is only an approximation. The question that one should ask is whether the conventional MHD shock theory is a good approximation to the set of dissipative MHD equations. The answer is no whenever there is a rotation of the transverse magnetic field from the upstream to the downstream regions. The evidence will be provided in the next section.

Our study of intermediate shocks was due to our interest in the dayside reconnection processes at the Earth's magnetopause, which had led us to consider MHD Riemann problems with nonzero normal component of magnetic field and with initial left- and right-state transverse magnetic fields having opposite signs. In the conventional view as described above, the solutions of the Riemann problems would require a 180° Alfvén wave (rotational discontinuity) to facilitate the change of the sign of the transverse magnetic field. Instead, intermediate shocks were obtained by Brio and Wu [9,12] in the numerical calculations with many numerical shock-capturing schemes, including an MHD upwind scheme, the Lax-Friedrichs scheme, the FCT, and the Lax-Wendroff scheme. Therefore, we had a strong conflict between the traditional view of the intermediate shocks and the numerical solutions. This started our investigation into intermediate shocks.

3. New Theory of MHD Shock Waves

In this section, we use the viscosity method approach and consider the dissipative MHD equations. The dissipation coefficients are assumed to be small but finite. The initial value problems defined by the dissipative MHD equations are presumably well-posed. Thus solutions

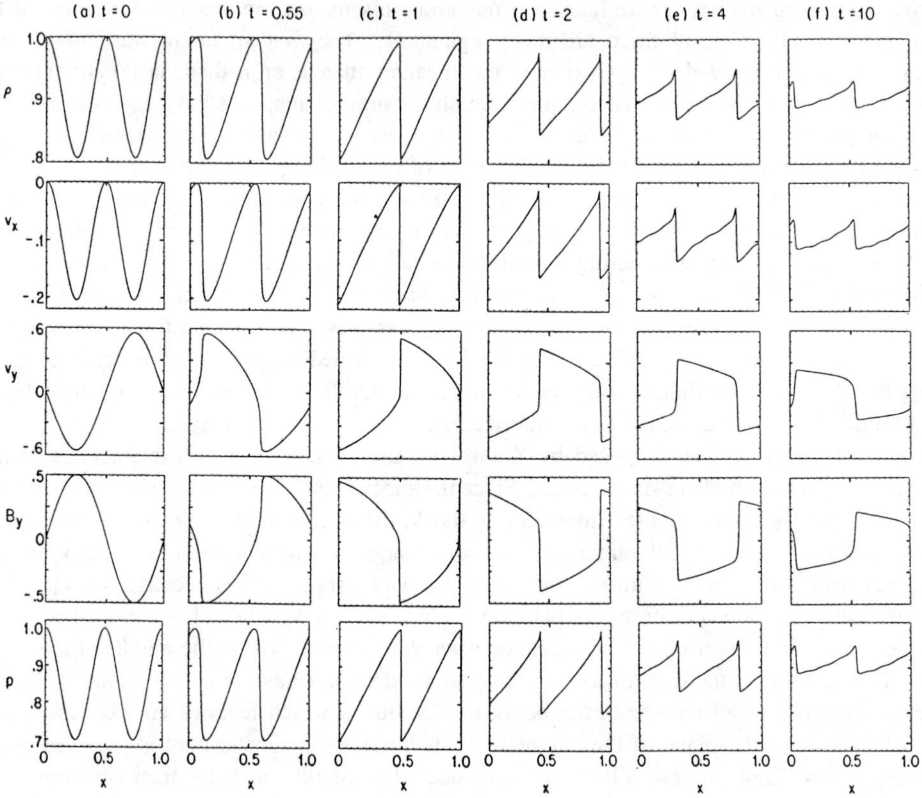

Figure 2: Formation of 2 → 4 intermediate shocks from a slow simple wave (from [4]).

of the problems remain smooth all the time and a shock is not a discontinuity, but a thin layer whose structure is described by the dissipative equations. In the following, we summarize some of the results from several previous papers [1-4].

(a) Formation of intermediate shocks

Case 1: Formation of 2 → 4 intermediate shock from a slow simple wave

In the first example, we present the evolution of a slow-mode simple wave solution in Figures 2 and 3. The initial condition is given by $B_y = 0.5\sin(2\pi x)$, $0 \le x \le 1$, with periodic boundary conditions. The other quantities were then numerically obtained from simple wave relations [1]. We have chosen such a simple initial distribution in which only one wave is involved initially so that one can easily identify the characteristics of the wave and the properties of the shocks formed. However, the result which shows that intermediate shocks can be formed through nonlinear steepening processes is quite general [4]. All quantities were normalized with respect to the following prescribed values at $B_y = 0$: $B_x = 1$, $\rho = 1$, $p = 1$, $\mathbf{v} = \mathbf{0}$, $B_z = 0$. Note that $v_z = 0$, $B_z = 0$; the problem is planar. We used $\gamma = 5/3$, thus the sound speed at $B_y = 0$ is $\sqrt{5/3}$. Since the sound speed is greater than the Alfvén speed at $B_y = 0$, the slow simple wave can cross $B_y = 0$.

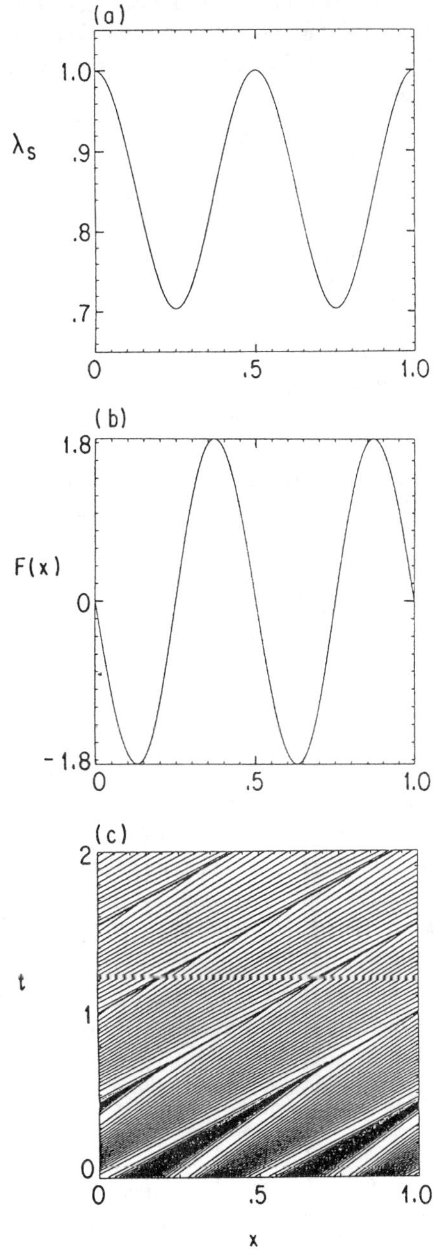

Figure 3: (a) $v_x + C_S$ versus x at $t = 0$; (b) a plot of the rate of change of $v_x + C_S$, $F(x)$, versus x at $t = 0$; (c) contours of the right-moving slow characteristic speed, $v_x + C_S$, in the (x, t) diagram for $2 \geq t \geq 0$.

For solving the dissipative MHD equations, we used an Eulerian uniform mesh containing 1600 to 6400 points and an explicit two-step Lax-Wendroff scheme [13]. The treatment of the dissipation terms was also explicit. Therefore the time step size was limited by the CFL condition for the flow and by the diffusion time scale. The convergence of the numerical results with respect to the size of the grid spacing has been checked [1].

The results was obtained with η = 0.0004, and $\mu = \nu = 0$. The column (a) of Figure 2 shows the initial state at t = 0. The subsequent columns for t = 0.55, 1, 2, 4, and 10 show that the initially smooth profile steepens to form $2 \to 4$ intermediate shocks. In Figure 3(c), the contours of the right-moving slow characteristic speed are plotted in the (x, t) diagram for $2 \geq t \geq 0$, Figure 3(a) plots the initial slow characteristic speed versus x, and Figure 3(b) plots the rate of change of the initial slow characteristic speed, $F(x)$. Before the formation of the shocks, all variables are constant on the characteristic curves and these curves are straight lines. From Figure 3(a), one sees that, initially, it is compressive in the range $0.25 \geq x \geq 0$ and $0.75 \geq x \geq 0.5$ where the speed is decreasing in x, or equivalently $F(x)$ is negative (Figure 3(b)). Similarly, it is expansive in the $0.5 \geq x \geq 0.25$ and $1.0 \geq x \geq 0.75$ regions where the speed is increasing in x. Consequently, the characteristics in the compressive regions begin to converge, while those in the expansive regions spread in time. Shocks begin to form when the converging characteristics intersect. Since $F(x)$ has a minimum -1.85 at both x = 0.130 and x = 0.630, the time of the first shock formation is expected to occur at t = 0.54 from the characteristics that are at x = 0.13 and x = 0.63 at t = 0. As shown in Figure 2(b), two shocks are formed at t = 0.55. Their characteristics can be traced back to the expected initial locations. For instance, from the B_y distribution in Figure 2(b), $|B_y| \approx 0.36$ at the shocks, which is equal to its value at x = 0.13 (0.63) at t = 0. The shocks are slow shocks since the value of B_y decreases with no change in its direction.

The results at t = 1 is shown in Figure 2(c). By fitting with the Rankine-Hugoniot jump relations for the shock at x = 0.495, a shock region is identified from x = 0.487 to x = 0.499. At x = 0.487, downstream of the shock, ρ, v_x, v_y, B_y, and p are 0.991, -0.00353, -0.0288, 0.0291 and 1.00, respectively; at x = 0.499, upstream of the shock, the values are 0.807, -0.211, 0.526, -0.473 and 0.702, respectively. From these two states, five values of shock speed are obtained from five jump relations. These values agree very well and the shock speed s is 0.907 ± 0.001. The fast, intermediate and slow characteristic speeds are 1.29, 1.00, and 1.00, respectively, for the downstream state, and 1.24, 0.902, and 0.709, respectively, for the upstream state. Thus both intermediate and slow characteristics are converging into the shock and the shock is a $2 \to 4$ intermediate shock. By this time, the characteristics from the compressive region in $0.25 \geq x \geq 0$ have converged into a single shock at x = 0.995, and those from $0.75 \geq x \geq 0.25$ into a single shock at x = 0.495 (Figure 3(c)).

Since the downstream value of B_y is so small that the shock is approximately a switch-off shock, the downstream intermediate and slow characteristic speeds are about the same and the upstream intermediate speed is about equal to the shock speed, as evidenced in the results. But the evolution will not stop at the switch-off shock, as expected in the conventional theory. Since the intermediate and slow characteristic speeds are faster in the downstream region and slower in the upstream region than the shock speed, their characteristics will continue to converge into the shock and form a stronger intermediate shock.

Shown in Figure 2(d) are the results at t = 2. Again, by fitting with the Rankine-Hugoniot jump relations, a shock region is identified from x = 0.41 to x = 0.43. At x = 0.41, downstream of the shock, ρ, v_x, v_y, B_y, and p are 0.9666,-0.02547, -0.1292, 0.1289 and 0.9780, respectively; at x = 0.43, upstream of the shock, the values are 0.8462, -0.1629, 0.4386, -0.4022 and 0.7772, respectively. From these two states, the shock speed s is found to be 0.940 ± 0.001. One can verify that in the shock frame the fluid velocities are superslow, super-Alfvénic and subfast ahead and subslow, sub-Alfvénic and subfast behind the shocks. These are 2 → 4 intermediate shocks.

The results at $t = 4$ and 10 are shown in Figures 2(e) and 2(f). They show that the sawtooth wave solutions with 2 → 4 intermediate shocks maintain at later times even though other waves begin to be present. At $t = \infty$, the solution will approach a constant state.

Case 2: Formation of 2 → 4 time-dependent intermediate shock

In Case 1, we have specialized to the planar situation in which B_z and v_z are zero everywhere. Therefore, in that case, there are no intermediate waves and the 2 → 4 intermediate shocks are due to the convergence of the slow waves. In Case 2 we include full MHD in which B_z and v_z are no longer zero. The initial condition is given by $B_z(x) = B_y(x)/2$, and $v_z(x) = 0$ with B_y and all other quantities being the same as in Case 1. The computation was again done with $\eta = 0.0004$. Figure 4 shows the results at $t = 0$, 0.5, 1, 1.5, and 2. Since the initial condition is not a pure simple wave, waves other than the slow wave are already present at the early times. However, the development is very similar to Case 1. For example, the B_y distributions of this case and Case 1 are not only qualitatively but quantitatively similar. On the other hand, there are new features in this calculation. Look at the shock at $x = 0.42$ at $t = 2$. The upstream state is given by $\rho = 0.853$, $v_x = -0.164$, $v_y = 0.429$, B_y=-0.390, $v_z = -0.0032$, $B_z = -0.210$, and $p = 0.791$. The fast, intermediate and slow characteristic speeds are 1.24, 0.919, and 0.791, respectively. The downstream state is given by $\rho = 0.964$, $v_x = -0.039$, $v_y = -0.145$, $B_y = 0.148$, $v_z = -0.149$, $B_z = -0.075$, and $p = 0.977$, and its fast, intermediate and slow characteristic speeds are 1.28, 0.979, and 0.962, respectively. The shock speed is 0.93. Therefore, the shock is a 2 → 4 intermediate shock. But the values of B_y/B_z at the upstream and the downstream of the shock are not equal, so the shock is not coplanar. Therefore it does not satisfy the Rankine-Hugoniot jump conditions. As a further consequence, the shock is time-dependent, since all stationary shocks have to satisfy the Rankine-Hugoniot jump relations.

As we will discuss later, it is important to have such time-dependent shocks. Because of the existence of the time-dependent non-coplanar intermediate shock solutions, the coplanar intermediate shocks based on the Rankine-Hugoniot jump conditions can have neighboring solutions when B_z is nonzero. In contrast, the conventional theory does not allow the time-dependent shock solutions, and the coplanar intermediate shocks become singular.

In [4], it was shown that all four types of intermediate shocks could be formed in a similar fashion by the wave steepening process.

(b) Shock Structure

Since intermediate shocks can be formed in dissipative MHD, they must have shock structures. Now we discuss their shock structure directly from the shock structure equations. Assuming steady state and that upstream state at $x = -\infty$ is spatially uniform, and integrating the dissipative MHD equations (1-4) once with respect to x, one gets the following MHD shock structure equations:

$$\{\rho v_x\} = 0 \tag{5}$$

$$\{\rho v_x^2 + p + \frac{B^2}{2} - B_x^2\} = (\mu + \frac{4}{3}\nu)\frac{dv_x}{dx} \tag{6}$$

$$\{\rho v_x v_y - B_x B_y\} = \nu \frac{dv_y}{dx} \tag{7}$$

$$\{\rho v_x v_z - B_x B_z\} = \nu \frac{dv_z}{dx} \tag{8}$$

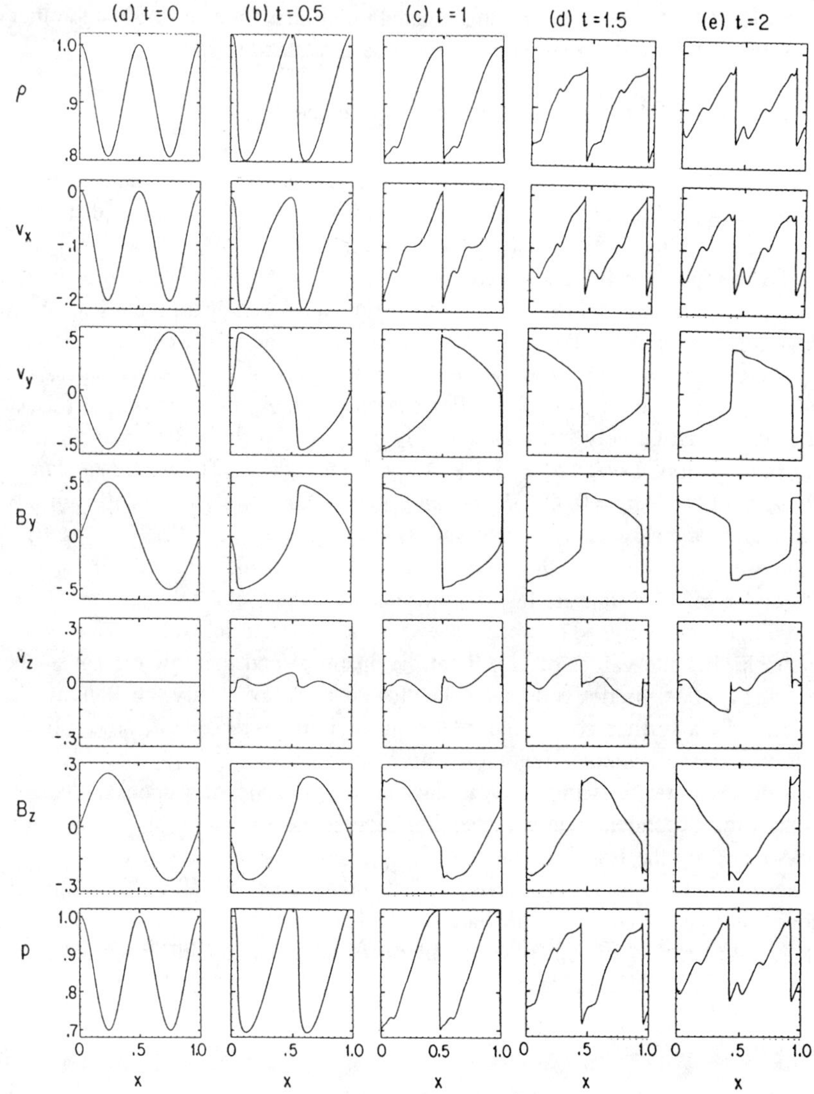

Figure 4: Formation of 2 → 4 time-dependent intermediate shocks (from [4]).

$$\{v_x B_y - v_y B_x\} = \eta \frac{dB_y}{dx} \tag{9}$$

$$\{v_x B_z - v_z B_x\} = \eta \frac{dB_z}{dx} \tag{10}$$

$$\{\rho v_x (\frac{v^2}{2} + \frac{\gamma}{(\gamma-1)}\frac{p}{\rho}) + B_y(v_x B_y - v_y B_x) + B_z(v_x B_z - v_z B_x)\} =$$

$$(\mu + \frac{4}{3}\nu)\frac{d}{dx}(\frac{v_x^2}{2}) + \nu\frac{d}{dx}(\frac{v_y^2+v_z^2}{2}) + \eta\frac{d}{dx}(\frac{B_y^2+B_z^2}{2}) + \kappa\frac{d}{dx}(\frac{p}{\rho}) \tag{11}$$

The brace notation is defined by $\{f\} = f(x) - f(-\infty)$ for any arbitrary function $f(x)$. Equation (5) means that ρv_x is constant through the shock layer. The equations (6-11) form a set of autonomous ordinary differential equations. Their solutions when all spatial derivatives are set to zero are singular points, and are the Rankine-Hugoniot relations.

The behavior of first- and second-order autonomous ordinary differential equations is simple and easy to analyze by local analysis. However solutions to equations of order three or more can be very complicated [14]. Here, we will limit our study of the MHD shock structure to two cases: one with η only and the other with η and μ. In addition, we will only consider some specific examples. Other situations can be found in [4].

In Case 1 of (a), we obtained a $2 \to 4$ intermediate shock at $t = 2$. From the Rankine-Hugoniot relations, one can obtain the other two singular points: state 1 and state 3. We will study shock structure for the shocks connecting these four states. For convenience, we will transform these shocks, which are due to the convergence of the $v_x + C_F$, $v_x + C_I$, and $v_x + C_S$ characteristics, to shocks with converging $v_x - C_F$, $v_x - C_I$, and $v_x - C_S$ characteristics (C_F, C_I and C_S denote fast, intermediate and slow speeds). Therefore, the upstream state of the shocks is at $x = -\infty$, and v_x decreases from state 1 to state 2, to state 3, and to state 4.

Besides discussing shock structure for the shocks connecting these four states, we also consider shock structure for shock states along a shock curve associated with the state 2, given above. The shock curve is defined as follows. It is known that one can determine the state behind a shock and its shock speed by specifying the state ahead and the jump of one quantity. Using state 2 for the upstream state, we can obtain the shock waves by resolving the Rankine-Hugoniot jump relations in terms of the downstream value of B_y. The upstream state (which is at $x = -\infty$) will be called the left state and will have a subscript l; the downstream state (which is at $x = \infty$) will be called the right state and will have a subscript r. Thus the left state is given by state 2 above. The dependence of the shock speed (s) and the fast, intermediate, and slow characteristic speeds, denoted by λ's, with respect to the parameter B_{yr} in the range from B_{yl} to $-B_{yl}$ is shown in Figure 5. All of these shocks, which form a shock curve, are compressive and satisfy the entropy condition.

In the range $B_{yl} \le B_{yr} \le 0$, the shocks are slow shocks. (Note B_{yl} is negative.) As is shown in the figure, the shocks have the following relations $\lambda_{Sl} \ge s \ge \lambda_{Sr}$, $s \ge \lambda_{Il} \ge \lambda_{Ir}$, and $s > \lambda_{Fr} \ge \lambda_{Fl}$. Thus the shock frame fluid velocities are subfast, sub-Alfvénic, and superslow ahead, and subfast, sub-Alfvénic, and subslow behind the shock. There is one pair of slow characteristics converging on the shock. When $B_{yr} = 0$, we have a slow, switch-off shock, in which $\lambda_{Il} = s$. Once B_{yr} becomes positive, i.e., B_{yr} has a opposite sign from B_{yl}, the shocks become intermediate shocks. Thus, $\lambda_{Il} > s > \lambda_{Ir}$. In addition, $\lambda_{Sl} > s \ge \lambda_{Sr}$ for $0.22 > B_{yr} \ge 0$; thus in this B_{yr} range, we have $2 \to 4$ intermediate shocks. At $B_{yr} = 0.22$, the right slow characteristic speed (λ_{Sr}) equals to the shock speed and the shock is $2 \to 3 = 4$. When $|B_{yl}| > B_{yr} > 0.22$, we have $2 \to 3$ shocks with converging intermediate characteristics. Finally, at $B_{yr} = -B_{yl}$, we have a 180° Alfvén wave. At $B_{yr} = 0.3046$, and 0.12868, the shock speed is 0, and they, respectively, correspond to state 3 and 4 defined above.

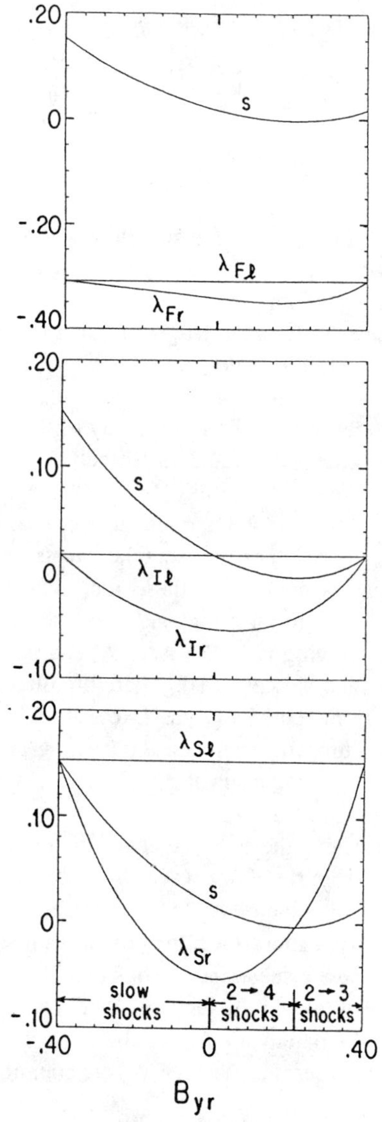

Figure 5: Shock speed (s) and fast, intermediate and slow characteristic speeds (λ_F, λ_I, λ_S) of left- and right-states along a shock curve (from [4]).

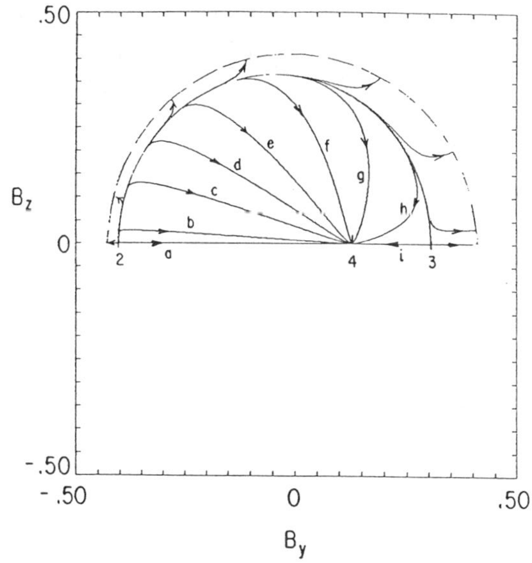

Figure 6: Trajectories for Case 1 in the non-planar situation. State 2 is an unstable node, state 3 is a saddle point, and state 4 is a stable node. The dashed curve represents a sonic curve along which v_x equals the sound speed. (From [4])

Case 1: η only

Let us begin with the case that there is only resistivity in the dissipation. We first consider the planar case that B_τ and v_τ vanish everywhere. The shock structure equations become a simple first-order autonomous system and can be represented by $dB_y/dx = f(B_y)$. However, $f(B_y)$ represents two different functions, one for supersonic flow and another for subsonic flow. This is because one obtains v_x in terms of B_y from solving a quadratic equation. One can easily show that one branch of the solution is supersonic, and the other subsonic.

States 2, 3, and 4, as defined above, are the singular points of the subsonic system; while state 1 is the only singular point of the supersonic system. One can show that, based on local analysis, state 1 is an unstable node, thus it diverges as x increases. On the other hand, for the subsonic branch, one can show that state 4 is a stable node, while states 2 and 3 are unstable nodes. Thus one predicts that shock structure exists between state 2 and state 4, and between state 3 and state 4, but there is no structure between state 2 and state 3. These can be verified by integrating directly the shock structure equation. Although the two branches of v_x can become sonic at the two limiting points, these two points are outside states 2 and 3 in the subsonic branch and are unstable. Therefore shock structure for $1 \to 2$, $1 \to 3$ and $1 \to 4$ shocks is impossible.

Next we consider the case where B_z and v_z can be nonzero. Here we have a second-order autonomous system of ordinary equations, which we express in terms of B_y and B_z. As above, there are two branches of solutions for v_x. The supersonic branch consists of a singular point, state 1. The subsonic branch consists of three singular points, corresponding to states 2, 3, and 4. One can check that the singular point at state 2 is an unstable node, state 3 is a saddle point, and state 4 is a stable node.

Since state 2 is an unstable node, the trajectories can leave it in all directions. The one that enters the state 3 saddle forms a $2 \to 3$ intermediate shock. Some may diverge to ∞. The rest

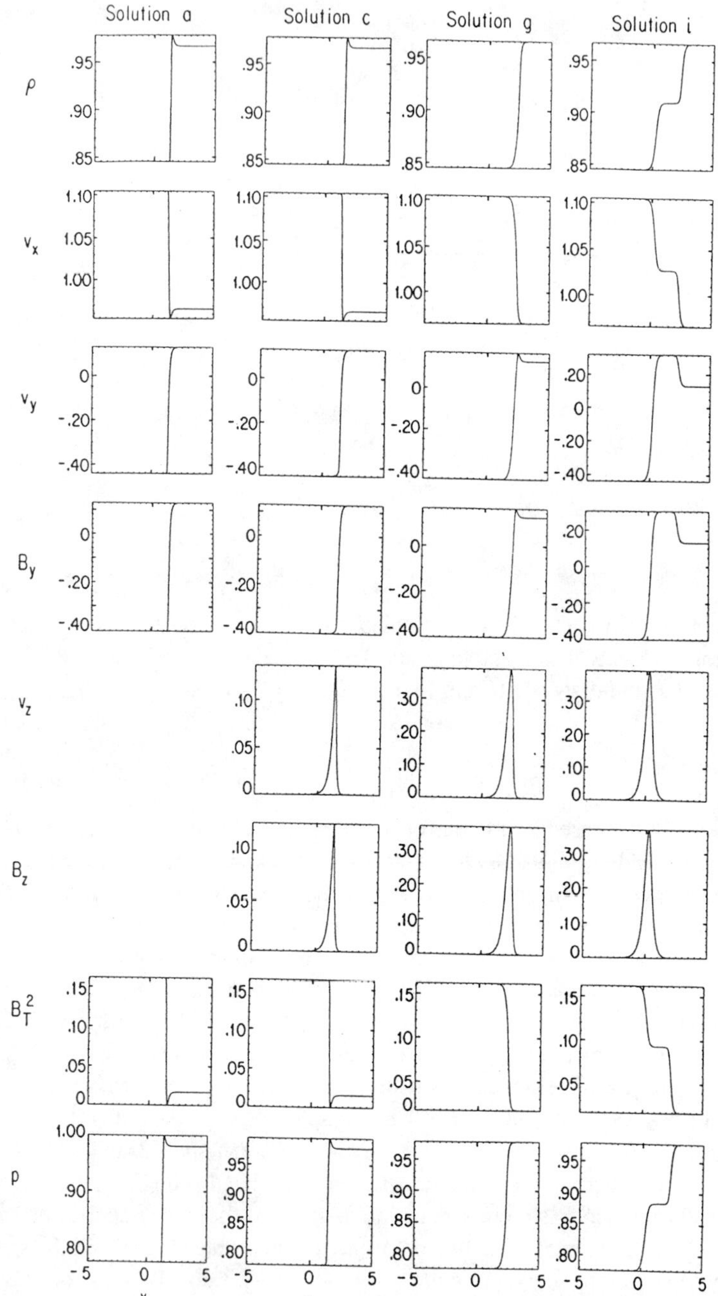

Figure 7: Shock structure solutions of a, c, g, and i of Figure 6 are plotted versus x for $\eta = 0.01$ (from [4]).

end up at state 4. Thus there is a family of $2 \to 4$ intermediate shock structure trajectories. These qualitative arguments are verified by the plot of the exact trajectories in Figure 6. From the state 3 saddle point, a trajectory can connect it to state 4 and form a $3 \to 4$ slow shock. Since at the state 3 saddle, the trajectories can enter only the $\pm B_z$ axes, the $2 \to 3$ intermediate shock structure has to be non-planar; while the trajectory can only leave along the B_y axis, and the $3 \to 4$ slow shock is planar. However the $2 \to 4$ intermediate shock can be either planar or non-planar. The reason for this is that the $2 \to 3$ intermediate shock is formed by the convergence of intermediate waves, which are non-planar, the $3 \to 4$ slow shock is due to the convergence of slow waves, which are planar, and the $2 \to 4$ intermediate shock can be formed by either intermediate waves, or by slow waves, or by both. Since the shock structure equations are invariant under the changes of B_z to $-B_z$, and v_z to $-v_z$, each solution with positive B_z has a mirror image with negative B_z.

The family of the $2 \to 4$ intermediate shock structure trajectories can be labelled by one parameter. Since the shock structure equations are entirely deterministic, the $2 \to 4$ trajectories can not intersect except at singular points. Then the curve C formed by the B_y axis and a given shock solution encloses a unique area A in the $B_y - B_z$ plot,

$$A = 1/2 \int_C (B_z dB_y - B_y dB_z). \tag{12}$$

Using (9-10), we can show that

$$A = -\frac{1}{2\eta} D \int_{-\infty}^{\infty} B_z dx \tag{13}$$

with constant $D = B_y(v_x - B_x^2/(\rho v_x))$ evaluated at $x = -\infty$ or $+\infty$, upstream or downstream of the shock. Let F_z denote the integral $\int_{-\infty}^{\infty} B_z dx$. Thus, for a given η, we can pick a $2 \to 4$ intermediate shock structure by specifying F_z. Note that F_z is proportional to η, and, since A is finite, for a given η, F_z is bounded, whose maximum value will be denoted by F_{zM}. From the expression for D, we note that $D \to 0$ when $v_x \to B_x/\sqrt{\rho}$ or $B_y \to 0$. Thus $F_{zM} \to \infty$ for switch-on and switch-off shocks, where $b_y = 0$, and rotational discontinuities, where $v_x = B_x/\sqrt{\rho}$.

The values of F_z for the shock structures $a - i$ shown in Figure 6 are 0, 0.012, 0.058, 0.102, 0.149, 0.206, 0.252, 0.290, and 0.301, respectively. In Figure 7, the shock structure solutions a, c, g, and i are plotted versus x for $\eta = 0.01$. (The solution a is the same as the one in Case 1 of (a) after a proper scaling in x, due to different values of η, is carried out.) For the solutions $a - e$, one sees that the shock structure includes a compression region and an expansive region. In the compression region, the pressure and the density increase, and $|B_t|$ decreases; while in the expansive region the pressure and the density decrease, and $|B_t|$ increases. As F_z increases, $|B_t|$ becomes monotonically decreasing, and the whole shock structure is totally compressive (see solutions $f - i$). When F_z is at its maximum (solution i), the $2 \to 4$ intermediate shock becomes a $2 \to 3$ intermediate shock plus a $3 \to 4$ slow shock.

In Figure 6 the sonic curve, along which v_x equals to the sound speed, is plotted in dashed lines. Since it is outside the region of the shock structure solutions, no shock solution can cross the sonic curve from states 2, 3, and 4. Therefore, there does not exist any shock structure from state 1 if only η is present.

Next we consider shock structure for the intermediate shocks along the shock curve of Figure 5. The solid curve in Figure 8 plots the calculated values of F_{zM} obtained by a numerical integration of the shock structure equations versus B_{yr}. The $2 \to 4$ intermediate shock may have any $F_z \leq F_{zM}$ in $0.22 \geq B_{yr} \geq 0$, and the $2 \to 3$ intermediate shock must have F_z precisely equal to F_{zM} in $|B_{yl}| > B_{yr} > 0.22$. The value of F_{zM} for the $2 \to 4$ intermediate shock is given by its corresponding $2 \to 3$ intermediate shock of the same singular point family. F_{zM} is minimum at $B_{yr} = 0.22$, and it tends to ∞ at both $B_{yr} = 0$, and $-B_{yl}$. The figure shows

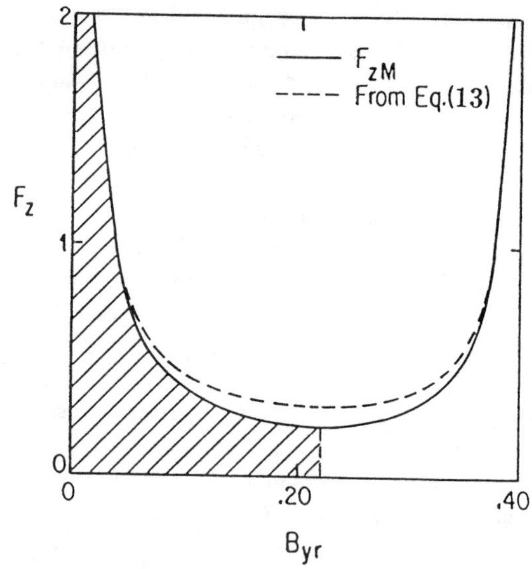

Figure 8: A plot of F_{zM} versus B_{yr} along the shock curve given in Figure 5. The $2 \to 4$ intermediate shock may have any $F_z \leq F_{zM}$ in $0.22 \geq B_{yr} \geq 0$ and the $2 \to 3$ intermediate shock must have F_z precisely equal to F_{zM} in $|B_{yl}| > B_{yr} > 0.22$. The dashed curve is from (13) with $A = \pi B_{yl}^2/2$. (From [4])

only the case with $F_z > 0$; there is an exactly similar curve for $F_z < 0$, due to the symmetry mentioned above. From (13) we can obtain an upper bound for F_z. The area A defined in (12) attains a maximum value of $\pi B_{yl}^2/2$ when $B_{yr} \to -B_{yl}$ (a 180° Alfvén wave). Using this value for A, we get F_z from (13), which is plotted as a dashed curve in Figure 8. As expected, the exact F_{zM} (solid curve) is close to the upper bound when $B_{yr} \to 0$ and $-B_{yl}$.

At $B_{yr} = -B_{yl}$, we have a 180° Alfvén wave; $F_z = \infty$ means that the width of a 180° Alfvén wave is infinitely long, or in other words, a 180° Alfvén wave (rotational discontinuity) does not have a finite-width structure in resistive MHD. On the other hand, at $B_{yr} = 0$, the shock can have any value of F_z. In one limit $F_z = 0$, we have a switch-off slow shock. In the other limit $F_z = \infty$, the shock can be considered as consisting of a broad (infinitely wide) rotational discontinuity attached by a switch-off shock, because the upstream intermediate characteristic speed equals the switch-off shock speed.

Note that, in the past, switch-off and switch-on shocks were considered singular entities. But in our treatment, it is a point in the admissible shock curve, and there is no need for special consideration.

Case 2: μ and η

In the previous case with η alone, there is no structure for the $1 \to 2$, $1 \to 3$, and $1 \to 4$ shocks, since state 1 is on the supersonic branch, while states 2, 3, and 4 are on the subsonic branch. It is well known that additional dissipation mechanism is required for supersonic to subsonic shocks to have structure [15]. Here we include the bulk viscosity μ. We will show that the $1 \to 4$ intermediate shock possess a two-parameter family of shock structure trajectories. An additional reason for studying this case is motivated by Germain [10], who had shown that when

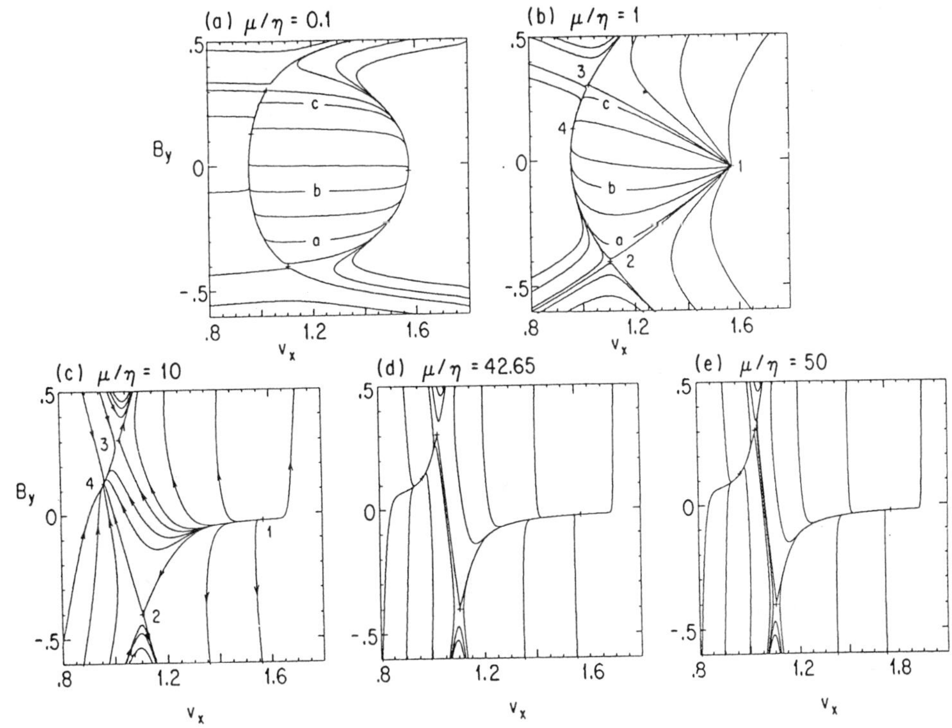

Figure 9: Trajectories for Case 2 in the planar situation for various ratios of μ/η: (a) $\mu/\eta = 0.1$, (b) $\mu/\eta = 1$, (c) $\mu/\eta = 10$, (d) $\mu/\eta = 42.65$, and (e) $\mu/\eta = 50$. (From [4])

μ is much larger than η or ν, intermediate shocks do not have structure. We shall discuss this situation.

By including η and μ, one has a two-dimensional autonomous system in a planar situation, which we write in terms of the variables v_x and B_y. Instead of having two separate branches as in Case 1, we now have a single system with all four states as its singular points. Figure 9 shows the computed trajectories of the system for various ratios of μ/η. From local linear analysis one can show that state 1 is an unstable node, states 2 and 3 are saddle points, and state 4 is a stable node. When μ/η is less than 42.65, there are structure solutions for the $1 \to 2$, $1 \to 3$, $1 \to 4$, $2 \to 4$ and $3 \to 4$ shocks, but not for the $2 \to 3$ intermediate shock. When the ratio is equal to 42.65, there are structure solutions for the $1 \to 2$, $3 \to 4$, and also $2 \to 3$ intermediate shocks, but not for the $1 \to 3$, $2 \to 4$, and $1 \to 4$ intermediate shocks. When the ratio is larger than 42.65, only the $1 \to 2$ fast shock and the $3 \to 4$ slow shock can exist. This dependence on the ratio of μ/η can be traced to the change in the directions of the eigenvectors of the two saddle points.

As shown in Figure 9(b) for the case μ/η=1, while the shock structure for the $1 \to 2$, $1 \to 3$, $2 \to 4$ and $3 \to 4$ shocks is unique in the planar situation, there is a one-parameter family of the $1 \to 4$ intermediate shock structure. When $B_z \neq 0$ and $v_z \neq 0$, one can show that the $1 \to 4$ intermediate shock structure can have two parameters, while $1 \to 3$ and $2 \to 4$ shocks can have one parameter [4].

The critical ratio of μ/η, above which there is no intermediate shock structure, depends on the properties of the shocks. In Figure 10, we show the critical ratio as a function of B_{yr} along the shock curve of Figure 5. The $2 \to 4$ intermediate shock can have shock structure as long

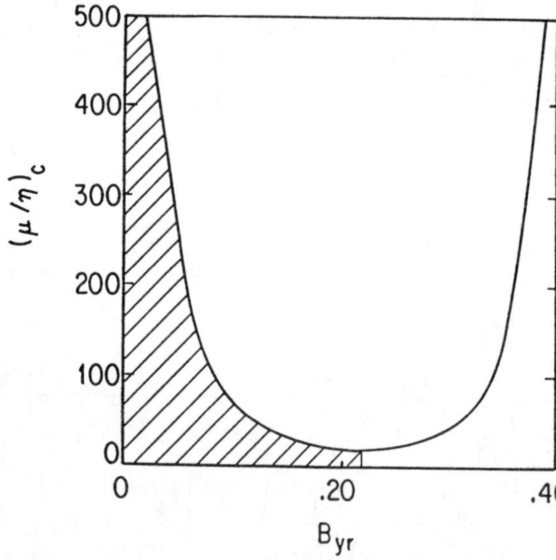

Figure 10: Critical ratio $(\mu/\eta)_C$ as a function of B_{yr} along the shock curve defined in Figure 5. The $2 \to 4$ intermediate shock in $0.22 \geq B_{yr} \geq 0$ can have shock structure if $\mu/\eta \leq (\mu/\eta)_C$; the $2 \to 3$ intermediate shock in $|B_{yl}| \geq B_{yr} > 0.22$ must have $\mu/\eta = (\mu/\eta)_C$ to have a planar structure. (From [4])

as μ/η is less than the critical ratio in $0.22 \geq B_{yr} \geq 0$; the $2 \to 3$ intermediate shock can have a planar structure only when μ/η exactly equals to the critical value. It shows that the critical ratio tends to ∞ as B_{yr} goes to 0 or $-B_{yl}$. Although we can not prove it rigorously, this result is reasonable because it makes a smooth transition to the slow shock, which can have structure for any value of μ.

In [4], the case with μ and ν was also considered. The intermediate shock structure has a similar dependence on μ/ν as on μ/η. The result that intermediate shocks do not have structure if μ is much larger than ν (η) agrees with the proof by Germain [10]. Because of this property, Germain had concluded that all intermediate shocks are not physical. However, we do not share his view and offer a different perspective, which can be best illustrated by the following numerical example. In [4], it was shown that the $2 \to 4$ intermediate shock with $B_{yr} = 0.201$ on the shock curve has shock structure if $\mu/\nu \leq 19.5$. Using its shock structure with $\mu = 0.19$ and $\nu = 0.01$ as an initial state, its time-development by solving the dissipative MHD equations for a different set of dissipation coefficients was considered. If the same μ and ν as the initial state were used, the same shock with the same shock structure would stay. If different values for μ and ν, and with $\mu/\nu \leq 19.5$, were used, we found, as expected, that the same shock, but with a different shock structure, would stay. On the other hand, if $\mu/\nu > 19.5$, the shock would evolve. Instead of the initial $2 \to 4$ shock, it evolved to a $2 \to 3$ intermediate shock followed by a $3 \to 4$ slow shock, in addition to fast and slow waves. The new $2 \to 3$ shock so obtained had a structure with the new μ/ν ratio. This example means that as μ/ν becomes larger than its critical value, the initial $2 \to 4$ intermediate shock can no longer exist, and it will evolve to a system that includes a new intermediate shock whose shock structure is consistent with the given ratio of μ/ν. Since the critical value of μ/ν can be infinitely large, it is expected that for a given finite value of μ/ν, one is able to find an intermediate shock to fit in the solution.

In sum, an intermediate shock can not exist if μ/ν is larger than its critical value. But for a given value of μ/ν, there are intermediate shocks, which can be part of solutions of the dissipative MHD equations. The above example also shows that a planar $2 \to 3$ shock, which is under-compressive, can indeed exist.

(c) Stability of intermediate shocks and evolutionary conditions.

In (a), we have shown that intermediate shocks can be formed from the wave steepening process. This means that these shocks must be stable. There is another (more formal) way of studying the stability problem. By solving the shock structure equations, one can obtain the structure of a shock. One can then carry out linear and nonlinear stability analysis of this shock by solving the time-dependent dissipative MHD equations. We have carried out several cases for intermediate shocks, and, as expected, the intermediate shocks are stable.

The evolutionary condition, as was formulated by Akhiezer et al [11] and by Taniuti [16], is a linear analysis. It is based on ideal MHD equations, and the shock is considered a discontinuity. By impinging a disturbance to the discontinuity, a set of outgoing waves will be emitted. In order to ensure a unique non-trivial solution of these outgoing waves, it was found that after including a free parameter associated with the shock location, the number of outgoing waves, n_{out}, should equal to $N - 1$, with $N = 7$ for the number of MHD equations. This is the so-called evolutionary condition. For $1 \to 4$ intermediate shocks, $n_{out} = 4$, for $1 \to 3$ and $2 \to 4$ intermediate shocks, $n_{out} = 5$, and for $2 \to 3$ intermediate shocks, $n_{out} = 6$. Therefore, $1 \to 4$, $1 \to 3$, and $2 \to 4$ intermediate shocks do not satisfy the evolutionary condition, and were hence disregarded. Although $n_{out} = 6$ for $2 \to 3$ intermediate shocks, they were also considered nonevolutionary due to the additional requirement: For $2 \to 3$ intermediate shocks, the linear perturbations can be separated into two linearly independent subspaces. The $2 \to 3$ intermediate shocks fail to satisfy the evolution condition for each separate subspace.

The difference between our analysis and the analysis of the evolutionary condition lies in the fact that we are using dissipative MHD equations. In our analysis, shocks are not discontinuities and there are free parameters in the $1 \to 4$, $1 \to 3$ and $2 \to 4$ shock structure. For instance, if a perturbation is impinging on a $1 \to 3$ intermediate shock, the perturbation could be carried away not only by outgoing waves but also by changing the shock structure. By adding the additional parameters associated with the shock structure to the number of outgoing waves we get exactly 6 unknowns for $1 \to 4$, $1 \to 3$ and $2 \to 4$ shocks. This then suggests that these shocks should be stable. Furthermore, the shock structure is generally three-dimensional (involving B_z and v_z) for $2 \to 3$ intermediate shocks. Thus, it is impossible to form two independent subspaces, and the additional evolutionary requirement of Taniuti [16] for this case is not valid.

(d) Riemann problem and dissipation coefficients

In this section, we consider an MHD Riemann problem, whose left and right states satisfy the jump conditions of a rotational discontinuity. For the rotational discontinuity with characteristic speeds $v_x \pm B_x/\sqrt{\rho}$, one has the following jump relations [8]:

$$[\mathbf{v}_t] = \mp sgn(B_x)[\mathbf{B}_t/\sqrt{\rho}] \tag{14}$$

$$[p] = [\rho] = [entropy] = [B^2] = [v_x] = 0 \tag{15}$$

Here, angle brackets denote a jump.

In the calculations, we have used the following initial distributions for $\mathbf{B}$ and $\mathbf{v}$

$$B_y(x) = B_t \cos(\theta(x)) \tag{16}$$

$$B_z(x) = B_t \sin(\theta(x)) \tag{17}$$

with $\theta(x) = (\theta_l + \theta_r)/2 - (\theta_l - \theta_r)\tanh(x)/2$ and B_t a constant. According to the wave relations, (14), $v_y(x) = B_y(x)/\sqrt{\rho_0}$ and $v_z(x) = B_z(x)/\sqrt{\rho_0}$. From (15), ρ, p, v_x, and B_x have constant initial distributions ρ_0, p_0, v_{x0}, and B_{x0}, respectively. B_{x0} is assumed positive. These distributions represent an intermediate wave moving with velocity $v_{x0} - B_x/\sqrt{\rho_0}$, which rotates the transverse magnetic field from θ_l to θ_r within a transition region of width 2. The smooth transition is required to assure the validity of the dissipation terms.

Figure 11 shows the result of such a calculation with the initial conditions given by (16) and (17) with $\theta_l = 0$, $\theta_r = -180°$, $p_0 = 1$, $\rho_0 = 1$, $B_t = 2$, $B_{x0} = 1$, $v_{x0} = 0$, and $\gamma = 5/3$. The conditions represent an intermediate wave moving to the left with speed $v_{x0} - B_{x0}/\sqrt{\rho_0}$, which is -1. We have used constant dissipation coefficients $\eta = \nu = 0.01$ and $\mu = 0$ in Figure 11 and $\eta = \nu = 0.05$ and $\mu = 0$ in Figure 12. The calculations were carried out with 14400 grid points covering the region $-180 \le x \le 180$ (thus the spatial spacing $\Delta x = 1/40$) and with the time step size $\Delta t = 30/4000$.

If there is no dissipation, the intermediate wave will just move to the left with constant velocity -1. However, when there is dissipation, the rotational discontinuity does not have structure (as was discussed in (b)) and it evolves into several waves. As shown in Figure 11 for the case $\eta = \nu = 0.01$ at $t = 60$, the solution consists of the following waves separated by constant states. The waves moving to the left are a fast rarefaction wave, denoted by FR in the figure, a $2 \to 3$ intermediate shock, denoted by IS, and a slow rarefaction wave, SR. The waves moving to the right include a contact discontinuity, C, a slow shock, SS, and a fast shock, FS. Note that v_z and B_z are nonzero only inside the intermediate shock. We have checked the solution and found it in very good agreement with the Riemann invariants across the rarefaction waves and the Rankine-Hugoniot jump conditions across the shock waves. By using the computational result as a guide, we have constructed an analytical solution of the Riemann problem with the same sequence of waves. The analytical solution (shown as thin curves in Figure 11) agrees with the numerical result very well [3].

In Figure 12, we show the result for the case $\eta = \nu = 0.05$ at $t = 60$. The solution is qualitatively similar, but quantitatively different, to the previous case. Again, an analytical solution was constructed and was in good agreement with the computational result [3].

This shows that there are many solutions to the above Riemann problem and that each solution depends on the dissipation coefficients. Thus this Riemann problem is not well defined in the usual sense. Now we show how the solution can be obtained from the consideration of shock structure. Instead of varying η (ν), the set of Riemann problems can be equally considered as a set of problems with various widths of the profile for fixed η (ν) coefficients. For example, a case with $\eta = \nu = 0.001$ and with width equal to 2 is the same as the case with $\eta = \nu = 0.01$ and with width equal to 20. Since there is B_z component in the profile, the integral of B_z depends linearly on the width of the profile. For a larger width, the B_z flux is larger. Now we assume that the B_z flux is only carried by the $v_x - C_I$ wave and not by the $v_x + C_I$ wave, which seems to be the case in the numerical results. A solution to the Riemann problem can then consist of a left-moving fast shock (or fast rarefaction), a $2 \to 3$ intermediate shock, a slow shock (rarefaction), and a right-moving entropy wave, a slow shock (rarefaction), and a fast shock (rarefaction). Since there is a parameter to specify for each wave, we get a six parameter family of states which can be connected to the initial upstream state. However, the shock structure of the $2 \to 3$ shock will be required to have the same integral of B_z, a constant of the system. Thus, we have five parameters. To arrive at a solution, the upstream state of the right-moving fast shock (rarefaction) should agree with the downstream state of the initial 180° wave. Since there are five values (ρ, v_x, v_y, B_y, and p) to specify and there are five parameters, it seems plausible to have a unique solution. Since the integral of B_z is closely related to the $2 \to 3$ shock, a different B_z flux will give a different $2 \to 3$ shock, and thus a different solution of the Riemann problem.

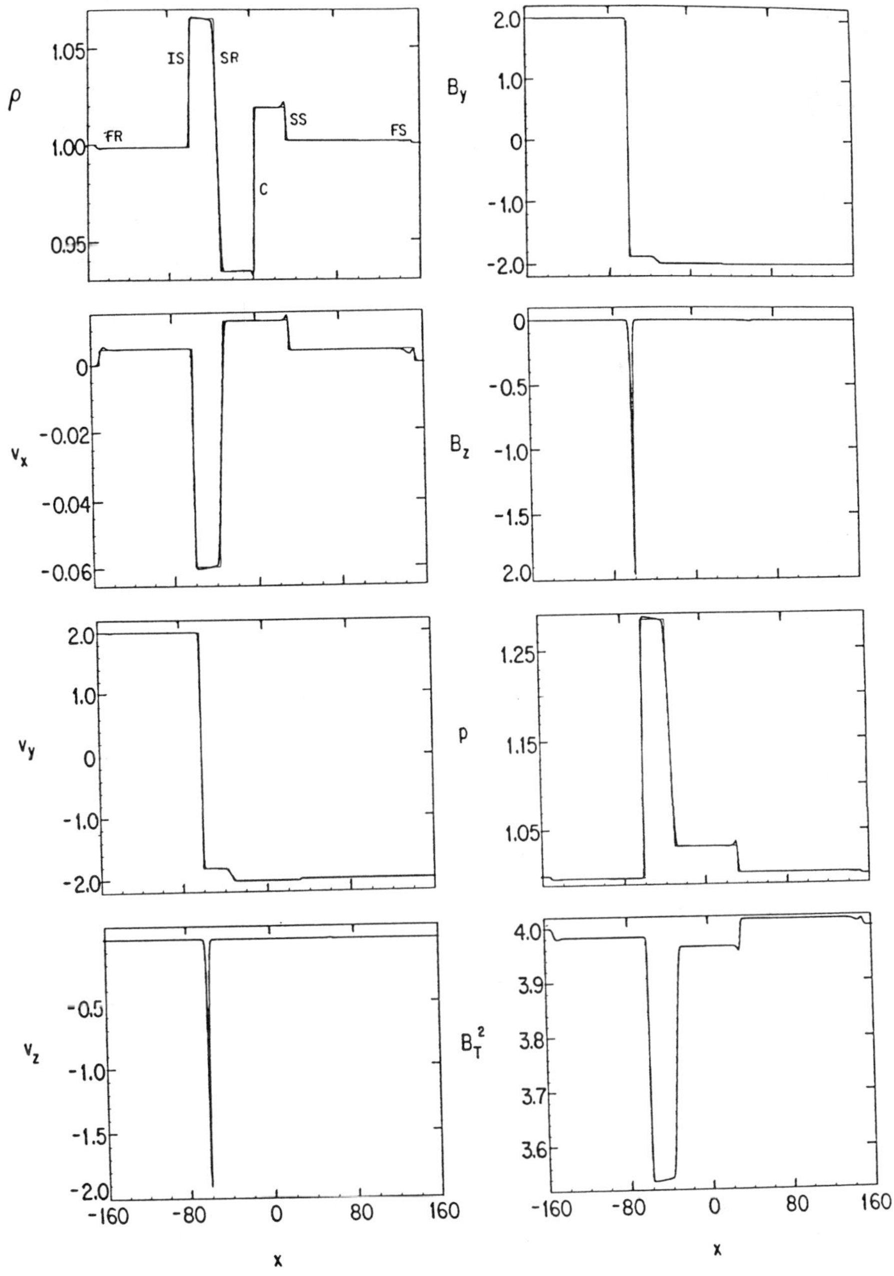

Figure 11: Solution at t = 60 for the coplanar Riemann problem with $\eta = \nu = 0.01$. The thin curves in the plots of ρ, v_x, v_y, B_y and p represent an analytical solution of the problem. B_T^2 is defined as $B_y^2 + B_z^2$. (From [3])

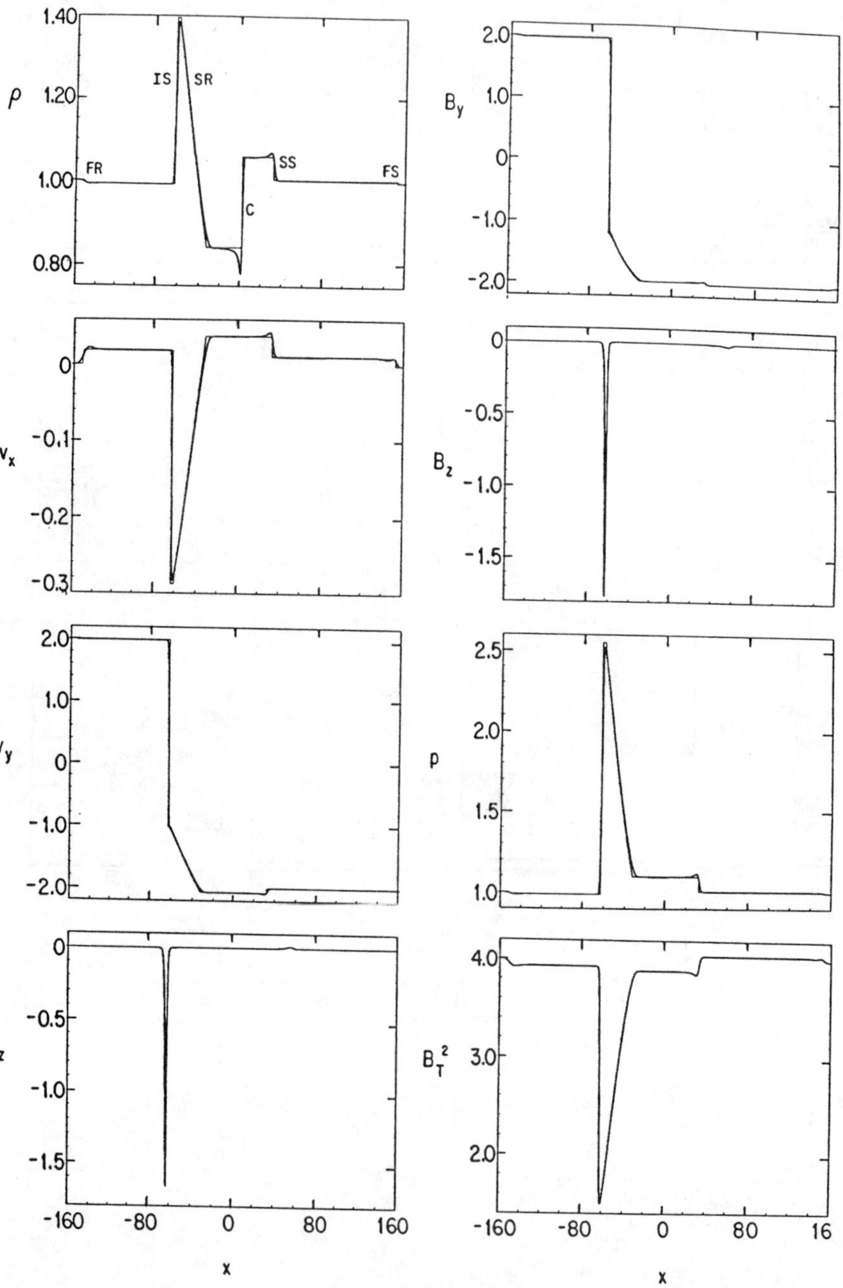

Figure 12: Solution at $t = 60$ for the coplanar Riemann problem with $\eta = \nu = 0.05$. The thin curves in the plots of ρ, v_x, v_y, B_y and p represent an analytical solution of the problem. B_T^2 is defined as $B_y^2 + B_z^2$. (From [3])

(e) Time-dependent shock solutions

In (d), we have considered a coplanar Riemann problem. It shows that B_y changes its sign across an intermediate shock. This is what one would expect to occur, because it is the only candidate that allows B_y to change sign and also possesses shock structure. One can not use a rotational discontinuity to rotate the magnetic field, because it does not have structure.

What will happen if we consider a non-coplanar Riemann problem, where the left and right magnetic fields are not on the same plane? On one hand, for the same reason as above, we can not use rotational discontinuities to rotate the field. On the other hand, according to the Rankine-Hugoniot jump conditions, all time-independent shocks, including intermediate shocks, are coplanar. To answer this question, the non-coplanar Riemann problems were solved numerically [3]. The results showed that the solutions of these non-coplanar problems are not self-similar and new entities, which we call time-dependent intermediate shocks, are formed. These new shocks are non-coplanar and so must be time-dependent. Furthermore, the fluid velocities of these shocks have a super-Alfvénic to sub-Alfvénic transition. The magnetic field rotates through these shocks. In the limit that it becomes coplanar, these shocks go to the usual intermediate shocks. When the Riemann problem is not coplanar, the solution will continue to evolve, the B_z flux will continue to increase inside the time-dependent intermediate shock, which will then approach a rotational discontinuity with an infinite width as time goes to infinity.

In the traditional theory, Kantrowitz and Petschek [7] predicted that an intermediate shock would disintegrate instantly once it interacts with an intermediate wave. But because of the existence of time-dependent intermediate shocks, a coplanar intermediate shock can have neighboring solutions even when the system becomes non-coplanar. Thus an intermediate shock does not need to disintegrate instantly, as was confirmed in a numerical experiment [2].

4. MKdVB Shock Waves

Our new theory of MHD shock waves indicates that it is essential to take into account the dissipation mechanism in considering MHD shocks. The dissipation mechanism affects not only the intermediate shock structure but also the global solutions when there is a rotation of the magnetic field. This relationship between evolution and structure for intermediate shocks also holds for the CKB (Cohen-Kulsrud Burgers) equation, which can be derived from the dissipative MHD equations in the limits of weak nonlinearity, weak dissipation, and small angles of propagation relative to the upstream magnetic field [5].

One may thus suggest that it is always necessary to consider the dissipation mechanism for shock waves in any system which possesses more than two stationary points for its shock structure equations and whose shock structure solutions depend on the dissipation mechanism. Here, we give another example to show this. We consider the shock waves for the Modified KdV-Burgers (MKdVB) equation, which is of the form

$$\frac{\partial b}{\partial t} + \alpha \frac{\partial}{\partial x}(b^3) = \eta \frac{\partial^2 b}{\partial x^2} + \mu \frac{\partial^3 b}{\partial x^3} \tag{18}$$

In this equation, there are both dissipative (η) and dispersive (μ) terms. Without the dissipative term, it is the usual MKdV equation. We will consider the case $\alpha > 0$; entirely analogous results may be obtained for $\alpha < 0$.

The stationary wave solution of (18) moving to the right with constant velocity V satisfies

$$\alpha \frac{\partial}{\partial x}(b^3) - V \frac{\partial b}{\partial x} = \eta \frac{\partial^2 b}{\partial x^2} + \mu \frac{\partial^3 b}{\partial x^3} \tag{19}$$

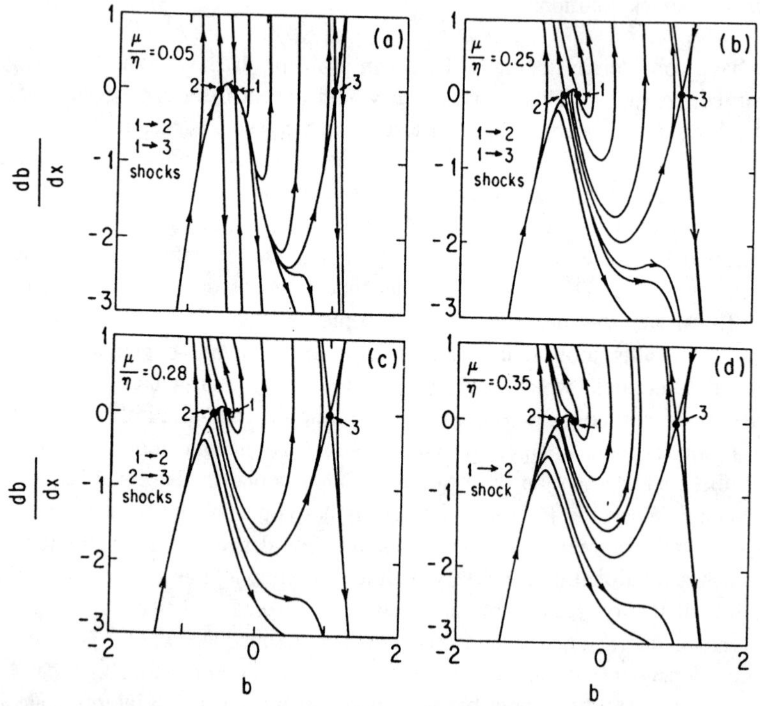

Figure 13: MKdVB shock trajectories using $\alpha = 1/4$, $\eta = 0.05$ and various ratios of μ/η for the case where the stationary state 1 is $b = -0.4$, state 2 is $b = -0.6$, and state 3 is $b = 1$.

Integrating once with respect to x gives

$$\alpha b^3 - Vb - F_0 = \eta \frac{\partial b}{\partial x} + \mu \frac{\partial^2 b}{\partial x^2} \tag{20}$$

where F is the constant of integration. Since both dissipation and dispersion are assumed important only within the shock layer, F_0 equals the flux $F = \alpha b^3 - Vb$ both upstream and downstream of the shock. In other words, the upstream and downstream shock states are stationary points of (20), which satisfy the jump relation

$$\alpha b^3 - Vb = F_0 \tag{21}$$

Depending on α, V, and F_0, (21) may have one or three real solutions. When it has only one, there are no shocks. When it has three solutions, it is easy to see that their sum must be zero. Thus two of the solutions have the same sign and the third has the opposite sign. We call the third solution a stationary point (state) of type 3 and the other stationary points (states) of types 1 and 2. We define state 1 to have a smaller magnitude than state 2. Clearly the magnitude of state 3 is the sum of state 1 and 2.

The MKdVB equation has a characteristic wave with speed $3\alpha b^2$. For positive α, this wave corresponds to the fast MHD wave and the stationary states 1, 2, and 3 correspond to those in MHD. There is no intermediate wave in the MKdVB equation, so $2 \to 3$ intermediate shocks do not have any converging pairs of characteristics and are under-compressive shocks. When α is negative, the stationary points of the MKdVB equation correspond to states 2, 3, and 4 in MHD. The shock classification of the MKdVB system is exactly the same as the CKB equations [5].

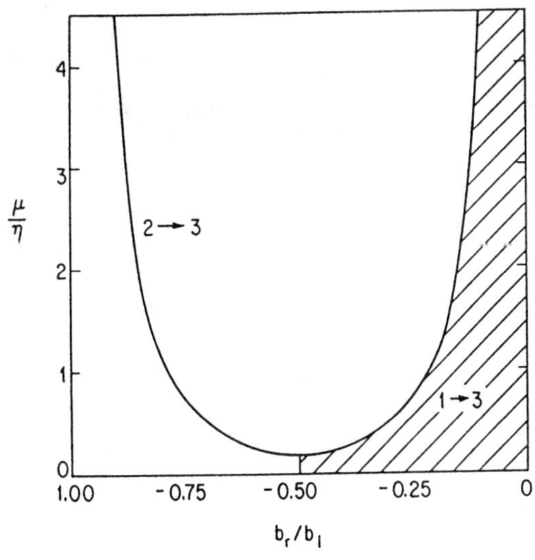

Figure 14: The dependence of the critical μ/η upon the ratio of the upstream to downstream b values. The $1 \rightarrow 3$ shocks can have shock structure solutions in the shaded area in $-1/2 \leq b_r/b_l \leq 0$; the $2 \rightarrow 3$ shocks can have solutions only when μ/η equal to its critical value in $-1 \leq b_r/b_l \leq -1/2$. The plot is for $\eta = 0.05$ and $\alpha = 1/4$.

The shock structure equation (20) is a second order autonomous equation. It can be rewritten as

$$\frac{\partial b}{\partial x} = \phi \tag{22}$$

$$\frac{\mu}{\eta}\frac{\partial \phi}{\partial x} = \frac{\alpha}{\eta}b^3 - \frac{V}{\eta}b - \frac{F_0}{\eta} - \phi \tag{23}$$

where $V = \alpha(b_l^2 + b_l b_r + b_r^2)$, with the subscripts l and r denoting downstream and upstream states, respectively. Figure 13 shows the numerically computed trajectories using $\alpha = 1/4$, $\eta = 0.05$ and various ratios of μ/η for a case where the stationary state 1 is $b = -0.4$, state 2 is $b = -0.6$, and state 3 is $b = 1$. From local linear analysis one can show that states 2 and 3 are always saddle points, while state 1 is a saddle point if $\mu/\eta < 0.18$ and a spiral point if $\mu/\eta > 0.18$. However the local analysis of the stationary points does not provide the global behavior of its solutions. As shown in the figure, the $1 \rightarrow 2$ shock always exists, independent of the μ/η ratio. On the other hand, the $1 \rightarrow 3$ shock can exist only when $\mu/\eta < 0.28$, and the $2 \rightarrow 3$ shock only exists at $\mu/\eta = 0.28$. When $\mu/\eta > 0.28$, neither $1 \rightarrow 3$ nor $2 \rightarrow 3$ can exist.

This μ/η dependence is true for all MKdVB shocks. In Figure 14, the dependence of the critical μ/η upon the ratio of the upstream to downstream b values is plotted. The $1 \rightarrow 3$ shocks can have shock structure solutions in the shaded area in $-1/2 \leq b_r/b_l \leq 0$; the $2 \rightarrow 3$ shocks can have solutions only when μ/η equal to its critical value in $-1 \leq b_r/b_l \leq -1/2$. The plot is for $\eta = 0.05$ and $\alpha = 1/4$. From (23), it is clear that a different value of η corresponds to a different value of α. Although the critical μ/η ratio depends on η, the qualitative behavior remains the same.

Now we can see how the solutions of the MKdVB Riemann problem depend on the ratio of μ/η. We will solve numerically for the time evolution of an initial profile, $b(x, 0)$, that varies between b_l at $x \rightarrow -\infty$ (downstream, to the left) and b_r at $x \rightarrow \infty$ (upstream, to the right). To

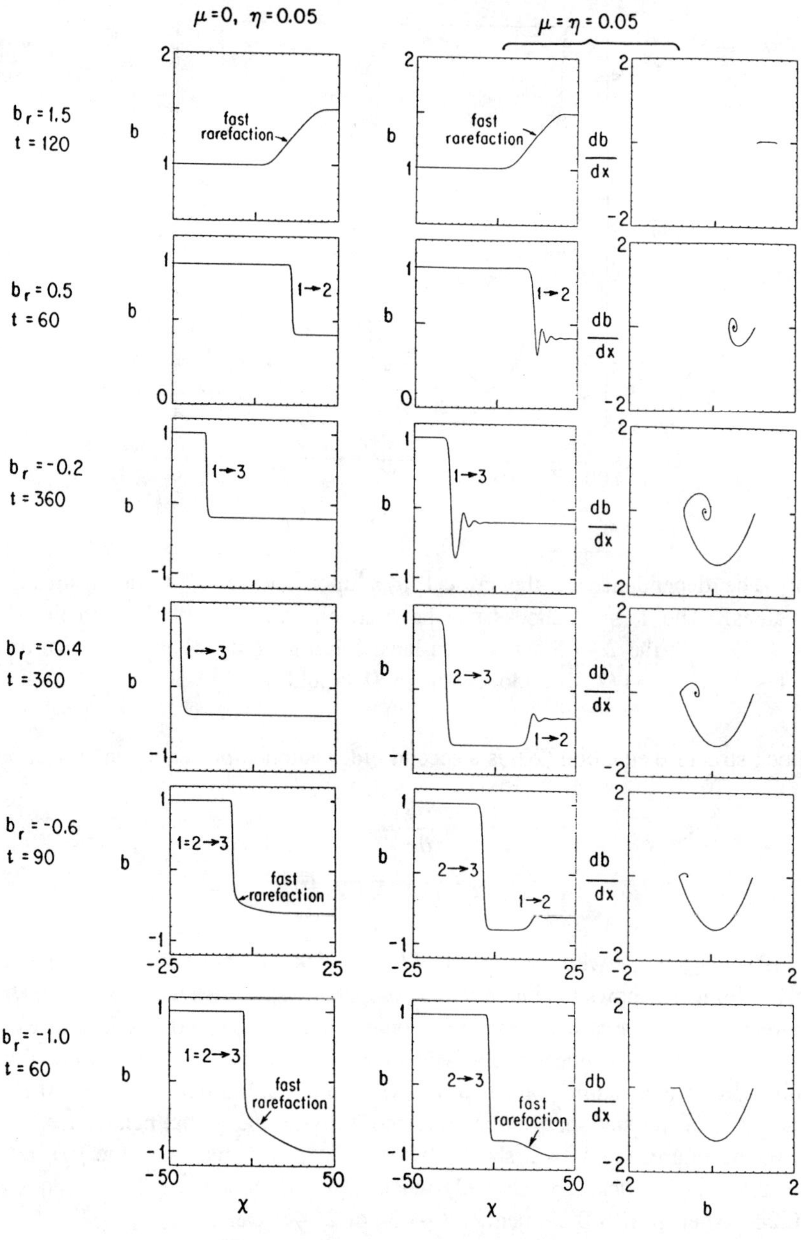

Figure 15: MKdVB Riemann problems. The left state is given by $b_l = 1$; the right state is given by several values. The results with $\mu = 0$ and $\eta = 0.05$ are on the left-hand panels; while the results with $\mu = \eta = 0.05$ are on the right-hand panels. This shows the solutions of the Riemann problems depend on the μ/η ratio.

be specific, we choose $\alpha = 1/4$, $\eta = 0.05$, and $\mu = 0$ or 0.05, use the following initial distribution

$$b(x) = b_t \cos(\theta(x)) \tag{24}$$

with $\theta(x) = (\theta_l + \theta_r)/2 - (\theta_l - \theta_r)\tanh(x)$ and $b_t(x) = (b_l + b_r)/2 - (b_l - b_r)\tanh(x)$, choose $b_l = 1$, $\theta_l = 0$, $\theta_r = 180°$ or $-180°$, $b_0 = 1$, and various values of b_r, and integrate

$$\frac{\partial b}{\partial t} + \alpha \frac{\partial}{\partial x}(b(b^2 - b_0^2)) = \eta \frac{\partial^2 b}{\partial x^2} + \mu \frac{\partial^3 b}{\partial x^3} \tag{25}$$

over the range $-25 \leq x \leq 25$ or $-50 \leq x \leq 50$. Figure 15 summarizes the results of the calculations. Each row shows the results for different values of b_r. The results for the case $\mu = 0$ and $\eta = 0.05$ are on the left-hand panels; the results, which include b - db/dx phase plots, for the case $\mu = \eta = 0.05$ are on the right-hand panels. Let us begin where b_l and b_r are of the same sign. If $b_r > b_l$, a fast rarefaction wave will be created. If $b_r < b_l$, a steady fast shock will be generated. Since the $1 \to 2$ fast shocks always have shock structure solutions, the results are independent of the μ/η ratio. This is confirmed in the top two rows of Figure 15. (Because the stationary state 1 becomes a spiral point, an upstream wave exists in the right-hand panels of the second row.)

In the case that b_l and b_r are of opposite signs, intermediate shocks are expected. From Figure 14, we see that when $\mu = 0$, $1 \to 3$ shocks in the range $-1/2 \leq b_r/b_l \leq 0$ can exist, but $2 \to 3$ shocks can not, except in the degenerate case $1 = 2 \to 3$, where states 1 and 2 are the same and $b_r/b_l = -0.5$. On the other hand, when $\mu = \eta = 0.05$, only $1 \to 3$ shocks in the range $-0.21 \leq b_r/b_l \leq 0$ and $2 \to 3$ shocks with $b_r/b_l = -0.79$ exactly can exist. These differences in the shock structure solutions determine the differences in the solutions of the Riemann problems. The third row of Figure 15 shows the results for $b_r/b_l = -0.2$. Since shock solutions exist for both $\mu = 0$ and $\mu = 0.05$ cases, we expect that the initial state to evolve into a single $1 \to 3$ shock, as the results confirm.

The next row of the figure is for $b_r/b_l = -0.4$. Since when $\mu = 0$ and $\eta = 0.05$, the $1 \to 3$ shocks with $b_r/b_l = -0.4$ exist, one expects the system to evolve into such a single $1 \to 3$ shock. However in the case where $\mu = \eta = 0.05$, the $1 \to 3$ shocks with $b_r/b_l = -0.4$ can not exist; the system has to evolve into other waves. The possible solutions are either a fast shock or a fast rarefaction wave plus a $2 \to 3$ shock with $b_r/b_l = -0.79$. (Since fast waves converge into $1 \to 3$ shocks, it is not possible to have a fast shock or rarefaction wave plus a $1 \to 3$ shock.) In the present case, the fast wave is a fast shock since b changes from -0.4 to -0.79. The numerical results confirm these predictions. Similarly, one can construct the solutions for the other two cases in the bottom two rows of Figure 15. For both $b_r/b_l = -0.6$ and $b_r/b_l = -1$, the system evolves into a fast rarefaction wave plus a $1 = 2 \to 3$ shock in the case $\mu = 0$. Since the upstream fast wave speed is the same as the $1 = 2 \to 3$ shock speed, the fast rarefaction wave is attached to the $1 = 2 \to 3$ shock. On the other hand, when $\mu = \eta = 0.05$, it evolves into a fast shock plus a $2 \to 3$ intermediate shock for the case $b_r/b_l = -0.6$, and into a fast rarefaction wave plus a $2 \to 3$ intermediate shock for the case $b_r/b_l = -1$. In sum, we show that, depending on the ratio of μ/η, the global solutions of the Riemann problem can be very different. We also show that the Riemann problems can be solved once the curve of critical μ/η of Figure 14 is known.

In the case where $\mu = \eta = 0.05$, we see the formation of a $2 \to 3$ shock, which, as was mentioned above, is under-compressive. The stability of under-compressive shocks is of current interest. Clearly, the MKdVB $2 \to 3$ shock is stable; we may reason that it is so because an MKdVB $2 \to 3$ shock uniquely solves the Riemann problem defined by the upstream and downstream states of the shock.

Acknowledgments

It is a pleasure to acknowledge useful discussions with Drs. M. Brio, C. K. Chu, F. V. Coroniti, B. Engquist, A. Harten, C. F. Kennel, L. C. Lee, T. P. Liu, and S. Osher.

References

1. C. C. Wu, *On MHD intermediate shocks*, Geophys. Res. Lett., 14, 668, 1987.

2. C. C. Wu, *The MHD intermediate shock interaction with an intermediate wave: Are intermediate shocks physical?*, J. Geophys. Res., 93, 987, 1988.

3. C. C. Wu, *Effects of dissipation on rotational discontinuities*, J. Geophys. Res., 93, 3969, 1988.

4. C. C. Wu, *Formation, structure, and stability of MHD intermediate shocks*, J. Geophys. Res., 95, 8149, 1990.

5. C. F. Kennel, R. D. Blandford, and C. C. Wu, *Structure and evolution of small amplitude intermediate shock waves*, Phys. Fluids B, 2, 253, 1990.

6. L. D. Landau, and E. M. Lifshitz, *Electrodynamics of Continuous Media*, Addison-Wesley, Reading, Mass., 1960.

7. A. R. Kantrowitz, and H. E. Petschek, *MHD characteristics and shock waves*, in *Plasma Physics in Theory and Application*, edited by W.B. Kunkel, pp. 148-206, McGraw-Hill, New York, 1966.

8. A. Jeffrey, and T. Taniuti, *Non-Linear Wave Propagation*, Academic, San Diego, Calif., 1964.

9. M. Brio, and C. C. Wu, *Characteristic fields for the equations of magnetohydrodynamics*, in *Nonstrictly Hyperbolic Conservation Laws*, edited by B. Keyfitz and H. C. Kranzer, pp. 19-23, American Mathematical Society, Providence, R. I., 1987.

10. P. Germain, *Shock waves and shock-wave structure in magneto-fluid dynamics*, Rev. Mod. Phys., 32, 951, 1960.

11. A. I. Akhiezer, G. J. Lubarski, and R. V. Polovin, *The stability of shock waves in magnetohydrodynamics*, Zh. Eksp. Teor. Fiz., 35, 731, 1958. (Sov. Phys. JETP, 8, 507, 1959.)

12. M. Brio, and C. C. Wu, *An upwind differencing scheme for the equations of magnetohydrodynamics*, J. Comput. Phys., 75, 400, 1988.

13. R. D. Richtmyer, and K. W. Morton, *Difference Methods for Initial-Value Problems*, 2nd ed., Interscience Publishers, New York, 1967.

14. J. E. Anderson, *Magnetohydrodynamic Shock Waves*, MIT. Press, Cambridge, Mass., 1963.

15. W. Marshall, *The structure of magnetohydrodynamic shock waves*, Proc. R. Soc. London, Ser. A, 233, 367, 1955.

16. T. Taniuti, *A note on the evolutionary condition on hydromagnetic shocks*, Prog. Theor. Phys., 28, 756, 1962.

CHAPTER 18

The Fluid-Dynamic Limit of the Broadwell Equation*

Zhouping Xin†

Abstract. This paper studies the asymptotic equivalence of the Broadwell model of the nonlinear Boltzmann equation to its corresponding Euler equation of compressible gas dynamics in the limit of small mean free path ϵ. We present two results here. First, we show that the fluid dynamical approximation is valid even if there are shocks in the fluid flow, although there are thin shock layers in which the convergence does not hold uniformly. More precisely, by assuming that the fluid flow is piecewise smooth with a finite number of noninteresting shocks and suitably small oscillations, we can show that there exist solutions to the Broadwell equations such that the Broadwell solutions converge to the fluid dynamical solutions away from the shocks at a rate of order ϵ. Our second result is that in the case that the fluid flow is a smooth rarefaction wave, the corresponding Broadwell solution exists globally in time, and the Broadwell solution converges to the rarefaction wave uniformly for all time except an initial layer as mean free path goes to zero.

1. Introduction. We study the validity of the fluid dynamical approximation for the Broadwell model of the nonlinear Boltzmann equation. The general Boltzmann equation of kinetic theory gives a statistical description of a gas of interacting particles. An important property of this equation is its asymptotic equivalence to the Euler or Navier-Stokes equations of the compressible gas dynamics, in the limit of small mean free path. We expect, away from initial layers, boundary layers, and shock layers, the Boltzmann solutions should relax to its equilibrium state (local Maxwellian state) in the small mean free path, and the gas should be governed by the macroscopic equations — the fluid equations. This asymptotic relationship can be seen formally by using the Hilbert expansion and Chapman-Enskog expansions, and the asymptotic equivalence has been justified rigorously only in the case that the fluid flows are smooth [1,2,13,14].

The full Boltzmann equation is a very difficult subject to study due to the nonlocal nonlinearity which describes a complicated mechanism for interacting of particles. A similar but much simpler model, the Broadwell model, was proposed by Broadwell [3] which describes a gas as composed of particles of only six speeds with a binary collision law and spatial variation in only one direction. This leads to a simpler model Boltzmann equation, called the Broadwell equation, and to nontrivial

*Received by the editors June 1990. This work was supported by NSF grants DMS 88-06731 and DMS90-02286..
†Courant Institute, New York University, 251 Mercer Street, New York, NY 10012..

fluid dynamic Euler equations. In one space dimension, this model can be written in the form [1]

$$(1.1)\qquad \begin{aligned} \frac{\partial}{\partial t}f_+ + c\frac{\partial}{\partial x}f_+ &= \frac{1}{\epsilon}Q(f,f) \\ \frac{\partial}{\partial t}f_0 &= -\frac{1}{2\epsilon}Q(f,f) \\ \frac{\partial}{\partial t}f_- - c\frac{\partial}{\partial x}f_- &= \frac{1}{\epsilon}Q(f,f) \end{aligned}$$

where $Q(f,f) = f_0^2 - f_+ f_-$, ϵ is the mean free time between collisions, the unknowns f_+, f_0 and f_- represent the mass densities for gas particles moving in the x-direction with constant speed c, 0, and $-c$ respectively. The terms on the left side of this system represent the streaming of particles; those on the right represent binary collision between particles. For a microscopic distributional function $f = (f_+, f_0, f_-)$, we define its macroscopic fluid moments as

$$(1.1)'\qquad \rho = f_+ + 4f_0 + f_- \ , \quad m = c(f_+ - f_-) \ , \quad u = \frac{m}{\rho}$$

which are hydrodynamical quantities; the mass density, momentum and fluid velocity respectively. Introducng another quantity $z = c^2(f_+ + f_-)$, we can rewrite the Broadwell equations (1.1) in terms of (ρ, m, z) as follows.

$$(1.2)\qquad \begin{aligned} \rho_t + m_x &= 0 \\ m_t + z_x &= 0 \\ z_t + c^2 m_x &= \frac{1}{\epsilon}\frac{1}{8c^2}\{(c^2\rho - z)^2 - 4(z^2 - c^2m^2)\}. \end{aligned}$$

The conditions for (ρ, m, z) to be a local Maxwellian are

$$\rho > 0 \ , \quad |u| < c \ , \quad \text{and} \quad z = \rho\sigma(u)$$

where

$$\sigma(u) = \frac{1}{3}c^2[2(1 + \frac{3u^2}{c^2})^{1/2} - 1] \ .$$

The corresponding Euler equations of fluid dynamics for the model gas are derived by assuming that the gas is in a state of equilibrium for the collision process, i.e., (ρ, m, z) is a local Maxwellian. Then (1.2) is reduced to the following closed 2×2 system

$$(1.3)\qquad \begin{aligned} \rho_t + m_x &= 0 \\ m_t + (\rho\sigma(m/\rho))_x &= 0 \ . \end{aligned}$$

These model Euler equations give an alternative description of the gas in terms of its macroscopic density ρ and momentum m. Caflisch [4] has shown that (1.3) is a strictly hyperbolic and genuinely nonlinear system, its two characteristic speeds are given by

$$\lambda_1(u) = 2\frac{u - \sigma(u)^{1/2}}{\frac{3}{c^2}\sigma(u) + 1} \ , \quad \text{and} \quad \lambda_2(u) = 2\frac{u + \sigma(u)^{1/2}}{\frac{3}{c^2}\sigma(u) + 1}$$

which admit the bounds

$$-c < \lambda_1(u) < 0 < \lambda_2(u) < c \quad \text{if} \quad |u| < c \ .$$

We will consider Cauchy problems for both (1.1) and (1.3). The asymptotic equivalence between Broadwell equations (1.1) and the Euler equations (1.3) at small mean free path has been proved by Caflisch and Papanicolaou [1] in any fixed finite time interval for which the fluid flow is smooth. More precisely, they proved that if the Euler equation (1.3) has a smooth solution which exists up to time $T < +\infty$, then there exists corresponding solution to Broadwell equation (1.1) in the same time interval such that the Broadwell solution converges to the local Maxwellian state determined by the fluid solution as the mean free path goes to zero. We will give two generalizations of this result. First, since (1.3) is strictly hyperbolic and genuinely nonlinear, its solutions develop singularities, in general, no matter how smooth the initial data are [15], so we would like to prove the validity of the fluid approximation in the case that there are shocks in the fluid flow. We have shown that, roughly speaking, if the model fluid dynamic equation (1.3) has a solution up to time $T < +\infty$ which is piecewise smooth with finitely many noninteracting shocks, then one can construct smooth solutions to the Broadwell equations (1.1), and the Broadwell solutions converge asymptotically to the fluid dynamical solution as the mean free path tends to zero, although there are thin shock layers in which the convergence does not hold uniformly. For precise statement, see Theorem 1 in §2. Second as is well known the basic quasilinear wave patterns for the Euler equation (1.3) are shock waves and rarefaction waves. It thus would be interesting to study the asymptotic behavior of the Broadwell solution toward the rarefaction waves of the fluid flow at small mean free path. We can prove that in the case that the fluid flow is a smooth rarefaction wave for the Euler equation (1.3), then the corresponding Broadwell solution exists globally in time, and away from an initial layer, the Broadwell solution converges to the local Maxwellian determined by the rarefaction wave uniformly with a rate independent of time. We refer to §3 for details.

The fluid dynamical limit in the presence of shocks will be studied in §2. We will give a precise statement of the convergence theorem, its proof consists of two main parts. The first part is a construction of a formal solution to the Broadwell equation by matching the truncated Hilbert expansion and shock layer expansion, the second part is a stability analysis for the approximate solution which yields the existence of a Broadwell solution and its convergence to the fluid dynamic solution. We will give an outline, details can be found in [5]. We study the convergence at small mean free path toward rarefaction waves in §3. The time uniform convergence theorem can be obtained by combining the finite time estimates of Caflisch and Papanicolaou [1] with a large time stability analysis of rarefaction waves. The precise statement of the convergence and the reformulation of the problem by appropriate scalings will be given in §3. The technical estimates will appear elsewhere.

2. Fluid Dynamical Limit in the Presence of Shocks. In this section, we will study the fluid dynamical limit of the Broadwell equations in the case that the fluid flow is piecewise smooth with finitely many noninteracting entropy satisfying shocks. We will concentrate on the case that the fluid flow contains only a single curved shock, the general case follows easily from our argument.

2.1. The main convergence theorem. We assume that the Euler equation has a distributional solution $(\rho_0, m_0)(x,t)$ on $R\times[0,T]$, with $0 < T < +\infty$, such that
i) The initial data of the fluid solution $(\rho_0, m_0)(x,t)$ satisfy

$$\rho_0(x,0) > 0\ , \quad |m_0(x,0)| < c\rho_0(x_0,0)\ , \quad x \in R\,. \tag{2.1}$$

ii) The fluid solution $(\rho_0, m_0)(x,t)$ contains a shock moving along a smooth curve $x = s(t)$ such that the fluid state $(\rho_0, m_0)(x,t)$ and its derivatives are uniformly smooth away from the shock but up to the shock, furthermore, we have the following bounds

$$\sum_{1\le\alpha\le 6}\int_0^T \not\int_{R} |\nabla_x^\alpha(\rho_0, m_0)(x,t)|^2 dx\, dt < +\infty \tag{2.2}$$

iii) Crossing the shock $x = s(t)$, it satisfies the Rankine-Hugoniot conditions

$$\begin{aligned} &\rho_0^\ell(u_0^\ell - \dot{s}) = \rho_0^r(u_0^r - \dot{s}) \\ &\rho_0^\ell(\sigma(u_0^\ell) - \dot{s}u_0^\ell) = \rho_0^r(\sigma(u_0^r) - \dot{s}u_0^r) \end{aligned} \qquad 0 \le t \le T\,, \tag{2.3}$$

and the Lax entropy condition [6]

$$\lambda_2^r < \dot{s} < \lambda_2^\ell \quad \text{or} \quad \lambda_1^r < \dot{s} < \lambda_1^\ell\,. \tag{2.4}$$

We set $z_0(x,t) = \rho_0(x,t)\sigma(u_0(x,t))$, then $(\rho_0, m_0, z_0)(x,t)$ is a local Maxwellian. The following theorem shows that there are solutions of the Broadwell equation (1.1) which are close to this local Maxwellian, in which the macroscopic quantities are governed by the Euler equation (1.3).

THEOREM 1. *For each fixed point $(\bar{\rho}, \bar{m}, \bar{z})$ $(\bar{z} = \bar{\rho}\sigma(\bar{m}/\bar{\rho}))$ in the state space, there exist positive constants β_0 and ϵ_0 so that if (ρ_0, m_0) is a solution to the Euler equation (1.3) as described above and the local Maxwellian $(\rho_0, m_0, z_0)(x,t)$ lies in the β_0-neighborhood of $(\bar{\rho}, \bar{m}, \bar{z})$, then for each ϵ with $0 < \epsilon \leq \epsilon_0$, a smooth solution $(\rho^\epsilon, m^\epsilon, z^\epsilon)(x,t)$ of the Broadwell equation (1.2) exists on $R\times[0,T]$, with*

$$(\rho^\epsilon, m^\epsilon, z^\epsilon) - (\rho_0, m_0, z_0) \in L^\infty([0,T], H^2)\ . \tag{2.5}$$

Moreover,

$$\sup_{\substack{0\leq t\leq T\\ |x-s(t)|\geq h}} \left|(\rho^\epsilon - \rho_0, m^\epsilon - m_0, z^\epsilon - z_0)(x,t)\right| \leq c_h\epsilon \quad \text{for any } h > 0 \tag{2.6}$$

and

$$\sup_{0\leq t\leq T} \int_{R} \left|(\rho^\epsilon - \rho_0, m^\epsilon - m_0, z^\epsilon - z_0)(x,t)\right|^2 dx \leq c_1\epsilon^\gamma \quad \text{for any } \gamma < 1\ . \tag{2.7}$$

Remarks.

1) The asymptotic structure of the Broadwell solution $(\rho^\epsilon, m^\epsilon, z^\epsilon)$ in Theorem 1 will be clear, since it is a perturbation of a formal solution which will be constructed explicitly by matching Hilbert expansion with shock layer expansions.

2) The initial data of the Broadwell solution in Theorem 1 are essentially those of Hilbert expansions and shock layer expansion. The matching of these solutions with an initial layer expansion is left for the future.

A sketch of the proof of Theorem 1 will occupy the rest of this section. Details can be found in [5]. The proof has two main parts. First, we will construct a smooth vector valued function $V^\epsilon(x,t) = (\rho^\epsilon, m^\epsilon, z^\epsilon)(x,t)$ such that the following are true.

i) $V^\epsilon(x,t)$ is a smooth approximate solution to the Broadwell equation (1.2) with an error of order ϵ^2. ii) $V^\epsilon(x,t)$ approximates the given fluid flow $(\rho_0, m_0, z_0)(x,t)$ away from the shock with an error of order ϵ. iii) Up to leading order, $V^\epsilon(x,t)$ has a smooth time dependent Broadwell shock profile of width ϵ in place of the shock discontinuity. V^ϵ will be constructed by matching the truncated Hilbert expansion and shock layer expansion in the next subsection. The second part of the proof is essentially a nonlinear stability analysis. We rigorously prove that there exists an exact solution to Broadwell equation in the neighborhood of V^ϵ and the solution converges to the given fluid flow. The main difficulty is that the linearized Broadwell operator has singularity of order $O(\frac{1}{\epsilon})$ due to the Broadwell profile. We will overcome this difficulty by using appropriate scalings and applying an energy method which makes full use of the inner structure of time dependent shock profile for the Broadwell equation [5,7].

We remark here that our approach above is motivated by the recent work of Goodman and Xin for viscous conservation laws in which we prove that a piecewise smooth, entropy satisfying Euler solution can be realized as the limit of Navier-Stokes solutions as dissipation goes to zero [8]. However, the fluid dynamical limit at small mean free path for nonlinear Boltzmann equation is more singular. Since the Broadwell equation (1.1) is purely hyperbolic, the dissipation involved in this system is much weaker than that in Navier-Stokes equations. Thus the stability analysis is more complicated.

2.2. Approximate Broadwell solution. In this subsection, we sketch how to construct an approximate Broadwell solution $V^\epsilon(x,t)$ described in §2.1 by the method of matched asymptotic expansion. The main idea is as follows. The outer solutions, which approximate the fluid flow $(\rho_0, m_0, z_0)(x,t)$ away from the shock, will be found by the Hilbert expansion, and the inner solutions, which approximate the shock discontinuity near the shock, can be found by shock layer expansion. The various order of outer and inner functions will be obtained simultaneously. Then one can form $V^\epsilon(x,t)$ by matching the outer and inner solutions on an appropriate "matching zone". We start with outer expansion.

(1) Hilbert Expansion with a Shock

Set $X^0(x,t) = (\rho^0, m^0, z^0)(x,t)$, and $\chi_i(x,t) = (\rho_i, m_i, z_i)(x,t)$, $i = 0, 1, 2, \ldots$. In a zone away from the shock $x = s(t)$, the Broadwell solution of (1.2) can be approximated by truncation of the formal power series

$$\chi^0(x,t) + \epsilon\chi_1(x,t) + \epsilon^2\chi_2(x,t) + \cdots , \quad x \neq s(t) . \tag{2.8}$$

Substituting (2.8) into (1.2) and equating the coefficients of powers of ϵ, we obtain after some manipulations that

$$\begin{aligned} &z^0 = \rho^0\sigma(m^0/\rho^0) \equiv \rho^0\sigma(u^0) \\ &\rho^0_t + m^0_x = 0 \quad x \neq s(t) , \\ &m^0_t + (\rho^0\sigma(u^0))_x = 0 , \end{aligned} \tag{2.9}$$

$$\begin{aligned} &z_1 = (\sigma(u^0) - u^0\sigma'(u^0))\rho_1 + \sigma'(u^0)m_1 - K(u^0)u^0_x \\ &\rho_{1t} + m_{1x} = 0 \\ &m_{1t} + \{(\sigma(u^0) - u^0\sigma'(u^0))\rho_1 + \sigma'(u^0)m_1\}_x = (K(u^0)u^0_x)_x \end{aligned} \tag{2.10}$$

$$\begin{aligned} &z_2 = (\sigma(u^0) - u^0\sigma'(u^0))\rho_2 + \sigma'(u^0)m_2 - \mu(x,t) \\ &\rho_{2t} + m_{2x} = 0 \\ &m_{2t} + \{(\sigma(u^0) - u^0\sigma'(u^0))\rho_2 + \sigma'(u^0)m_2\}_x = \mu_x \end{aligned} \tag{2.11}$$

where K and μ are known functions [5].

The outer functions $\chi_i(x,t)$ may be discontinuous at the shock $x = s(t)$, but are expected to be smooth away from the shock uniformly up to $x = s(t)$. The leading order equation is the only nonlinear equation which is exactly the model Euler equation (1.3). So we can take $\chi^0(x,t)$ to be the local Maxwellian χ_0 determined by the given fluid flow, i.e.,

$$\chi^0(x,t) = \chi_0(x,t) = (\rho_0, m_0, z_0)(x,t) , \quad x \neq s(t) . \tag{2.12}$$

All the other equations for higher order outer functions $\chi_i(x,t)$ are inhomogeneous linear systems with the same principal part which is the linearized operator of the nonlinear Euler equation (1.3) around the given fluid state $(\rho_0, m_0)(x,t)$. So they are linear hyperbolic systems with characteristic speeds $\lambda_1(u_0) < 0$, and $\lambda_2(u_0) > 0$. In order to obtain the higher order outer functions, we need to solve initial boundary value problems for these 2×2 linear hyperbolic systems on both regions

$$\Omega_- = \{(x,t),\ x < s(t)\ ,\ 0 \leq t \leq T\} , \qquad \text{and } \Omega_+ = \{(x,t)\ ,\ x > s(t)\ ,\ 0 \leq t \leq T\} .$$

By the standard theory for mixed problems for linear hyperbolic system [9,10], and the Lax entropy condition (2.4) (for definiteness, we assume $\lambda_2^r < \dot{s} < \lambda_2^\ell$). We see that in order to have smooth solutions of (2.10) and (2.11) on Ω_+, one needs only to impose appropriate initial conditions. However, to have smooth solutions on Ω_-, additional boundary conditoins on the shock curve $x = s(t)$ will be needed. The required boundary conditions will come from matching with shock layer expansions.

(2) Shock Layer Expansions

Near the shock, the Broadwell solution will be represented by an inner expansion of the form:

$$X_0(\xi,t) + \epsilon X_1(\xi,t) + \epsilon^2 X_2(\xi,t) + \cdots , \tag{2.13}$$

where $X_i = (P_i, M_i, Z_i)$, and ξ is the stretched variable given by

$$\xi = \frac{x - s(t)}{\epsilon} + \delta(t,\epsilon) , \tag{2.14}$$

in which $\delta(t,\epsilon)$ is the perturbation of the shock position which is assumed to have an expansion of the form

$$\delta(t,\epsilon) = \delta_0(t) + \epsilon\delta_1(t) + \epsilon^2\delta_2(t) + \cdots \tag{2.15}$$

After substituting (2.13)-(2.15) into (1.2) and matching powers of ϵ, we arrive at

$$\begin{aligned} &-\dot{s}\frac{\partial}{\partial\xi}P_0 + \frac{\partial}{\partial\xi}M_0 = 0 \\ &-\dot{s}\frac{\partial}{\partial\xi}M_0 + \frac{\partial}{\partial\xi}Z_0 = 0 \\ &-\dot{s}\frac{\partial}{\partial\xi}Z_0 + c^2\frac{\partial}{\partial\xi}M_0 = \frac{1}{8c^2}\left[(c^2P_0 - Z_0)^2 - 4(Z_0^2 - c^2M_0^2)\right] , \end{aligned} \tag{2.15'}$$

$$\begin{aligned} &-\dot{s}\frac{\partial}{\partial\xi}P_1 + \frac{\partial}{\partial\xi}M_1 + \dot{\delta}_0(t)\frac{\partial}{\partial\xi}P_0 + \frac{\partial}{\partial t}P_0 = 0 \\ &-\dot{s}\frac{\partial}{\partial\xi}M_1 + \frac{\partial}{\partial xi}Z_1 + \dot{\delta}_0(t)\frac{\partial}{\partial\xi}M_0 + \frac{\partial}{\partial t}M_0 = 0 \\ &-\dot{s}\frac{\partial}{\partial\xi}Z_1 + c^2\frac{\partial}{\partial\xi}M_1 + \dot{\delta}_0(t)\frac{\partial}{\partial\xi}Z_0 + \frac{\partial}{\partial t}Z_0 \\ &= \frac{1}{8c^2}\left[2(c^2P_0 - Z_0)(c^2P_1 - Z_1) - 4(2Z_0Z_1 - 2c^2M_0M_1)\right] , \end{aligned} \tag{2.16}$$

etc.

The inner expansion is supposed to be valid in a small zone of order ϵ around $x = s(t)$. We need boundary conditions at $\xi = \pm\infty$ to solve above systems of ordinary differential equations. The conditions will come from matching conditions.

(3) Matching conditions. The Hilbert expansion and shock layer expansion are expected to be valid in the "matching zone", in which both $|\xi| \to \infty$ and $0 < |x - s(t)|$ small are valid. Thus they must agree there. This yields the following so called "matching conditions" as $\xi \to \pm\infty$,

$$X_0(\xi,t) = \chi_0(s(t)\pm,t) + o(1) , \tag{2.17}$$

$$X_1(\xi,t) = \chi_1(s(t)\pm,t) + (\xi - \delta_0)\frac{\partial\chi_0}{\partial x}(s(t)\pm,t) + o(1) , \tag{2.18}$$

$$\begin{aligned} X_2(\xi,t) &= \chi_2(s(t)\pm,t) - \delta_1\frac{\partial\chi_0}{\partial x}(s(t)\pm,t) \\ &\quad + (\xi - \delta_0)\frac{\partial\chi_1}{\partial x}(s(t)\pm,t) + \frac{1}{2}(\xi - \delta_0)^2\frac{\partial\chi_2}{\partial x}(s(t)\pm,t) + o(1) . \end{aligned} \tag{2.19}$$

We note that the matching conditions require that inner functions have algebraic growth rates at both infinities.

Now we are in the position to determine various order of inner and outer functions and δ's. First, by assumption, $\chi_0(s(t)\pm,t)$ are local Maxwellians and satisfy the Lax entropy conditions, so it follows that systsem (2.15) with boundary condition (2.17) has a smooth traveling wave solution $X_0 = (P_0, M_0, Z_0)(\xi,t)$ which is unique up to phase shift. X_0 can be constructed explicitly and has the following properties:

$$\begin{aligned} &\dot{s}P_{0\xi} < 0 , \quad M_{0\xi} < 0 , \quad \dot{s}\,Z_{0\xi} < 0 , \quad \forall\xi \in R \text{ and} \\ &X_0(\xi,t) = \chi_0(s(t)\pm,t) + O(1)\,exp\{-\alpha|\xi|\} \text{ as } \xi \to \pm\infty , \text{ for some } \alpha > 0 \end{aligned} \tag{2.20}$$

These properties play an essential role in our stability analysis and the construction of high order corrections. Next, we determine $\chi_1(x,t)$, $X_1(\xi,t)$ and $\delta_0(t)$ simultaneously. The existence of a smooth solution to (2.16) with linear growth at both infinities is a consequence of entropy conditions.

The solvability condition for the matching condition (2.18) yields an ordinary differential equation for $\delta_0(t)$ in the form [5]

$$(2.21)\qquad (\dot{s}-\lambda_1^\ell)\frac{d}{dt}\{(\rho_0^\ell-\rho_0^r)\delta_0\}+\ddot{s}\{(\rho_0^\ell-\rho_0^r)\delta_0\}=O(t)$$

with some known function $O(t)$. (2.21) can be solved easily, then matching condition (2.18) yields the necessary conditions on $x = s(t)$ so that one can obtain smooth $\chi_1(x,t)$ over Ω_+ and Ω_- respectively. This in turn determines the inner function $X_1(\xi,t)$. In fact, one can show [5]

LEMMA 2. *$\chi_1(x,t)$, $X_1(\xi,t)$ and $\delta_0(t)$ can be be determined simultaneously such that (i) $\chi_1(x,t)$ and its derivatives are uniformly continuous up to $x = s(t)$ and*

$$(2.22)\qquad \sum_{|\alpha|\le 2}\int\!\!\!\int |\partial_x^\alpha \chi_1(x,t)|^2 dx\,dt < +\infty$$

(ii) $\delta_0(t)$ and $X_1(\xi,t)$ are smooth, and there exists $\alpha_0 > 0$ such that

$$(2.23)\qquad X_1(\xi,t)=\chi_1(s(t)\pm,t)+(\xi-\delta_0)\chi_{0x}(s(t)\pm,t)+O(1)\,exp\{-\alpha_0|\xi|\}\,,\quad \xi\to\pm\infty\,.$$

Higher order functions can be constructed in a similar way. In particular, second order outer and inner functions $\chi_2(x,t)$ and $X_2(\xi,t)$ and δ_1 can be detertmined simultaneously with similar properties as in Lemma 2. Now we set

$$O(x,t)=\chi_0(x,t)+\epsilon\chi_1(x,t)+\epsilon^2\chi_2(x,t)\,,\quad x\neq s(t)\,,$$

and

$$I(x,t)=X_0(\frac{x-s(t)}{\epsilon}+\delta_0+\epsilon\delta_1,t)+\epsilon X_1(\frac{x-s(t)}{\epsilon}+\delta_0+\epsilon\delta_1,t)+\epsilon^2X_2(\frac{x-s(t)}{\epsilon}+\delta_0+\epsilon\delta_1,t)\,.$$

Let $\chi(y)$ be a $C_0^\infty(R)$ function such that $0\le\chi(y)\le 1$, $\chi(y)=1$ for $|y|<1$ and $\chi(y)=0$ for $|y|\ge 2$, and α be a constant in $(\frac{2}{3}\,,1)$. We define a smooth function as follows,

$$(2.24)\qquad V^\epsilon(x,t)=\chi(\frac{x-s(t)}{\epsilon^\alpha})I(x,t)+(1-\chi(\frac{x-s(t)}{\epsilon^\alpha}))\,O(x,t)+d(x,t)\,,$$

where $d(x,t)=(d_1,d_2,d_3)(x,t)$ is a higher order correction term which is chosen so that the following proposition holds.

PROPOSITION 3. *The smooth function $V^\epsilon(x,t)$ constructed in (2.24) satisfies*

$$(2.25)\qquad \begin{cases}\rho_t^\epsilon+m_x^\epsilon=0\\ m_t^\epsilon+z_x^\epsilon=0\\ z_t^\epsilon+c^2m_x^\epsilon-q(v^\epsilon,v^\epsilon)=R(x,t)\,,\end{cases}$$

where $q(f,f)=\frac{1}{\epsilon}\frac{1}{8c^2}\{(c^2f_1-f_3)^2-4(f_3^2-c^2f_2^2)\}$, and the error term $R(x,t)$ admits following bounds, for $0\le\ell\le 2$.

$$(2.26)\qquad \frac{\partial^\ell}{\partial x^\ell}R(x,t)=O(1)\epsilon^{(2-\ell)\alpha}\,,\quad \int_0^T\int_R\Big|\frac{\partial^\ell}{\partial x^\ell}R\Big|^2dx\,dt\le O(1)\,\epsilon^{(2-\ell)2\alpha}\,.$$

It can be shown such a $d(x,t)$ exists and it is a linear hyperbolic wave [5]. We note that Proposition 3 asserts that $V^\epsilon(x,t)$ is the desired approximate solution to Broadwell equation (1.2). This finishes the construction of the formal approximate Broadwell solution.

2.3. Stability analysis. We now show that there exist exact solutions $U^\epsilon(x,t)$ to the Broadwell equation (1.2) in a neighborhood of the approximate solution $V^\epsilon(x,t)$ for suitably small ϵ, and the asymptotic behavior of the Broadwell solution $U^\epsilon(x,t)$ is governed by $V^\epsilon(x,t)$. We first derive an error equation. Then, after appropriate scalings and reformulation, the error equation can be estimated by using an energy analysis of Kawashima and Matsumura which makes full use of the inner structure of the Broadwell profile [5,7].

Consider the Broadwell solution $U^\epsilon(x,t)$ on $R\times[0,T]$ with initial data $U^\epsilon(x,0)=V^\epsilon(x,0)$. We decompose the solution as

$$U^\epsilon(x,t)=V^\epsilon(x,t)+\overline{V}(x,t)\equiv V^\epsilon(x,t)+(\overline{\phi},\overline{\psi},\overline{\omega})(x,t)\ ,\quad (x,t)\in R\times[0,T]\ . \tag{2.27}$$

Then, by using the defining relations for V^ϵ, we obtain after some manipulations the following error equation.

$$\begin{aligned}&\bar\phi_t+\bar\psi_x=0\\ &\bar\psi_t+\bar\omega_x=0\\ &A^\epsilon\bar\phi+m^\epsilon\bar\psi-B^\epsilon\bar\omega=\epsilon(\bar\omega_t+c^2\bar\psi_x)+\epsilon R(x,t)-\Gamma(\bar\phi,\bar\psi,\bar\omega)\\ &\bar V(x,0)=0\ ,\end{aligned} \tag{2.28}$$

where

$$A^\epsilon=\frac{1}{4}(c^2\rho^\epsilon-z^\epsilon)\ ,\quad B^\epsilon=\frac{1}{4c^2}(c^2\rho^\epsilon+3z^\epsilon)\ ,\quad P(\cdot)=\epsilon q(\cdot,\cdot)\ .$$

To exploit the structure of the Broadwell shock profile, we need to integrate the equations for the hydrodynamical moments in (2.28) once. Thus we set

$$\bar\phi=\bar{\bar\phi}_x\ ,\quad \bar\psi=\bar{\bar\psi}_x\ ,\bar\omega=\bar{\bar\omega}\ . \tag{2.29}$$

Substituting (2.29) into (2.28) and integrating the first two equations yields

$$\begin{aligned}&\bar{\bar\phi}_t+\bar{\bar\psi}_x=0\\ &B^\epsilon\bar{\bar\psi}_t+m^\epsilon\bar{\bar\psi}_x+A^\epsilon\bar{\bar\phi}_x=\epsilon(c^2\bar{\bar\psi}_{xx}-\bar{\bar\psi}_{tt})+\epsilon R(x,t)-\Gamma(\bar{\bar\phi}_t,\bar{\bar\phi}_x,-\bar{\bar\psi}_t)\\ &\bar{\bar\phi}(x,0)=\bar{\bar\psi}(x,0)=\bar{\bar\psi}_t(x,0)=0\ .\end{aligned} \tag{2.30}$$

Formula (2.30) can be further simplified by introducing the following scalings.

$$\bar{\bar\phi}(x,t)=\epsilon\phi(y,\tau)\ ,\quad \bar{\bar\psi}(x,t)=\epsilon\psi(y,\tau)\ ,\quad y=\frac{x-s(t)}{\epsilon}\ ,\quad \tau=\frac{t}{\epsilon}\ .$$

We then rewrite (2.30) as

$$\begin{aligned}&\phi_\tau-\dot s\phi_y+\psi_y=0\\ &(\psi_\tau-\dot s\psi_y)_\tau-\dot s(\psi_\tau-\dot s\psi_y)_y-c^2\psi_{yy}+A^\epsilon\phi_y+(m^\epsilon-\dot sB^\epsilon)\psi_y+B^\epsilon\psi_\tau\\ &\qquad=\epsilon R(\epsilon y+\dot s(\epsilon\tau),\epsilon\tau)-\Gamma(\phi_y,\psi_y,-(\psi_\tau-\dot s\psi_y))\\ &\phi(0,y)=\psi(0,y)=\phi_\tau(y,0)=0\ .\end{aligned} \tag{2.31}$$

Our purpose is to show that for ϵ sufficiently small, (2.31) has a smooth "small" solution up to $\tau=T/\epsilon$. Since (2.31) is a strictly hyperbolic system, so the local existence is standard. In order to get the existence up to $\tau=T/\epsilon$, we need the following a priori estimate.

PROPOSITION 4. *(a priori estimate). Suppose (2.31) has a solution (ϕ,ψ) in*

$$X(0,\tau_0)\equiv\{(\phi,\psi):|(\phi,\psi)\in C^0(0,\tau_0;H^2)\ ,\psi_\tau\in C^0(0,\tau_0;H^1)\} \tag{2.32}$$

for some $\tau_0\in(0,T/\epsilon)$. Then there exist positive constants ϵ_0, β_0, k_0, and K_0 independent of τ_0 and ϵ such that if $0<\epsilon\le\epsilon_0$, $|(\rho_0,m_0,z_0)-(\bar\rho,\bar m,\bar z)|_{L_\infty}\le\beta_0$ and

$$\sup_{0\le s\le\tau_0}(\|(\phi,\psi)\|_2+\|\psi_\tau(\cdot)\|_1)\le k_0\ ,$$

it holds that

$$\sup_{0\le\tau\le\tau_0} \left(\|(\phi,\psi)(\tau)\|_2^2 + \|\psi_\tau(\tau)\|_1^2\right) + \int_0^{\tau_0} \left(\|(\phi_y,\psi_y)(\tau)\|_1^2 + \|\psi_\tau(\cdot,\tau)\|_1^2\right) d\tau \le K_0\epsilon^{4\alpha-1} . \tag{2.33}$$

The proof of this proposition is very technical, which is an energy analysis making use of the structure of V^ϵ, in particular (2.20). We refer to [5] for details.

Now we can finish the proof of Theorem 1. Combining the standard local existence of solutions to (2.31) in X (see (2.32)) and Proposition 4, we conclude, by using the standard continuous induction argument for hyperbolic systems that there exists a unique solution (ϕ,ψ) to (2.31) with $(\phi,\psi)\in X(0,T/\epsilon)$. Furthermore (2.33) holds with τ_0 replaced by T/ϵ. It follows from this and the structure of V^ϵ that there exists a smooth Broadwell solution U^ϵ to (1.2) on $\boldsymbol{R}\times[0,T]$ such that (2.5) is satisfied. To get the desired asymptotic behavior, first we have from (2.27),(2.29) and (2.33) that

$$\begin{aligned}\sup_{0\le t\le T}\|U^\epsilon - V^\epsilon(\cdot,t)\|^2 &= \sup_{0\le t\le T}\|(\bar{\bar{\phi}}_x,\bar{\bar{\psi}}_x,\bar{\bar{\omega}}\|^2 = \epsilon^2\sup_{0\le t\le T}\|(\phi_x,\psi_x,\psi_\tau)\|^2\\ &= \epsilon\sup_{0\le t\le T}\|(\phi_y,\psi_y,\psi_\tau)\|^2 \le K_0\epsilon^{4\alpha} .\end{aligned}$$

On the other hand, by construction of V^ϵ, we have

$$\sup_{0\le t\le T}\|V^\epsilon - \chi_0\| \le O(1)\epsilon^\alpha$$

Thus (2.7) follows. Formula (2.6) follows from (2.33) and Sobolev's inequality in a similar way. The proof of Theorem 1 is complete.

3. Fluid Dynamical Limit to Rarefaction Waves.

The rarefaction wave is one of the basic nonlinear wave patterns for the model Euler equation (1.3). As a direct application of the result by Caflisch and Papanicolaou [1]: For any given fixed finite time interval, we have strong convergence of Broadwell solutions to a smooth rarefaction wave at small mean free path. Using the inner geometrical structure of the rarefaction wave, we are able to obtain time uniform convergence to a smooth rarefaction for the Broadwell solution as the mean free path goes to zero. This result also sheds light on the fluid dynamical limit problem in the case that fluid flow contains centered rarefaction waves. The study of centered rarefaction waves will appear elsewhere.

3.1. The convergence theorem. We consider the Cauchy problem for the Broadwell equation (1.1) with initial data

$$f(x,0) \equiv (f_+,f_0,f_-)(x,0) = \mathring{f}(x) , \tag{3.1}$$

where each component of $\mathring{f}(x)$ is a strictly positive function, and $lim_{x\to\pm\infty}\mathring{f}(x)$ are Maxwellians. Let $\rho_0(x)$ and $m_0(x)$ be the fluid moments associated with this $\mathring{f}(x)$ (cf. (1.1)′), and $u_0(x) = m_0(x)/\rho_0(x)$. Then it follows from the positiveness of $\mathring{f}(x)$ that

$$\rho_0(x) > 0 , \quad |u_0(x)| < c . \tag{3.2}$$

Assume that the initial fluid velocity, $u_0(x)$, satisfies the following conditions.
i) $\frac{\partial}{\partial x}u_0(x) > 0$ for all $x\in R^1$.
ii) $u_0(x)$ approaches $u_0(\pm\infty) = u_{0\pm}$ sufficiently fast as $x\to\pm\infty$.
iii) The decay rates of $\frac{\partial^\ell}{\partial x^\ell}u_0(x)$ as $x\to\pm\infty$ are not less than those of $\frac{\partial u_0}{\partial x}$ $(\ell = 2,3)$.
We remark that condition i) guarantees the existence of the rarefaction wave for (1.3). Under the above assumptions, the Cauchy problem for the model Euler equation (1.3) with initial data

$$\rho(x,0) = \rho_0(x) , \quad u(x,0) = u_0(x) \tag{3.3}$$

has a global (in time) smooth rarefaction wave solution $(P,U)(x,t)$ which can be determined by the following relations

$$\begin{aligned} &\lambda_i(U) = W_i(x,t) \\ &P(x,t) = \rho_{0-}\, exp\{\int_{u_{0-}}^{U(x,t)} (\lambda_i(s)-s)^{-1}\, ds\}\ , \end{aligned} \qquad i=1,2\ , \tag{3.4}$$

in which $\rho_{0-} = lim_{x\to-\infty}\rho_0(x)$, $u_{0-} = lim_{x\to-\infty} u_0(x)$, and $W_i(x,t)$ is the solution of the following Cauchy problem.

$$\begin{aligned} &W_{it} + W_i W_{ix} = 0 \\ &W_i(x,0) = \lambda_i(u_0(x))\ . \end{aligned} \tag{3.5}$$

The existence of $(P,U)(x,t)$ is a consequence of the fact [4] that for $|u| < c$,

$$\begin{aligned} &u^2 < \sigma(u) < c^2\ , \quad d\lambda_i(u)/du > 0 \quad (i=1,2)\ , \\ &-c < \lambda_1(u) < min(u,0) \le max(u,0) < \lambda_2(u) < c\ . \end{aligned} \tag{3.6}$$

In particular, (3.6) implies that the system (1.3) is strictly hyperbolic and whose charactersitic fields are both genuinely nonlinear.

For the smooth rarefaction wave $(P,U)(x,t)$, the local Maxwellian associated with this state is given by $f^r(x,t) = (f_+^r, f_0^r, f_-^r)(x,t)$, where

$$\begin{aligned} f_+^r &= \frac{P}{2c^2}(\sigma(U)+cU) \\ f_0^r &= \frac{P}{4c^2}(c^2-\sigma(U)) \\ f_-^r &= \frac{P}{2c^2}(\sigma(U)-cU)\ . \end{aligned} \tag{3.7}$$

Our following theorem asserts that the local Maxwellian f^r governs the asymptotic behavior at small mean free path of the solutions to Broadwell equations (1.1) with initial data (3.1).

THEOREM 5. *Let $\rho_0(x)$ and $m_0(x)$ be the fluid moments associated with the initial microscopic density fucntion $\mathring{f}(x)$ for the Broadwell equation (1.1). Suppose that $\rho_0(x)$ and $m_0(x)$ satisfy the conditions as described above so that the fluid equation (1.3) with initial data (3.3) has a global smooth rarefaction wave solution $(P,U)(x,t)$. Let $f^r(x,t)$ be the local Maxwellian associated with the rarefaction wave $(P,U)(x,t)$ by (3.7).*

Then there exists an $\epsilon_0 > 0$ such that for each $\epsilon \in (0,\epsilon_0]$ Broadwell equation (1.1) with initial data (3.1) has a unique solution f^ϵ such that

$$\sup_{\delta\le t<+\infty} \|f^\epsilon(u) - f^r(t)\|^2 \le c_\delta \epsilon^{1/2}\ . \tag{3.9}$$

In particular, we have

$$\sup_{\delta\le t<+\infty} |f^\epsilon(t) - f^r(t)|_{L_\infty} \le C'_\delta \epsilon^{1/4}\ . \tag{3.10}$$

Remark 1. As in [1], we can get rid of the restriction $t \ge \delta$ if the initial data $\mathring{f}(x)$ is a local Maxwellian.

Remark 2. Although we have a global in time existence of bounded smooth soluions to (1.1) and (3.1), we do not have requirements on the strength of the rarefaction wave except the existence of the rarefaction wave. This is in contrast to Theorem 1.

For the proof of Theorem 5, we are going to combine the finite time result of Caflisch and Papanicolaou with a large time estimate which can be obtained by an energy method making use of the geometrical structure of the rarefaction wave. In the next subsection, we derive an error equation by decomposing the Broadwell solution as a perturbation of a Chapman-Enskog ansatz and using appropriate scalings. The main energy estimate is outlined in §3.3.

3.2. Some preliminaries. In the following, without loss of generality, we always assume that $c = 1$. We begin with listing some properties of the smooth rarefaction wave $(P, U)(x, t)$ which we will use later.

LEMMA 6. *The rarefaction wave $(P, U)(x, t)$ constructed in (3.4) and (3.5) satisfies the following (i) (P, U) is a smooth global in time solution of (1.3), i.e.*

$$\begin{aligned} P_t + (PU)_x &= 0 \\ (PU)_t + (P\sigma(U))_x &= 0 \end{aligned} \tag{3.11}$$

(ii) $\frac{\partial}{\partial x}U(x,t) > 0$, $\frac{\partial}{\partial x}\lambda_i(U) > 0$ $(i = 1, 2)$, $\frac{\partial P}{\partial x} < 0$ $(i = 1)$, $\frac{\partial P}{\partial x} > 0$ $(i = 2)$.
(iii) There exists a positive constant η_0 such that

$$|U| \le 1 - \eta_0 \ , \quad |\lambda_i(U)| \ge \eta_0 \ , \quad |\lambda_i(U) - U| \ge \eta_0 \ , \quad 1 - \eta_0 \ge \sigma(U) \ge |U| + \eta_0$$

(iv) For any $p \in [1, \infty]$, there exists a positive constant C suchthat for $t \ge 0$

$$\begin{aligned} &\|(U_x, U_t)\|_{L_p} \le c(1+t)^{-1+1/p} \ , \\ &\|(U_{xx}, U_{tx}, U_{xxx}, U_{txx})\|_{L_p} \le c(1+t)^{-1} \ . \end{aligned} \tag{3.12}$$

Proof. (i) and (ii) follow from (3.4)-(3.6) and the assumption $\frac{\partial}{\partial x}u_0(x) > 0$. (iii) follows from (3.2), (3.6) and (ii). To prove (iv), it suffices to obtain similar estimate on the solution W_i of (3.5). However W_i can be estimated by explicit solution formula and the assumptions on $u_0(x)$. #

Now suppose $f^\epsilon(x,t)$ is a smooth solution to (1.1) and (3.1). We decompose the Broadwell solution f^ϵ as

$$f^\epsilon(x,t) = \bar{f}(x,t) + g(x,t) \ , \tag{3.13}$$

where $\bar{f}(x,t)$ is the Chapman-Enskog ansatz

$$\bar{f}(x,t) = f^r(x,t) + \epsilon\,\theta(x,t) \ , \tag{3.14}$$

in which $\theta(x,t) = (\theta_+, \theta_0, \theta_-)(x,t)$ to be determined. Substituting (3.13) into the Broadwell equation (1.1), and using the defining relation for f^r ((3.7) and (3.11)), we obtain (recall $c = 1$) after some computations that

$$\begin{aligned} \frac{\partial g_+}{\partial t} + \frac{\partial g_+}{\partial x} + \frac{1}{\epsilon}Lg - \frac{1}{\epsilon}Q(g,g) &= \frac{1}{2}\{2[2f_0^r\theta_0 - (\theta_+ f_-^r + \theta_- f_+^r)] - [P\sigma(U))_t + (PU)_x]\} \\ &\quad + \epsilon[(\theta_0^2 - \theta_+\theta_-) - (\frac{\partial}{\partial t}\theta_+ + \frac{\partial}{\partial x}\theta_+)] \\ \frac{\partial g_0}{\partial t} - \frac{1}{2\epsilon}Lg + \frac{1}{2\epsilon}Q(g,g) &= -\frac{1}{4}\{2[2f_0^r\theta_0 - (\theta_+ f_-^r + \theta_- f_+^r)] \\ &\quad - [(P\sigma(U))_t + (PU)_x]\} - \epsilon\frac{\partial}{\partial t}\theta_0 \\ \frac{\partial g_-}{\partial t} - \frac{\partial g_-}{\partial x} + \frac{1}{\epsilon}Lg - \frac{1}{\epsilon}Q(g,g) &= \frac{1}{2}\{2[2f_0^r\theta_0 - (\theta_+ f_-^r + \theta_- f_+^r)] - [(P\sigma(U))_t + (PU)_x]\} \\ &\quad + \epsilon[(\theta_0^2 - \theta_+\theta_-) - (\frac{\partial}{\partial t}\theta_- - \frac{\partial}{\partial x}\theta_-)] \ , \end{aligned} \tag{3.15}$$

where $Lg = -2Q(\bar{f}, g) = \bar{f}_- g_+ + \bar{f}_+ g_- - 2\bar{f}_0 g_0$. Now, we choose $\theta(x,t)$ such that the right-hand sides above contain no order $O(1)$ terms. This requires

$$2[2f_0^r\theta_0 - (\theta_+ f_-^r + \theta_- f_+^r)] - [(P\sigma(U))_t + (PU)_x] = 0 \ .$$

One particular choice is

$$2\theta_+ = 2\theta_- = -4\theta_0 = e$$
$$(f_+^r + f_0^r + f_-^r)e = -(P\sigma(U))_t - (PU)_x$$

A direct computation yields

$$e(x,t) = -\mu(U)U_x \ , \tag{3.16}$$

where $\mu(U) = 16(1-\sigma(U))/(1+\sigma(U))^3$. It follows that the equations for $g(x,t)$ take the form

$$g_{+t} + g_{+x} + \frac{1}{\epsilon}Lg = \frac{1}{\epsilon}Q(g,g) - \epsilon\{\frac{e_t + e_x}{2} + \frac{3e^2}{16}\}$$
$$g_{0t} - \frac{1}{2\epsilon}Lg = -\frac{1}{2\epsilon}Q(g,g) + \epsilon\{\frac{e_t}{4} + \frac{3e^2}{32}\}$$
$$g_{-t} - g_{-x} + \frac{1}{\epsilon}Lg = \frac{1}{\epsilon}Q(g,g) - \epsilon\{\frac{e_t - e_x}{2} + \frac{3e^2}{16}\} \ .$$

We transform the above system by the following scaling.

$$g(x,t) = \bar{g}(y,\tau) \ , \quad y = \frac{1}{\epsilon} \ , \quad \text{and} \quad \tau = \frac{t}{\epsilon} \ . \tag{3.17}$$

We then obtain the following error equation (dropping the bars)

$$\begin{aligned} g_{+\tau} + g_{+y} + Lg &= Q(g,g) - \epsilon^2(\frac{e_t + e_x}{2} + \frac{3e^2}{16}) \\ g_{0\tau} - \frac{1}{2}Lg &= -\frac{1}{2}Q(g,g) + \epsilon^2(\frac{e_t}{4} + \frac{3e^2}{32}) \\ g_{-\tau} - g_{-y} + Lg &= Q(g,g) - \epsilon^2(\frac{e_t - e_x}{2} + \frac{3e^2}{16}) \\ g(y,0) = \overset{\circ}{g}(y) &\equiv \overset{\circ}{f}(\epsilon y, 0) - f^r(\epsilon y, 0) - \epsilon\theta(\epsilon y, 0) \ . \end{aligned} \tag{3.18}$$

Instead of Theorem 5, we shall show that problem (3.18) has a unique "small", smooth, and global in time solution. First, we note that by reexamining the proof of Caflisch and Papanicolaou in [1], we have easily

PROPOSITION 7. (Caflisch and Papanciolaou). *There exists a positive constgant ϵ such that for each $\epsilon \in (0, \epsilon_1]$, the problem (3.18) has the unique solution*

$$g \in C^0(0, \tau_0; H_1) \cap C^1(0, \tau_0; L_2)$$

satisfying

$$\begin{aligned} \sup_{\delta \le \tau \le \tau_0} \|g(\tau)\|_{L_2(R_y)} &\le c_\delta \epsilon^{1/2} \ , \\ \sup_{\delta \le \tau \le \tau_0} \|g_y(t)\|_{L_2(R_y)} &\le c_\delta \epsilon^{3/2} \ , \end{aligned} \tag{3.19}$$

where $\tau_0 = \frac{1}{\epsilon}$, and c_δ is positive constant independent of ϵ.

We note that in the proof of Proposition 7, we have used the structure of $e(x,t)$ (3.16). Now in order to extend the solution $g(y,\tau)$ in Proposition 7 beyond the time τ_0, we need the following a priori large time estimates.

PROPOSITION 8. (*a priori estimate*). *Let $g(y,\tau)$ be the solutions of (3.18) so that $g \in C^0(0, \tau_1; H_1) \cap C^1(0, \tau_1; L_2)$ for some time $\tau_1 > \tau_0$. Then there exist positive constants ϵ_2 ($\le \epsilon_1$), δ_1 and C_2 independent of ϵ and τ_1 such that if*

$$N(\tau_1) = \sup_{\tau_0 \le \tau \le \tau_1} \|g(\tau)\|_1 < \delta_1 \ , \quad \textit{and} \quad \epsilon \le \epsilon_2 \tag{3.20}$$

then the following inequality holds

$$\sup_{\tau_0\le\tau\le\tau_1} \|g(\tau)\|_1^2 + \int_{\tau_0}^{\tau_1} \|g_y(\tau)\|^2 d\tau \le C_2(\|g(\tau_0)\|_1^2 + \epsilon^{1/2}) . \tag{3.21}$$

Since (3.18) is a semilinear hyperbolic system, it is standard to obtain the global in time exisence and estimate of the solution as long as we can prove Proposition 8. The proof of Proposition 8 will occupy the rest of this section.

3.3. Main energy estimate. We now prove Proposition 8 by using a variation of the energy method first used by Kawashima in [11] and then by Matsumura in [12]. The basic idea behind the detailed energy analysis is to design an appropriate energy function based on the fact that the linearized collision operator is nonnegative and the crucial property that rarefaction waves for the Euler equation are expansive. We begin with the basic L_2 estimate. As in [12], we set for $i = +, 0, -$,

$$\begin{aligned} E_0(\bar f_i, g_i) &= (\bar f_i + g_i)\,log(\bar f_i + g_i) - \bar f_i\, log\, \bar f_i - (1 + log\, \bar f_i) g_i ,\\ E_1(\bar f_i, g_i) &= log(\bar f_i + g_i) - log(\bar f_i) ,\\ E_2(\bar f_i, g_i) &= log(\bar f_i + g_i) - log\, \bar f_i - g_i/\bar f_i .\end{aligned}$$

For suitably small $N(\tau_1)$ and ϵ, it follows from (3.7), (3.14) and Lemma 6 that

$$\nu|g_i|^2 \le E_0(\bar f_i, g_i) \le c_3|g_i|^2 , \quad |E_1(\bar f_i, g_i)| \le c_3|g_i| , \quad |E_2(\bar f_i, g_i)| \le c_3|g_i|^2 , \quad i = +, 0, - , \tag{3.22}$$

and

$$\begin{aligned} &(Q(g,g) - Lg)\{log[(\bar f_0 + g_0)^2 \bar f_+ \bar f_-] - log[(\bar f_+ + g_+)(\bar f_- + g_-)\bar f_0^2]\}\\ &\qquad \ge \nu_1|Q(g,g) - Lg|^2 - c_3|c|^2|g|^2 ,\end{aligned} \tag{3.23}$$

where we have used Sobolev's inequality, the Cauchy inequality and fact that $\bar f_0^2 - \bar f_+ \bar f_- = O(1)\epsilon e$. Similarly, using (ii) of Lemma 6 and the equations satisfied by $\bar f(x,t)$, one has after some computations that

$$\begin{aligned} &- E_2(\bar f_+, g_+)(\bar f_{+\tau} + \bar f_{+y}) - 4E_2(\bar f_0, g_0)\bar f_{0\tau} - E_2(\bar f_-, g_-)(\bar f_{-\tau} - \bar f_{-y})\\ &= +\frac{P}{8}(1 + 3\sigma(U))\epsilon\mu U_x\{g_+^2/(2\bar f_+^2) + g_-^2/(2\bar f_-^2) - g_0^2/(\bar f_0^2)\} + O(1)\{\epsilon U_x|g|^3\\ &\qquad + \epsilon^2|g|^2|(e_t, e_x, e^2)|\}\\ &= \frac{P}{8}(1 + 3\sigma(U))3\mu U_x\{g_+^2/(2\bar f_+^2) + g_-^2/(2\bar f_-^2) - \frac{1}{\bar f_0^2}(\frac{\bar f_- g_+}{2\bar f_0} + \frac{\bar f_+ g_-}{2\bar f_0})^2\}\\ &\qquad + O(1)\epsilon U_x|g||Q(g,g) - Lg| + O(1)\{\epsilon U_x|g|^3 + \epsilon^2|g|^2|(e_t, e_x, e^2)|\}\\ &\ge \epsilon\nu_2 U_x(\bar f_- g_+ - \bar f_+ g_-)^2 - c_3\beta^{-1}\epsilon^2 U_x^2|g|^2 - \beta|Q(g,g) - Lg|^2 - c_3\{\epsilon U_x|g|^3\\ &\qquad + \epsilon^2|g|^2|(e_t, e_x, e^2)|\} ,\end{aligned} \tag{3.24}$$

where $\beta > 0$ to be determined, and ν's and c_3 are some appropriate positive constants. Now in (3.18), we multiply the first and third equations by $E_1(\bar f_+, g_+)$ and $E_1(\bar f_-, g_-)$ respectively, and the second equation by $4E_1(\bar f_0, g_0)$, sum up and integrate the resulting equation over $R\times[\tau_0, \tau_1]$ to get after using (3.22)-(3.24)

$$\begin{aligned} &\|f(\tau)\|^2 + \int_{\tau_0}^{\tau} (\|U_y^{1/2}(\bar f_- g_+ - \bar f_+ g_-)\|^2 + \|Q(g,g) - Lg\|)\, ds\\ &\qquad \le c_4\{\|g(\tau_0)\|^2 + \epsilon\int_{\tau_0}^{\tau}\int |U_x|\,|g|^2 dy\, ds + \epsilon^2\int_{\tau_0}^{\tau}\int |g|\,|(e_t, e_x, e^2)|\, dy\, ds\}\end{aligned} \tag{3.25}$$

provided $N(\tau_1)$ and ϵ are suitably small, where $c_4 > 0$ is a constant.

Next, we estimate the derivatives of $g(y,\tau)$. To this end, we differentiate (3.18) with respect to y and multiply the first, second and third equation by $\bar{f}_-g_{+y}$, $4\bar{f}_0g_{0y}$ and $\bar{f}_+g_{-y}$ respectively. Summing them up and integrating over $R^1\times[\tau_0,\tau]$, we have after using the facts

$$(Lg)_yL(g_y)\geq\frac{1}{2}(L(g_y))^2-c_5|g|^2\epsilon^2U_x^2\ ,\quad |\bar{f}_y|=O(1)\epsilon\quad\text{and}\quad|\bar{f}_\tau|=O(1)\epsilon\ ,$$

that

$$\|g_y(\tau)\|^2+\int_{\tau_0}^{\tau}\|L(g_y)\|^2ds\leq c_6\{\|g_y(\tau_0)\|^2+(N(\tau_1)+\epsilon)\int_{\tau_0}^{\tau}\|g_y(s)\|^2ds$$
$$+\int_{\tau_0}^{\tau}\int_{R}[\epsilon^2|g|^2U_x^2+|g_y|\epsilon^2|(e_{ty},e_{xy},ee_y)|]\,dy\,ds\}\ . \tag{3.26}$$

To bound $\int_{\tau_0}^{\tau}\|g_y(s)\|^2ds$, we estimate $\int_{\tau_0}^{\tau}\|\bar{f}_-g_{+y}+\bar{f}_+g_{-y}\|^2(s)\,ds$ and $\int_{\tau_0}^{\tau}\|\bar{f}_-g_{+y}-\bar{f}_+g_{-y}\|^2(s)\,ds$ respectively. First, we rewrite (3.18) as

$$\bar{f}_-g_{+\tau}-\bar{f}_+g_{-\tau}+\bar{f}_-g_{+y}+\bar{f}_+g_{-y}=(\bar{f}_--\bar{f}_+)(Q(g,g)-Lg)+O(1)\epsilon^2|(e_t,e_x,e^2)|\ , \tag{3.27}$$
$$\bar{f}_-g_{+\tau}+\bar{f}_+g_{-\tau}+\bar{f}_-g_{+y}-\bar{f}_+g_{-y}=(\bar{f}_-+\bar{f}_+)(Q(g,g)-Lg)+O(1)\epsilon^2|(e_t,e_x,e^2)|\ , \tag{3.28}$$
$$\bar{f}_-g_{+\tau}+\bar{f}_+g_{-\tau}-2\bar{f}_0g_{0\tau}+\bar{f}_-g_{+y}-\bar{f}_+g_{-y}=(\bar{f}_++\bar{f}_0+\bar{f}_-)(Q(g,g)+O(1)\epsilon^2|(e_t,e_x,e^2)|\ . \tag{3.29}$$

Multiplying (3.27) and (3.28) by $\bar{f}_-g_{+y}+\bar{f}_+g_{-y}$ and $\bar{f}_-g_{+y}-\bar{f}_+g_{-y}$ repsectively, and integrating them over $R\times[\tau_0,\tau]$, we have

$$\int_{\tau_0}^{\tau}\|(\bar{f}_-g_{+y}+\bar{f}_+g_{-y})\|^2ds\leq C_7\{\|g(\tau_0)\|_1^2+\|g(\tau)\|_1^2+\int_{\tau_0}^{\tau}(\|(\bar{f}_-g_{+y}-\bar{f}_+g_{-y})\|^2(s)$$
$$+\|Q(g,g)-Lg\|^2(s)+\||g|(\bar{f}_\tau,\bar{f}_y)\|^2+\epsilon^4\|(e_t,e_x,e^2)\|_{L_2(R_y)}^2)ds\}\ , \tag{3.30}$$

and

$$\int_{\tau_0}^{\tau}\|\bar{f}_-g_{+y}-\bar{f}_+g_{-y}\|^2(s)\,ds\leq c_8\{\|g(\tau_0)\|_1^2+\|g(\tau)\|_1^2+\int_{\tau_0}^{\tau}[\beta\|\bar{f}_-g_{+y}+\bar{f}_+g_{-y}\|^2(s)$$
$$(1+\beta^{-1})(\|Lg_y\|^2+\||g|(\bar{f}_\tau,\bar{f}_y)\|^2)+\|Q(g,g)-Lg\|^2]\,ds$$
$$+\int_{\tau_0}^{\tau}\int\epsilon^4|(e_t,e_x,e^2)|^2\,dy\,ds\}\ . \tag{3.31}$$

where $\beta>0$ is to be determined.

We conclude from (3.25), (3.26), and (3.30)-(3.31) that

LEMMA 9. *There exist positive constants δ_2, ϵ_3 and c_9 such that if*

$$N(\tau_1)\leq\delta_2\quad\text{and}\quad 0<\epsilon\leq\epsilon_3\ ,$$

then for all $\tau\in[\tau_0,\tau_1]$,

$$\|g(\tau)\|_1^2+\int_{\tau_0}^{\tau}(\|g_y(s)\|^2+\|U_y^{1/2}(\bar{f}_-g_+-\bar{f}_+g_-)\|^2(s)+\|Q(g,g)-Lg\|^2(s))\,ds$$
$$\leq c_9\{\|g(\tau_0)\|_1^2+\int_{\tau_0}^{\tau}\int_{R}(\epsilon U_x|g|^3+\epsilon^2|g|\,|(e_t,e_x,e^2)|+|g|^2|(\bar{f}_\tau,\bar{f}_y,U_y)|^2$$
$$+\epsilon^4|(e_t,e_x,e^2)|^2+\epsilon^4|(e_{ty},e_{xy},ee_y)|^2)\,dy\,ds\}\ . \tag{3.32}$$

We now show that Lemma 9 implies Proposition 8. To this end, we estimate each term on the right-hand side of (3.32). First, by using Sobolev's inequality, Young's inequality and Lemma 6, we obtain

$$\tag{3.33} \begin{aligned} c_9 \int_{\tau_0}^{\tau} \int_{\boldsymbol{R}} \epsilon U_x |g|^3 dy\, ds &\leq O(1)\epsilon \int_{\tau_0}^{\tau} \|U_x\|_{L_\infty} \|g\|^{1/2} \cdot \|g_y\|^{1/2} \|g\|^2\, ds \\ &\leq \frac{1}{16} \int_{\tau_0}^{\tau} \|g_y(s)\|^2 ds + O(1)\epsilon^{4/3} \int_{\tau_0}^{\tau} \|U_x\|_{L_\infty}^{4/3} \|g\|^{10/3}\, ds \\ &\leq \frac{1}{16} \int_{\tau_0}^{\tau} \|g_y(s)\|^2 ds + \big(O(1)\epsilon^{1/3} \int_{\tau_0}^{\tau} \|U_x\|_{L_\infty}^{4/3} d(\epsilon s)\big) \sup_{\tau_0 \leq s \leq \tau} \|g(s)\|^2 \\ &\leq \frac{1}{16} \int_{\tau_0}^{\tau} \|g_y(s)\|^2 ds + O(1)\epsilon^{1/3} \sup_{\tau_0 \leq \tau \leq \tau_1} \|g(s)\| \,. \end{aligned}$$

Next, we have

$$\begin{aligned} C_9 \int_{\tau_0}^{\tau} \int_{\boldsymbol{R}} \epsilon^2 |g|\, |e_t|\, dy\, ds &\leq O(1)\epsilon \int_{\tau_0}^{\tau} \|g\|^{1/2} \|g_y\|^{1/2} \|e_t\|_{L_1(\boldsymbol{R}_x)} ds \\ &\leq \frac{1}{16} \int_{\tau_0}^{\tau} \|g_y(s)\|^2 ds + O(1)\epsilon^{4/3} \int_{\tau_0}^{\tau} \|e_t\|_{L_1(x)}^{4/3} \|g\|^{2/3} ds \\ &\leq \frac{1}{16} \int_{\tau_0}^{\tau} \|g_y(s)\|^2 ds + \frac{1}{16} \sup_{\tau_0 \leq s \leq \tau} \|g(s)\|^2 + O(1)\epsilon^{1/2} \Big(\epsilon \int_{\tau_0}^{\tau} \|e_t\|_{L_1(x)}^{4/3}\, ds\Big)^{3/2} . \end{aligned}$$

Since $e_t = O(1)\big(|U_t|\, |U_x| + |U_{xt}|\big)$, we get from Lemma 6 that

$$\int_{\tau_0}^{\tau} \epsilon \|e_t\|_{L_1(x)}^{4/3}\, ds = O(1) \,.$$

Thus

$$\tag{3.34} c_9 \int_{\tau_0}^{\tau} \int_{\boldsymbol{R}} \epsilon^2 |g|\, |e_t|\, dy\, ds \leq \frac{1}{16} \int_{\tau_0}^{\tau} \|g_y(s)\|^2 ds + \frac{1}{16} \sup_{\tau_0 \leq \tau \leq \tau_1} \|g(\tau)\|^2 + O(1)\epsilon^{1/2} .$$

Similarly, one has

$$\tag{3.35} \begin{aligned} &c_9 \int_{\tau_0}^{\tau} \int_{\boldsymbol{R}} \epsilon^2 |g|\, |(e_x, e^2)|\, dy\, ds \\ &\quad \leq \frac{1}{16} \int_{\tau_0}^{\tau} \|g_y(s)\|^2 ds + \frac{1}{16} \sup_{\tau_0 \leq \tau \leq \tau_1} \|g(\tau)\|^2 + O(1)\epsilon^{1/2} , \end{aligned}$$

and

$$\tag{3.36} \begin{aligned} c_9 \int_{\tau_0}^{\tau} \int_{\boldsymbol{R}} |g|^2 |(\bar{f}_\tau, \bar{f}_y, U_y)|^2 dy\, ds &\leq O(1)\epsilon^2 \int_{\tau_0}^{\tau} \int_{\boldsymbol{R}} |g|^2 \big(|U_t|^2 + |U_x|^2\big)\, dy\, ds \\ &\leq O(1)\epsilon^2 \sup_{\tau_0 \leq \tau \leq \tau_1} \|g(\tau)\|^2 \int_{\tau_0}^{\tau} \big(\|U_t\|_{L_\infty}^2 + \|U_x\|_{L_\infty}^2\big)\, ds \\ &\leq O(1)\epsilon \sup_{\tau_0 \leq \tau \leq \tau_1} \|g(\tau)\|^2 . \end{aligned}$$

Finally, we obtain from Lemma 6 that

$$\tag{3.37} c_9 \int_{\tau_0}^{\tau} \int_{\boldsymbol{R}} \big(\epsilon^4 |(e_t, e_x, e^2)|^2 + \epsilon^4 |(e_{ty}, e_{xy}, e e_y)|^2\big)\, dy\, ds \leq O(1)\epsilon^2 .$$

It follows from (3.32)-(3.37) that

$$\begin{aligned} &\|g(\tau)\|_1^2 + \int_{\tau_0}^{\tau} \big(\|g_y(s)\|^2 + \|U_y^{1/2}(\bar{f}_- g_+ - \bar{f}_+ g_-)\|^2(s) + \|Q(g,g) - Lg\|^2(s)\big)\, ds \\ &\quad \leq \Big(\frac{1}{4} + O(1)\epsilon^{1/3}\Big) \sup_{\tau_0 \leq \tau \leq \tau_1} \|g(\tau)\|^2 + \frac{1}{4} \int_{\tau_0}^{\tau} \|g_y(s)\|^2 ds + O(1)\epsilon^{1/2} + c_9 \|g(\tau_0)\|_1^2 . \end{aligned}$$

We conclude from the above that

PROPOSITION 10. *There exist positive constants δ_3, ϵ_4 and c_{10} such that if*

$$N(\tau_1) \le \delta_3 , \quad \text{and} \quad 0 < \epsilon \le \epsilon_4 ,$$

then we have

$$\sup_{\tau_0 \le \tau \le \tau_1} \|g(\tau)\|_1^2 + \int_{\tau_0}^{\tau_1} (\|g_y(s)\|^2 + \|U_y^{1/2}(\bar{f}_- g_+ - \bar{f}_+ g_-)(s)\|^2 + \|(Q(g,g) - Lg)(s)\|^2)\, ds \le c_{10}\{\|g(\tau_0)\|_1 + \epsilon^{1/2}\} . \tag{3.38}$$

Now Proposition 8 follows from Proposition 10. By standard continuous induction argument, we conclude from Propositions 7 and 8 that the problem (3.18) has a unique smooth global in time solution $g(y,\tau) \in C^0(0,+\infty; H_1) \cap C^1(0,+\infty; L_2)$ such that

$$\sup_{0<\delta\le\tau<+\infty} \|g(\tau)\|_1^2 \le c_{11}\epsilon^{1/2} . \tag{3.39}$$

Consequently, we have Theorem 5. This completes the proof of Theorem 5.

REFERENCES

[1] R. E. Caflisch, and G. C. Papanicolaou, *The fluid dynamical limit of a nonlinear model Boltzmann equation*, Comm. Pure Appl. Math., 32 (1979), pp. 589-616.

[2] R. E. Caflisch, *The fluid dynamic limit of the nonlinear Boltzmann equation*, Comm. Pure Appl. Math., 33 (1980), pp. 651-666.

[3] J. E. Broadwell, *Shock structure in a simple discrete velocity gas*, Phys. Fluids, 7 (1964), pp. 1243-1247.

[4] R. E. Caflisch, *Navier-Stokes and Boltzmann shock profiles for a model of gas dynamics*, Comm. Pure Appl. Math., 32 (1979), pp. 521-554.

[5] Z. Xin, *The fluid-dynamic limit of the Broadwell model of the nonlinear Boltzmann equation in the presence of shocks*, to appear in Comm. Pure Appl. Math.

[6] P. D. Lax, *Hyperbolic systems of conservation laws, II*, Comm. Pure Appl. Math., 10 (1957), pp. 575-566.

[7] S. Kawashima, and A. Matsumura, *Asymptotic stability of traveling wave solutions of a system of one-dimensional gas motion*, Comm. Math. Phys., 101 (1985),pp. 97-127.

[8] J. Goodman and Z. Xin, *Viscous limit to piecewise smooth solution of system of conservation laws*, to appear.

[9] H. O. Kreiss, *Initial boundary value problem for hyperbolic systems*, Comm. Pure Appl. Math., 1970, pp. 277-295.

[10] J. Rauch, *L_2 is a continuable initial condition for Kreiss' mixed problem*, Comm. Pure Appl. Math., 25 (1972), pp. 265-285.

[11] S. Kawashima, *Global existence and stability of solutions for discrete velocity models of the Boltzmann equation*, Lecture Notes in Numer. Appl. Anal. 6, Recent Topics in Nonlinear PDE, Hiroshima, 1983, pp. 59-85.

[12] A. Matzumura, *Asymptotics toward rarefaction wave of solutions of the Broadwell model of a discrete velocity gas*, Japan J. Appl. Math., 4 (1987), pp. 489-502.

[13] H. Grad, *Singular and nonuniform limits of solutions of the Boltzmann equations*, SIAM-AMS Proccedings I, Transport Theory, 1969, R.I., pp. 296-308.

[14] R. Ellis and M. Pinsky, *The first and second fluid approximation to the linearized Boltzmann equation*, J. Math. Pure Appl., 54(1975), pp. 125-126.

[15] P. D. Lax, *Development of singularities of solutions of nonlinear hyperbolic partial differential equations*, J. Math. Phys., 5 (1964), pp. 611-613.